Exploiting Wastes in Concrete

Proceedings of the International Seminar
held at the University of Dundee, Scotland, UK
on 7 September 1999

Edited by

Ravindra K. Dhir
Director, Concrete Technology Unit
University of Dundee

and

Trevor G. Jappy
Research/Teaching Fellow, Concrete Technology Unit
University of Dundee

Published by Thomas Telford Publishing, Thomas Telford Limited, 1 Heron Quay, London E14 4JD.

URL: http://www.t-telford.co.uk

Distributors for Thomas Telford books are
USA: ASCE Press, 1801 Alexander Bell Drive, Reston, VA 20191-4400, USA
Japan: Maruzen Co. Ltd, Book Department, 3–10 Nihonbashi 2-chome, Chuo-ku, Tokyo 103
Australia: DA Books and Journals, 648 Whitehorse Road, Mitcham 3132, Victoria

First published 1999

The full list of titles from the 1999 International Congress 'Creating with Concrete' and available from Thomas Telford is as follows

- *Creating with concrete*
- *Radical design and concrete practices*
- *Role of interfaces in concrete*
- *Controlling concrete degradation*
- *Extending performance of concrete structures*
- *Exploiting wastes in concrete*
- *Modern concrete materials: binders, additions and admixtures*
- *Utilizing ready-mixed concrete and mortar*
- *Innovation in concrete structures: design and construction*
- *Specialist techniques and materials in concrete construction*
- *Concrete durability and repair technology*

A catalogue record for this book is available from the British Library

ISBN: 0 7277 2821 0

© The authors, except where otherwise stated

All rights, including translation, reserved. Except for fair copying, no part of this publication may be reproduced, stored in a retrieval system or transmitted in any form or by any means, electronic, mechanical, photocopying or otherwise, without the prior written permission of the Books Publisher, Thomas Telford Publishing, Thomas Telford Ltd, 1 Heron Quay, London E14 4JD.

This book is published on the understanding that the authors are solely responsible for the statements made and opinions expressed in it and that its publication does not necessarily imply that such statements and/or opinions are or reflect the views or opinions of the publishers or of the conference organizers.

Printed and bound in Great Britain by MPG Books, Bodmin, Cornwall

PREFACE

Concrete is the key material for Mankind to create the built environment, the requirements for which are both demanding in terms of technical performance and economy and yet greatly varied from architectural masterpieces to the simplest of utilities. This presents the greatest challenge and the question is how best to advance concrete and create imaginatively.

In response, the Concrete Technology Unit (CTU) of the University of Dundee organised this Congress following on from its established series of events, namely, Concrete in the Service of Mankind in 1996, Concrete 2000: Economic and Durable Concrete Construction Through Excellence in 1993 and Protection of Concrete in 1990.

Under the theme of Creating with Concrete, the Congress consisted of five Seminars: (i) Radical Design and Concrete Practices, (ii) Role of Interfaces in Concrete, (iii) Controlling Concrete Degradation, (iv) Extending Performance of Concrete Structures and (v) Exploiting Wastes in Concrete, and five Conferences: (i) Modern Concrete Materials: Binders, Additions and Admixtures, (ii) Utilising Ready-Mixed Concrete and Mortar, (iii) Innovation in Concrete Structures: Design and Construction, (iv) Specialist Techniques and Materials for Concrete and Construction and (v) Concrete Durability and Repair Technology. In all, a total of 421 papers were presented from 67 countries.

The Opening Addresses were given by Mr Henry McLeish, MP, MSP, Minister for Enterprise and Lifelong Learning, Scotland, Dr Ian Graham-Bryce, Principal and Vice Chancellor of Dundee University, Mrs Helen Wright, Lord Provost, City of Dundee and Professor Peter Hewlett, President of the Concrete Society. This was followed by four Opening Papers by leading international experts; Dr Bryant Mather, US Army Corps of Engineers, Professor Charles F Hendriks, Delft University of Technology, Netherlands, Dr Bjørn Jensen, Danish Technological Institute, Dr Oliver Kornadt, Philipp Holzmann AG, Germany, Professor Jurek Tolloczko, Concrete Society, UK, Mr Michael Téménidès, CIMBÉTON, France and Professor Yves Malier, Ecole Normale Superieure de Cachan, France. The Closing Address was given by Professor John Morris, University of the Witwatersrand, South Africa.

The support of 20 International Professional Institutions and 31 sponsors was a major contribution to the success of the Congress. An extensive Trade Fair, participated in by 50 organisations, formed an integral part of the Congress. The work of the Congress was an immense undertaking and all of those involved are gratefully acknowledged, in particular, the members of the Organising Committee for managing the event from start to finish; members of the International Advisory and National Technical Committees for advising on the selection and reviewing of papers; the Authors and the Chairmen of Technical Sessions for their invaluable contributions to the proceedings.

All of the proceedings have been prepared directly from the camera-ready manuscripts submitted by the authors and editing has been restricted to minor changes where it was considered absolutely necessary.

Dundee
September 1999

Ravindra K Dhir
Chairman, Congress Organising Committee

INTRODUCTION

Materials which would otherwise be regarded as waste products in modern society can have value as components in cementitious materials, as binders or fillers. At the end of the twentieth century, the need to dispose of waste materials sustainably has necessitated radical developments and has encouraged lateral thinking to establish novel uses of waste materials as useable commodities in the concrete construction industry.

Waste products from the electricity, quarrying, steelmaking, demolition, wood, textile and paper processing industries may be utilised, through blending with Portland cement.

This use of waste is not without its difficulties. Firstly, the waste is often not in a form directly suitable for use in concrete and material processing is usually required. Secondly, there may be small quantities of deliterious, or even harmful, components which must be effectively encapsulated. Concrete may well be the ideal medium for this, even if this renders the material unsuitable for construction purposes.

The Proceedings of this Seminar; *Exploiting Wastes in Concrete*, dealt with the waste product materials utilised in concrete applications under two distinct themes: (i) Maximising Use: conditioned/lagooned PFA, sewage-sludge ash, rock flour, steelmaking slags, recycled concrete, masonry and rubble; (ii) Alternative Options: wood aggregates, recycled paper waste, textile materials, aggregate from spoil heaps, waste clay brick, rice husks and vitrified soil materials. The Seminar was opened with a Leader Paper, followed by a Keynote Paper for each theme, presented by the foremost experts in their respective fields. A total of 34 papers have been compiled into these Proceedings.

Dundee
September 1999

Ravindra K Dhir
Trevor G Jappy

ORGANISING COMMITTEE
Concrete Technology Unit

Professor R K Dhir, OBE (Chairman)

Dr M R Jones (Secretary)

Mr M D Newlands (Joint Secretary)

Professor P C Hewlett
British Board of Agrément

Dr N A Henderson
Mott MacDonald Ltd

Professor V K Rigopoulou
National Technical University of Athens, Greece

Dr S Y N Chan
Hong Kong Polytechnic University

Dr N Y Ho
L & M Structural Systems, Singapore

Dr M J McCarthy

Dr M C Limbachiya

Dr T D Dyer

Dr K A Paine

Dr T G Jappy

Mr P A J Tittle

Mr J C Knights

Mr S R Scott (Unit Assistant)

Miss A M Duncan (Unit Secretary)

INTERNATIONAL ADVISORY COMMITTEE

Dr H M Z-Al-Abideen
Deputy Minister
Ministry of Public Works and Housing, Saudi Arabia

Professor M S Akman
Emeritus Professor of Civil Engineering
Istanbul Technical University, Turkey

Dr R Amtsbüchler
Manager-Technical Services
Blue Circle Ltd (South Africa), South Africa

Professor C Andradé
Director
Institute of Construction Sciences, Spain

Professor J M J M Bijen
Director
INTRON B.V., The Netherlands

Professor A M Brandt
Head of Section
Polish Academy of Sciences, Poland

Dr J-M Chandelle
Managing Director
CEMBUREAU, Belgium

Professor P Helene
Head of Civil Construction Engineering Department
University of Sao Paulo, Brazil

Dr G C Hoff
Senior Engineering Consultant
Mobil Technology Company, USA

Professor I Holand
Senior Research Engineer
SINTEF, Norway

Professor B C Jensen
Director
Carl Bro, Denmark

Professor S Mirza
Professor of Civil Engineering and Applied Mechanics
McGill University, Canada

Professor S Nagataki
Professor of Civil Engineering and Architecture
Niigata University, Japan

Professor H Okamura
Vice President
Kochi University of Technology, Japan

Professor E A e Oliveira
Director
Laboratório Nacional de Eng Civil, Portugal

Professor J-P Ollivier
Director of LMDC-INSA
LMDC, France

INTERNATIONAL ADVISORY COMMITTEE (CONTINUED)

Professor R Park
Professor of Civil Engineering
University of Canterbury, New Zealand

Mr S A Reddi
Managing Director
Gammon India Limited, India

Professor H-W Reinhardt
Head of Construction Materials Institute
University of Stuttgart, Germany

Professor R Rivera-Villarreal
Chief of Concrete Technology Department
Ciudad Universitaria, Mexico

Professor A Samarin
Consultant
Sustainable Development Technological Sciences and Engineering, Australia

Professor A E Sarja
Research Professor
Technical Research Centre of Finland, Finland

Professor S P Shah
Walter P Murphy Professor or Civil Engineering
Northwestern University, USA

Professor H Sommer
Head of Research Institute
VÖZ, Austria

Professor I Soroka
Professor of Civil Engineering
National Building Research Institute, Israel

Professor M Tang
Research Professor
Nanjing University of Chemical Technology, China

Professor T Tassios
Professor
National Technical University of Athens, Greece

Professor K Tuutti
Vice President
Skanska Technik AB, Sweden

Professor T Vogel
Professor of Structural Engineering
Swiss Federal Institute of Technology ETH, Switzerland

Professor F H Wittmann
Head of Building Materials Laboratory
Swiss Federal Institute of Technology ETH, Switzerland

Professor A V Zabegayev
Head of Department RC Structures
Moscow State University of Civil Engineering, Russia

NATIONAL TECHNICAL COMMITTEE

Mr P Barber
Manager of the Scheme, The Quality Scheme for Ready Mixed Concrete

Professor A W Beeby
Professor of Structural Design, University of Leeds

Mr B V Brown
Divisional Technical Executive, Readymix (UK) Ltd.

Dr T W Broyd
Technology Development Director, W S Atkins Ltd.

Professor J H Bungey
Professor of Civil Engineering, University of Liverpool

Dr P S Chana
Director, CRIC, Imperial College of Science, Technology & Medicine

Professor J L Clarke
Principal Engineer, The Concrete Society

Dr P C Das
Group Manager, Structures Management, Highways Agency

Dr S B Desai, OBE
Principal Civil Engineer, Department of the Environment, Transport and the Regions

Professor R K Dhir, OBE (Chairman)
Director, Concrete Technology Unit, University of Dundee

Mr C R Ecob
Director Special Services Division, Mott MacDonald Ltd.

Professor F P Glasser
University of Aberdeen

Professor T A Harrison
Technical Consultant, Quarry Products Association

Professor P C Hewlett
Director, British Board of Agrément

Professor J Innes
Director of Roads, Scottish Office

NATIONAL TECHNICAL COMMITTEE (CONTINUED)

Mr K A L Johnson
Director, AMEC Civil Engineering Ltd.

Dr M R Jones
Senior Lecturer, Concrete Technology Unit, University of Dundee

Mr P Livesey
National Technical Services Manager, Castle Cement Ltd.

Professor A E Long
Director of School, Queens University of Belfast

Professor P S Mangat
Head of Research, Sheffield Hallam University

Mr G Masterton
Director, Babtie Group Ltd.

Professor G C Mays
Director of Civil Engineering, Cranfield University

Mr L H McCurrich
Technology Development Consultant, Fosroc Construction

Professor R S Narayanan
Partner, SB Tietz & Partners Consulting Engineers

Dr P J Nixon
Head, Centre for Concrete Construction, Building Research Establishment Ltd.

Dr W F Price
Senior Associate, Messrs Sandberg

Professor G Somerville, OBE
Director of Engineering, British Cement Association

Professor D C Spooner
Director, Materials and Standards, British Cement Association

Dr H P J Taylor
Director, Tarmac Precast Concrete Ltd.

Mr M Walker
Technical Manager, The Concrete Society

Dr R J Woodward
Senior Project Manager, Transport Research Laboratory

SUPPORTING INSTITUTIONS

American Concrete Institute, USA

American Society of Civil Engineers, USA

Australian Concrete Institute

Concrete Association of Finland

Concrete Society of Southern Africa

Concrete Society, UK

Danish Concrete Society, Denmark

Fédération de l'Industrie du Beton, France

German Concrete Association (DBV)

Hong Kong Institution of Engineers

Indian Concrete Institute

Institute of Concrete Technology, UK

Institution of Civil Engineers, UK

Instituto Brasileiro Do Concreto, Brazil

Japan Concrete Institute

Netherlands Concrete Society

New Zealand Concrete Society

Norwegian Concrete Association

Singapore Concrete Institute

Spanish Association for Structural Concrete

Swedish Concrete Association

SPONSORING ORGANISATIONS WITH EXHIBITION

AMEC Civil Engineering Ltd.

Babtie Group Ltd.

Bardon Aggregates

Blue Circle Cement

Blyth & Blyth

British Board of Agrément

British Cement Association

Building Research Establishment

Castle Cement Ltd.

Cementitious Slag Makers Association

CIMBÉTON, France

Du Pont de Nemours International S.A., Switzerland

ECC International Ltd.

Elkem Ltd. (Materials)

Fosroc International Ltd.

Grace Construction Products

HERACLES General Cement Co., Greece

John Doyle Group

Lafarge Aluminates

L M Scofield Europe Ltd.

Minelco Ltd.

Mott MacDonald Ltd.

O'Rourke Group

Ove Arup and Partners

SPONSORING ORGANISATIONS WITH EXHIBITION (CONTINUED)

Readymix (UK) Ltd.

Rugby Cement

Scottish Enterprise Tayside

Sika Ltd.

SKW - MBT Construction Chemicals

Thomas Telford Publishing Ltd.

United Kingdom Quality Ash Association

W A Fairhurst & Partners

ADDITIONAL EXHIBITORS

Christison Scientific Equipment Ltd.

CMS Pozament Limited

The Concrete Society

David Ball Group plc.

E & FN Spon

Flexcrete Ltd.

Germann Instruments A/S, Denmark

Natural Cement Distribution Limited

Palladian Publications Ltd.

Quality Scheme for Ready Mixed Concrete

UK Certification Authority for Reinforcing Steel

Wacker-Chemie GmbH, Germany

Wexham Developments

CONTENTS

Preface	iii
Introduction	iv
Organising Committee	v
International Advisory Committee	vi
National Technical Committee	viii
Supporting Institutions	x
Sponsoring Organisations With Exhibition	xi
Additional Exhibitors	xii

Leader Paper
Wastes in concrete: Converting liabilities into assets — 1
A Samarin, University of Wollongong/Private Practice, Australia

THEME 1   MAXIMISING USE

Keynote Paper
Use of industrial by-products in cement-based materials — 23
T R Naik and R N Kraus, University of Wisconsin, United States of America

Studies on the effective use of concrete sludge - sudge water and pulverized dry sludge — 37
Y Sato, C Kiyohara, L Chia-Ming, Y Takeda, S Taguchi and T Yahushiji

Experimental basic aspects for reusing sewage sludge ash (SSA) in concrete production — 47
J Monzó, J Payá, J and M V Borrachero

The pozzolanic nature of ponded ash — 57
T L Robl, J C Groppo and A Hobbs

Admixture compatibility and mixer requirements for conditioned PFA concrete — 67
M J McCarthy, P A J Tittle and R K Dhir

Flowable rockdust - Cement slurry as backfill material — 81
B K Baguant

Structural lightweight concretes with fly ash pelletized coarse aggregate - mix proportioning — 95
T S Nagaraj and T Ishikawa

Fine-grained cementless concrete containing slag from foundry   101
S I Pavlenko and V I Malyshkin

Performance of steel slag aggregate concretes   109
M Maslehuddin, M Shameem, M Ibrahim and M N Khan

Properties of self-compacting concrete with slag fine aggregates   121
M Shoya, S Sugita, Y Tsukinaga, M Aba and K Tokuhasi

Electro-surface properties of aggregates from waste products and their influence on the quality of fine grained concrete   131
V A Matviyenko, S M Tolchin and N M Zaichenko

The use of recycled concrete and masonry aggregates in concrete: Improving the quality and purity of the aggregates   139
J Desmyter, J van Dessel and S Blockmans

Recycled aggregates from old concrete highway pavements   151
W Fleischer and M Ruby

Early age properties of recycled aggregate concrete   163
F T Olorunsogo

Processed concrete rubble for the reuse as aggregates   171
G Mellmann, U Meinhold and M Maultzsch

Strength and elasticity of brick and artificial aggregate concrete   179
M Zakaria

Properties and performance of recycled cementitious mortars   189
T Rad and D G Bonner

Separation as a requirement for a high level recycling of building rubble   199
A Mueller and F Splittgerber

THEME 2 ALTERNATIVE OPTIONS

Keynote Paper
Secondary materials: A contribution to sustainability?   209
J M J M Bijen, INTRON B.V., Netherlands

Simple treatments to reduce the sensitivity to water of clayey concretes lightened by wood aggregates   217
A Ledhem, A Bouguerra, R M Dheilly and M Queneudec

The use of fly ash in the compound wood-cement 227
F Z Mimoune, M Mimoune and M Laquerbe

Shrinkage and creep of recycled paper waste concrete 233
R P West, C A Ryan and A Thompson

The effect of moisture content and temperature 243
on thermal conductivity of lightweight environmental concrete
M L Benmalek, A Bali, A Bouguerra, M S Goual and M Queneudec

Fine-grained concrete containing aggregate from spoil of open cuts 251
S I Pavlenko

Waste clay brick - A European study of its effectiveness 261
as a cement replacement material
S Wild, A Gailius, H Hansen and J Szwabowski

Alkali silica resistance ground brick mortars 275
A Gailius and I Girniene

Durability of concrete containing rice husk as an additive 283
P R S Speare, K Eleftheriou and S Siludom

Properties of Portland cement mortars containing FBC fly ash 291
N Ghafoori and S Kassel

Characterisation of filler sandcretes with rice husk ashes additions - 299
Study applied to Senegal
I K Cisse and M Laquerbe

Rice husk ash: A filler for sand concrete 309
I K Cisse, R Jauberthie, M Temimi and J P Camps

Assessing the properties of mortars containing 319
municipal solid waste incineration fly ash
S Rémond, P Pimienta, N Rodrigues and J P Bournazel

Elaboration of a MSWI fly ash solidification stabilisation process: 327
Use of statistical design of experiments
S Morrel-Braymand, P Clastres and A Pellequer

Utilization of vitrified contaminated soil materials in concrete mixtures 337
S N Amirkhanian

Index of Authors **347**

Subject Index **349**

LEADER PAPER

WASTES IN CONCRETE: CONVERTING LIABILITIES INTO ASSETS

A Samarin
University of Wollongong
Australia

ABSTRACT. The paper presents a general overview of the current trends in the use of by-products or wastes in various types of concrete. A potential of using processed and unprocessed industrial by-products and domestic wastes as ingredients in concrete is evaluated. In particular, the feasibility of manufacturing hydraulic cements, non-hydraulic cements, cement *extenders*, admixtures and additives in concrete, unprocessed, processed and manufactured aggregate in concrete, and the use of processed or unprocessed by-products and wastes as concrete reinforcement is assessed. An attempt is made to predict possible future developments in the utilisation of industrial by-products as raw materials in cement and concrete due to ever increasing cost of waste disposal and due to the stricter environmental regulations. The impact of Kyoto Protocol, stipulating new levels of abatement of greenhouse gas emissions by the year 2012, on utilisation of cement *extenders* in concrete, and on the use of wastes as raw materials and fuel supplements in cement manufacturing, as well as the impact of the new technological developments on the use of wastes in concrete is evaluated.

Keywords: Broad definition of concrete, Industrial and domestic wastes, Processed and unprocessed wastes, Wastes as binders, Wastes as aggregates, New applications for wastes, Effects on the greenhouse gas emissions, Kyoto protocol, Economics of wastes utilisation, Environmental pollution, Sustainable development.

Dr Aleksander Samarin is a Private Consultant and Adviser in the Sustainable Development of Technological Sciences and Engineering and a Professorial Fellow at the School of Civil, Mining and Environmental Engineering, University of Wollongong, Australia. He is a Fellow of the Australian Academy of Technological Sciences and Engineering and a Vice-Chairman of the NSW Division of this Academy. Professor Samarin conducted R&D work into building and construction materials, energy and environment protection in many parts of the world, published widely, receiving several awards for his contribution to science and technology, and has taken a number of patents.

INTRODUCTION

In a very broad sense concrete can be defined as an artificial building and construction material consisting of relatively inert aggregate, held together with either inorganic or organic binder, or even by a hybrid of these two types.

Aggregate can also be either inorganic or organic in nature, it can be naturally occurring (processed or unprocessed), or manufactured. The reactivity of aggregate can also vary. Depending on the nature of reactivity of aggregate the effects can range from beneficial to highly destructive to concrete strength, dimensional stability and durability.

Industrial or domestic wastes, processed or unprocessed, can become cost-effective ingredients of concrete, defined in this way.

The processed or unprocessed industrial by-products or wastes can be used as raw materials in cement manufacturing, as components of concrete binders, or even as binders in their own right. They can also be used as aggregates, as a portion of aggregate, or as ingredients in manufactured aggregates.

Some wastes can be used as chemical admixtures and additives, which can alter and enhance selected, desirable properties of fresh or hardened concrete.

The successful use of industrial by-products or wastes in concrete depends on a number of factors. Of these, an application and nature of the end use of building and construction materials will dictate required properties of concrete, and thus will become a pre-requisite in deciding on whether or not the waste can be ultimately incorporated into this concrete. Desirable properties of concretes used for structural applications, or for block making, or in pavement construction, or for decorative use, etc., would obviously differ quite considerably from one another.

However, the economical factors would ultimately determine, if potentially beneficial waste can be used as an ingredient in concrete. These economical factors can differ considerably from one country to another, and particularly between developed and developing countries. They may also vary considerably for different locations in a given country, and are generally influenced by the cost of waste disposal, by the cost of transportation of waste to a manufacturing site and by the existing environmental regulations.

We can define waste as not readily avoidable by-product, for which there is no economical demand and for which disposal is required [1].

The first rule of waste management is to prevent its production in the first place, if it is at all possible. If not, we must attempt to generate an economical demand for the waste, for example by creating beneficial conditions for its recycling. Utilisation of waste, resulting in a cost-benefit to the end user is one way of creating a demand for a by-product. The benefits may include the elimination of quite tangible expenditure, such as cost of disposal, as well as some less corporeal advantages, such as abatement of pollution and better control of environmental contamination.

RECYCLED CONCRETE

Fundamental Requirements

After the second world war, when many European cities were reduced to rubble and there was a general shortage of cement and aggregate manufacturing facilities, the use of recycled concrete in building and road construction industries was reasonably wide spread.

The main problem with the use of recycled concrete, which existed in the late 1940-s, and which is still the principal source of annoyance in the present day use of recycled materials, is the difficulty of separating useful components of rubble from the contaminants, such as timber, glass, plastic, etc., which are often harmful to the properties of the end product [2].

The cost of "purification" of recycled rubble versus a tolerance to the impurities of concrete made with the recycled materials, generally determines in the end its potential suitability of use.

Rejuvenated Cement

Most of the current effort in recycling concrete is directed towards the re-use of coarse aggregate. The fine fraction of rubble, obtained from a hydraulic cement concrete, contains cement grains, which almost without exception are only partially hydrated. These cement particles can be "rejuvenated" by means of mechano-chemical activation and thus become either *cement extenders*, or even binders in their own right [3].

Recycled Aggregate in Road Construction

The most widely used form of concrete recycling is possibly in road construction. Bituminous concrete is particularly tolerant to the inclusions of all the components of old pavements, the main potential problem being badly oxidised bitumen. Addition of sufficient amount of fresh binder can overcome the problem. Special machinery has been developed for recycling of flexible pavements, mostly in the developed countries, although recycling of bituminous concrete is also practiced in the developing countries [4].

In my personal opinion *hybrid* or semi-rigid /semi-flexible pavements are a particularly good type of construction made from a recycled aggregate concrete [5].

Rigid pavements made with hydraulic cement concretes can be unreinforced, reinforced, or post-tensioned.

Failure of a rigid pavement can be either structural, or due to material deterioration in concrete. Structural failure often results from the settlement of a base course, from an inadequate provision of thermal expansion /contraction joints, or simply due to the traffic fatigue. The integrity of recycled concrete components in these cases is usually quite good.

When failure is due to the chemical or physical distress of concrete, such as alkali-aggregate reaction, frost damage, etc., the suitability of recycled aggregate for the new constructions work must be circumspect.

There has been a significant progress in the technological development of recycling methods of rigid pavements during the past decade [6], [7].

Generally the recycling work, using specially designed machinery, is done *in situ*. The old pavement is broken up, the rubble is crushed and screened. If the pavement was reinforced, steel rods are removed by electro-magnets or by mechanical means. The suitability if recycled material is assessed in a field laboratory, allowing design and construction of the new pavement, with the appropriate addition of new raw materials, to proceed without delay.

Recycled Aggregate in Building and Construction

Use of recycled aggregate in non-structural, non load-bearing concrete presents similar problems to those of aggregate used in pavements.

Requirements for structural concrete can vary significantly, depending on the type of structure, and within that structure they do change from one structural member to another. Similar structures in different environment conditions would behave in a dissimilar way, and hence would have specific provisions for the properties and the composition of concrete [8].

An impressive volume of research and development work is conducted in many parts of the world in this area, and there is no lack of published data on recycled aggregate in technical books, journals and conference proceedings [9], [10], [11], [12], [13], [14], [15], [16], [17].

Structural Standards do differ from one country to another, as they reflect environmental and economical conditions for the distinctive and sometimes unique parts of the world.

Thus, we can not expect to end up with universal standards or manuals for the use of recycled aggregate in structural concrete. Each specific application would have its own solution.

AGGREGATE DERIVED FROM WASTE

Unprocessed Waste as Aggregate in Concrete

A wide range of unprocessed waste, with variable degrees of success has been used as aggregate in concrete.

Probably the most widely researched is the use of blast furnace slag aggregate. The technology is well known and field results over thirty or even forty years, particularly in pavements, are available. Properties of blast furnace slag do vary considerably, and with the optimisation of smelting technology, have undergone considerable changes in the past thirty years or so. Australian blast furnace slags used to be glassy and dense, but with the use of fluxes, such as serpentine, to improve the furnace efficiency, became porous and even at times dendritic. However, most blast furnace slag aggregates are compatible with hydraulic cements, although free lime component of slag may cause dimensional instability in concrete, if slag is not properly weathered before use. The R&D work is continuing, and up to date reports are available [18], [19], [21].

Less widely used is steel slag, although it was reported that several steel slag aggregates performed well in Portland cement concrete [20], [21]. More R&D work, particularly into long term durability and also in the use of steel slag aggregate in semi-rigid pavements is required, in my opinion.

Some preliminary assessment of the behaviour of ferro-chromium and silica-manganese slags in concrete was also carried out [21].

Many building and industrial wastes can be used in special application concretes. Demolition waste, provided it is separated into the ingredients of similar properties can be used successfully in most non-structural and even in some structural concretes [22], [23]. The process of waste segregation, however, may prove difficult and therefore expensive [24], [25], [2].

The use of crushed brick or a mixture of crushed brick and of Portland cement concrete rubble as an aggregate in structural concrete was reported to be promising [26], [27]. Long term properties, which differ from conventional concretes, should be accounted for in the design. Modulus of elasticity of brick aggregate concrete may also differ from the Structural Code values, for a particular grade of concrete.

Crushed glass, from wine, beer or soft drink bottles, can be used as fine aggregate in concrete [28]. It is important however, to have sufficient content of either ultra fine glass particles, or of fly ash or other good quality pozzolans, to prevent alkali-aggregate expansion in larger glass particles.

Lightweight concretes can be produced with unprocessed waste materials, such as expanded polystyrene granules [29], [30]. There are several patented lightweight construction systems, incorporating expanded polystyrene concrete panels, which are now more than ten years old. A classical example is *Stracke Ges.m.b.H.* system in Vienna, Austria. Using special concrete mixer, lightweight Portland cement concrete with polyester aggregate of between 300 and 350 kg/m^3 is formed into blocks or panels, which can be sawn, drilled and nailed. These are then used as a permanent formwork in buildings, providing excellent thermal and acoustic properties to a reinforced structure, cast with high workability concrete.

Cork granules can be used in some of the above applications in place of expanded polystyrene chips [31].

Sawdust was used in different parts of the world, as aggregate for concrete [32]. Again, number of patents have been taken in these countries. Most describe the type of sawdust used, some special additives, preventing adverse reactions between the wood particles and Portland cement paste, mix proportions, and also mixing and curing techniques.

Typical sawdust concrete has a design strength of 15 MPa, density of approximately 1000 kg/m^3 and in some cases it can be sawn, milled and nailed. However, long term durability of this material, particularly in a wet-dry environment, needs additional investigation in my opinion.

An attempt to use shredded rubber as an aggregate in Portland cement concrete was generally not very successful [33]. Considerable reduction in strength was reported, even at relatively low rates of replacement of conventional aggregate with rubber. In my own, unpublished work it was found, that spherical rubber particles mixed into concrete, had an identical effect to the air voids of similar dimensions and total volume on the concrete properties. Use of rubber aggregate in semi-rigid concrete, however, can be quite promising.

There is an extremely wide range of different unprocessed wastes, currently being evaluated as potential aggregate in concrete. These include cane bagasse, wood ash [34], china clay waste, slate processing waste [35], paper waste, etc. These may be suitable for either temporary or low cost housing, particularly in the developing countries [36], [37], [38], [39].

Concrete Aggregate Manufactured from Waste

Work on the development of a process for conversion of pulverised fuel ash into a sintered lightweight aggregate - *Lytag* - for concrete, was carried out at the Building Research Establishment in Garston, U.K. since the early 1950-s, at the request of Central Electricity Generating Board. The R&D work was completed by the late 1950-s.

In the Lytag process, fly ash with a carbon content of between 7 and 10 percent is fed onto pan pelletisers, where the rotation action converts damp ash into spheroidal pellets. The pellets are then passed to the sinter strand, where they are fused at temperature of approximately 1300° C. Carbon in fly ash is ignited, producing gas and bloating the pellets, which, upon cooling, solidify to strong and lightweight particles. The first production plant of some 100,000 t/a was constructed in 1961 at Northfleet, U.K. [40].

Using Lytag, concrete with an air dry density of about 1,800 kg/m^3 can be obtained, with a maximum ultimate compressive strength of up to 70 MPa in individual, cube specimens. Thus, theoretically, characteristic strength of up to 50 MPa is achievable with special design techniques. In practice, it seldom exceeds 40 MPa. Long term behaviour and modulus of elasticity, however, can vary from the Structural Code values, for a given grade of concrete.

Lytag aggregate has been used also as a filtration material. There were several reports describing the manufacturing process and markets for Lytag in some detail [41].

There were number of attempts in the late 1980-s to convert household and industrial garbage into concrete aggregate. Commercial success of this concept depends *inter alia* on the price differential between the cost of aggregate production and the cost of garbage disposal.

Australian company, *Neutralysis* spent twelve million Aust. dollars over a period of six years developing the technology of manufacturing concrete aggregate from a blend of household and industrial waste. A pilot plant near Brisbane, Queensland, was built in 1991 [42].

The process starts with primary and secondary pulverisation of garbage blend accompanied by magnetic separation of scrap steel for recycling.

The garbage then enters the processing stage where it is mixed with the combustible liquid industrial waste and clay, extruded into pellets and dried. The pellets are fired in three successive kiln stages using excess energy given off during the process. In the final vetrification stage all the organic harmful materials are destroyed and inorganic hazardous materials are encapsulated. The aggregate was of the quality suitable for non-structural concrete, in concrete block manufacturing and as a road base material. The surplus energy can be used to generate electricity, which is then sold into the power grid.

Due to the relatively low cost of garbage disposal in Australia, the technology was sold to a sister company in the United States, where the waste disposal costs were considerably higher.

Another process, which converts the residue from incinerated toxic waste into pellets, was developed by a Singapore company, *Miltox*. Miltox, a joint venture between Singaporean, Malaysian and Australian interests built a pilot plant with an output of approximately 400 kilograms an hour in 1994. It seems, that long term research into the durability and environmental effects of the product is required, before it can be safely used in building and construction materials.

POZZOLANS AND OTHER PORTLAND CEMENT EXTENDERS

Applications in Concrete Technology

The use of fly ash, ground granulated blast furnace slag, silica fume, rice husks ash and other industrial or natural mineral by-products as extenders of Portland cement is very thoroughly researched and thousands, if not tens of thousand of papers, reports and books have been written on the subject. Wide spread industrial use of these materials have been almost universal for several decades. Number of International Conferences on the use of Fly Ash, Silica Fume, Slag and other Mineral By-products in Concrete were organised by ACI/CANMET and were held in 1983 (Canada), 1986 (Spain), 1989 (Norway), 1992 (Turkey), and 1995 (USA), and the Proceedings were published by the American Concrete Institute as the Special Publication (SP) series. Many other conference proceedings, including conferences at Dundee, contain valuable information on the use of these materials. To go into the detailed analysis of all the published data is outside the scope of this paper.

There are however several very important points, arising from my own R&D work with pozzolanic materials, which I would like to emphasise. When fly ash was first used in Australia as a Portland cement extender in the 1960-s, I was often asked by structural engineers and builders *"What happens when fly ash is mixed into concrete?"* The question of course is meaningless. First of all we must have clear understanding which concrete properties we wish to enhance, and which we are prepared to sacrifice to a certain degree, when using fly ash. The end result will depend not only on the mineralogical and chemical composition of fly ash, but also on its grading, particle shape and impurities, and, most importantly, on the compatibility of physical, chemical and mineralogical properties of fly ash with the corresponding properties of Portland cement and sand used in the concrete mix [43], [44].

Exactly the same rule applies to the granulated blast furnace slag, rice husks, and other natural and artificial pozzolans. Silica fume, due to its very high surface area and fineness is particularly sensitive to the above factors.

Thus, there are no universal relationships providing optimum mix designs with Portland cement extenders. These mix designs must be derived for each specific application of concrete, and for each set of the available raw materials.

Impact on Greenhouse Gas Emissions

To produce one tonne of Portland cement clinker in a rotary kiln, some 100 kg. of fuel oil is used for the electric energy generation. To make Portland cement, clinker has to be ground to a fineness in excess of 300 m^2/ kg. Grinding requires between 35 and 40 kilowatt hours (i.e. 126 - 144 MJ) of electric energy per tonne of clinker, or about 10 kg of fuel oil. Apart from oil, pulverised coal and natural gas, or dual fuels, such as oil-coal, oil-natural gas, and coal-gas are commonly used for clinker making. An average CO_2 emission factor in the fossil fuel combustion process is of the order of 0.22 kg of carbon dioxide for every kilogram of fuel. The electrical energy requirement to produce one tonne of cement is of the order of 440 kWh or approximately 1,580 MJ [3]. Calcination of one tonne of $CaCO_3$ generates 440 kilograms of CO_2, according to the stoichiometry of this chemical reaction. Energy demand for the calcination of one tonne of clinker is about 490 kWh, or approximately 1,750 MJ.

Consequently, for every *tonne* of Portland cement produced, cement plants generate approximately *one tonne* of carbon dioxide - one of the main greenhouse gases!

This figure varies, depending on the type of kiln and the fuel used. For a *wet* process, common in the developing countries, it is near 1.1 tones of CO_2 per tonne of cement; for a more energy efficient *dry* process it is of the order of 0.9 tonnes of carbon dioxide per tonne of cement, but a *tonne per tonne* is a good average.

As a result, one cubic metre of good structural concrete, made with 320 kg/m^3 of Portland cement only, will generate some 0.41 tonnes of CO_2, emitted into the atmosphere. If 30% of Portland cement is replaced with fly ash, this figure is reduced to about 0.29 tonnes of CO_2, that is approximately 30% reduction of green house gas emissions, neglecting of course the carbon dioxide produced during the combustion of coal. With very reactive, highly pozzolanic and ultra-fine fly ash much higher Portland cement replacement (or replacement-addition) rates are possible, without reduction of the ultimate strength of concrete. In my experience, the reduction of CO_2 emissions to below 0.2 tonnes per cubic metre of concrete is possible, with only a slight effect on the rate of concrete strength gain, say if 90 in place of 28 day Characteristic Strength criteria is adopted [40].

Much higher Portland cement replacement rates are achievable with ground granulated blast furnace slag, and with *triple* or even *quadruple* blends, particularly if silica fume and High Range Water Reducing Admixtures are used.

Depending on the source, grinding of slag usually, but not always, requires slightly more than 40 kilowatt hours per tonne of ground material.

Even so, ground granulated slag manufacturing results in less that 11% of energy required to produce one tonne of Portland cement.

Optimising the composition of blended cements and curing methods, and, when possible allowing 90 day Characteristic Strength criteria in Structural Design, it should be possible to reduce the CO_2 emissions to less than 0.1 of a tonne per cubic metre of concrete.

HYDRAULIC CEMENTS MANUFACTURED USING INDUSTRIAL BY-PRODUCTS

Alternative Fuels

In addition, or even in place of the fossil fuels mentioned above [3], waste - derived fuels can be used in Portland cement manufacturing. Instead of natural gas, coke oven, pyrolysis and landfill gases are being used. Mineral, hydraulic, industrial oils, distillation residues and halogen-free spend solvents are also viable alternative liquid fuels. Pulverised coal can be supplemented and in some cases possibly replaced with tar, petroleum coke, sawdust, dried sewage sludge, some selected plastics, agricultural residue and car tyres. Steel-belted tyres in some cases may assist in the iron deficiency or a raw feed [3].

In 1995 the thermal energy consumption from alternative fuels in the European cement industry was about 10% [45]. In most cases replacement of fossil fuels with wastes reduces carbon dioxide emissions from clinker and cement manufacturing, the maximum reduction reported is of the order of 40%. However, in each particular case the energy efficiency of waste combustion in cement kilns and the quality, and particularly the consistency of manufactured cement should be closely monitored.

Waste introduced into the burning or calcining zone is combusted with a flame and gas temperatures approximately in the range of between 1,900° C and 2,100° C. This is sufficient to destroy all the hazardous organic components of waste. However, presence of trace elements, and particularly of heavy metals should be monitored [46], as well as the potential variations in the emissions to atmosphere [47].

Apart from dust, cement kilns emit into the atmosphere nitrogen oxide, sulphur dioxide, organic carbon, cadmium, thallium, beryllium, mercury and traces of arsenic, cobalt, nickel and lead.

Utilisation of Industrial Wastes in Manufacturing of Hydraulic Cements

Industrial wastes can also be used as a raw feed in manufacturing new types of hydraulic cements, which often results in the reduction of energy required for clinker formation, and thus in the lower emissions of CO_2 [3]. Since the late 1950-s theoretical foundations of new types of binders, produced by chemical or mechano-chemical activation of natural siliceous materials or industrial siliceous wastes were developed in several East European countries.

Chemical activation with caustic alkalis and with salts of alkaline metals produced so called *soilcement*. Sodium metasilicates or water glass were used to obtain new hydraulic cements. The hardening and the strength gaining processes in these binders can be enhanced and accelerated by a heat treatment, with the best results reported at the temperatures in the range of between 900°C and 1000°C [48]. As this process requires no calcination, and with the significantly lower processing temperature, as compared with the temperature demand during the Portland clinker formation, which is of the order of 1,400°C, it results in better than 50% reduction in the emissions of carbon dioxide, in contrast to even the most modern Portland cement plants.

By optimising the technology of manufacturing slag-alkaline cements, one day compressive strength of more than 50 MPa was achieved in a concrete based on this binder. One month compressive strength was of the order of 160 MPa. Currently, there is wide spread reported use of slag-alkaline cements in the countries, which were part of the old Eastern block [49], [50], [51], [52].

Another family of inorganic binders, closely related to slag-alkaline cements, which also challenge Portland cements, are geopolymers. These are mineral polymers which are based on inorganic silico-aluminates. One of several possible hardening mechanisms of alumino-silicate oxides involves chemical reaction with alkalis and alkali-polysilicates, yielding polymeric Si - O - Al bonds [53], [54]. An inorganic polymerisation reaction results in a three-dimensional structure, similar to that of zeolites.

Geopolymers are usually produced by blending three ingredients - calcined alumino-silicates, alkali-disilicates and granulated blast furnace slag or fly ash. Geopolymeric concretes were reported to reach compressive strengths of 20 MPa in 4 hours, and up to 100 MPa in one month [55]. Geopolymers were commercially introduced into Eastern Europe and the United States in the 1980-s, and are used for most construction purposes and also for the immobilisation of hazardous wastes [56].

As the geopolymers do not use $CaCO_3$ as a raw feed component, and also because chemical reaction of formation of alumino-silicates takes place at the temperatures of approximately 750°C, it is claimed that production of geopolymers results in between 0.1 tonne and 0.2 tonnes of CO_2, emitted for every tonne of this mineral binder [57].

Yet another alternative to Portland cement are the new types of binders, produced by mechano-chemical activation of seemingly inert siliceous materials, such as fly ash or silica sand. This process requires no calcination of clinker, and also no high temperatures for binder formation, and thus produces no carbon dioxide from either the combustion of fuel, or from the conversion of $CaCO_3$, essential in the Portland cement clinker manufacturing.

These cements are known as silica-water-suspension-binders. They are produced in specially designed and patented high intensity planetary mills. It was reported, that the best results were obtained after the concretes based on silica-water-suspension-binders were cured at temperatures of between 40°C and 120°C. Even allowing for these high curing temperatures, up to 90% reduction of carbon dioxide emissions, as compared with Portland cement production, can be achieved with these binders [3], [58].

WASTES AS REINFORCEMENT AND AS SPECIAL ADDITIVES IN CONCRETE

Waste as Continuous and Fibre Reinforcement and Additives for Self-Curing Concrete

Possibly the most common and one of the oldest in use of all the organic continuous reinforcing materials in Portland cement concrete is bamboo. It has large percentage of fibre, which is high in tensile strength, and it also has very good strength to weight ratio. Bamboo waste, as well as specially harvested plants have been used as concrete reinforcement in many developing, tropical and sub-tropical countries. Bamboo is generally used with low modulus concrete and requires special provisions for the end anchorage [59]. It is recommended, that bamboo main reinforcement in Portland cement concrete should be in the range of between 3.5% and 4.5%. Under the serviceability loads the reinforcement strains are then relatively small, and there is a large safety factor between the rupture and the serviceability loads [60]. The main weakness of bamboo, as indeed of any organic material, is its high water absorption potential, which ultimately leads to swelling and shrinkage. Yet another handicap of bamboo reinforcement is its relatively fast rate of decay in concrete [61]. To overcome the problems influencing long term durability, bamboo is often pre-treated, initially using sand-blasting to remove the outer skin, with subsequent soaking in the liquid sulphur. To prevent insect attacks, bamboo reinforcement can also be soaked in diesel oil [60], [61]. Field tests indicate, that over periods of up to 10 years in service, durability of bamboo reinforced concrete subjected to a mild environmental exposure is usually not a significant problem.

Many organic fibres, including wastes, are used as concrete reinforcement. These can be broadly classified as natural, which can be either of vegetable or animal origin, or synthetic [62]. Most natural or even synthetic fibres of organic origin have relatively low Modulus of Elasticity. The most common use for these fibres is to restrain plastic shrinkage cracking of fresh concrete. In hardened, structural Portland cement concrete the strains in these fibres are usually too big to effectively control drying shrinkage, thermal and creep induced cracking [63]. There are, of course, very strong synthetic fibres, which are manufactured specifically for the purpose of reinforcing brittle materials, such as concrete, but these are certainly not wastes. There are, still, some natural fibres of reasonably high Elastic Modulus, such as coir (4 to 6 GPa), jute (17 to 18 GPa), and sisal (35 to 60 GPa) [62], [64], [65]. However, in many developing countries, and particularly in the low-cost and temporary housing projects, bagasse fibre [66], coconut husk fibre [67], as well as many other natural organic fibres [62] have been successfully used for these special applications.

Some of the industrial wastes, such as relatively short steel wire off-cuts, provided they are of reasonably uniform size, can be used in structural Portland cement concrete, in place of the specially manufactured steel fibre reinforcement [68]. It is most important, that the off-cuts are free of impurities harmful to concrete. Thus, they have to be properly cleaned of oil and grease, usually with the caustic soda. In one of my unpublished R&D projects, greasy steel off-cuts actually caused reduction in the tensile strength and increase in the creep of concrete, reinforced with these fibres.

It is known, that special admixtures, which help to retain moisture inside the fresh Portland cement concrete, and then to slowly release it, assisting the long term process of cement hydration in hardened concrete, can improve its strength and durability.

Some waste materials, such as water saturated porous slag aggregate, can also assist with the self-curing properties of concrete. There are several student projects at the University of Wollongong, in which water saturated fibres from fly ash filter bags are used in concrete, to assess their effectiveness as the self-curing agents.

Wastes as Organic Additives and Binders

Some organic waste materials can be used as binder-extenders in bituminous concrete [69], or as binders in the paving materials in their own right [70]. Wood lignins were successfully used as extenders of bitumen, and lignins as well as waste oils as binders for the wearing courses in the low cost and temporary (military) pavements. Owen dry powder-like lignin is mixed into bitumen at about 135°C. At lignin additions of some 30% by weight of bitumen, the hardness, softening point and viscosity of the binder is increased. The blend has higher resistance to the fatigue failure, with the resistance to moisture and freeze-thaw damage virtually unchanged [69]. In yet another project a blend of epoxy resins, coal tar and phenol was used as a binder for special application concrete [71].

In Canada, among several other countries, there were periods of considerable over supply of sulphur. Although not strictly speaking a waste in accordance with the adopted definition, sulphur, during the low-price periods, requires some new methods of utilisation. This lead to the development and reasonably wide spread use of sulphur concrete [72], [73]. The sulphur concrete, very much like polymer impregnated concrete [74], is a specialised product. Its compressive and flexural strengths are of the order of 100 and 10 MPa respectively. The process technology is relatively simple and can be commercially viable, if the cost of sulphur is sufficiently low. Sulphur concretes are used when good chemical resistance, high impermeability, and good mechanical properties are required [73].

WASTES AND CONCRETE - POTENTIAL AND REAL ENVIRONMENTAL BENEFITS

Utilisation of Concrete for the Effective Waste Disposal

Concrete, under the broad definition of this term, can be an excellent material for the encapsulation of hazardous wastes. Fundamentally, two types of encapsulation are possible. One is a *condensed waste storage* for those wastes, which can be later re-processed and recycled, and the second is *dispersed waste disposal system*, in which low concentration of hazardous waste does not produce harmful effects in the encapsulating concrete, which therefore can be used as building and construction material [74]. The composition of Portland cement concrete can be optimised, using mineral and chemical admixtures, which in themselves can be waste products, for the encapsulation of each particular hazardous waste [1], [46], [74], [75].

However, the best results in most cases can be obtained with concretes based on hydraulic cements other than Portland, or on the organic cements and, sometimes with concretes based on the hybrid of two of the above types [1], [8], [74].

Slag-alkaline cements [50], [76], [77], geopolymers [55], [56], [78] and silica-water-suspension binders can be particularly effective for the long term immobilisation of hazardous wastes in concrete.

Reduction of Environmental Pollution and Abatement of Greenhouse Gas Emissions

Most of the existing waste management systems attempt to either eliminate processes which generate waste, or, as often is the case, if this is not possible, to find some commercial use for the waste of given physical and chemical characteristics and composition. There is of course a third alternative, which can often prove to be the most economically viable solution to the process of waste management. It is usually possible to modify the process which generates waste, without any detriment to its efficiency and possibly even enhancing this process, in such a way that the properties of waste will change, converting waste into a value added product. One example of this philosophy is the concept of *Total Fly Ash Management* [40]. It is theoretically and economically possible to select or modify the composition of pulverised coal in such a way, that the fly ash produced in power generating plants would have desirable characteristics for particular commercial applications. These properties would include an improvement in the pozzolanic activity of fly ash, intended for use in concrete, or provide a desirable carbon content for aggregate manufacturing, or a dramatic increase in the content of cenospheres in fly ash. Elimination of ash disposal dams and landfill sites, which are the potential cause for the contamination of land and water with chemical leachates, should improve the environment. Reduction in energy of manufacturing hydraulic cements and concretes, which are made from or contain fly ash, does assist the abatement of Greenhouse Gas emissions. *Total Fly Ash Management*, of course, is only one of many concepts, which can be just as economically viable.

The Greenhouse Gases which were included in the Kyoto agreement are - carbon dioxide, methane, nitrous oxide, hydrofluorocarbons, perfluorocarbons, and sulphur hexafluoride. Unlike the main gases that make up 99.97 per cent of the air - nitrogen, oxygen and argon, which are either diatomic or monatomic, Greenhouse Gases have three or more atoms. These gases can absorb infra-red radiation because their molecules do naturally vibrate at frequencies the same as the infra-red. It should be remembered that water vapour is also a Greenhouse Gas, which at present we have no means of controlling. The need to comply with the Kyoto Protocol would impose significant pressure on the cement and concrete industries world wide to reduce Greenhouse Gas emissions. Use of wastes as raw materials for the traditional and new cements and also as fuel-extenders should provide these industries with the sufficient means to meet the adopted targets.

CONCLUSIONS

Under the "business as usual" scenario, the International Energy Agency (IEA) predicts the world Primary Energy Demand to increase from approximately 8,000 million tonnes of oil equivalent (MTOE) in 1995 to over 13,000 MTOE by the year 2020 [79]. In 1997-1998 oil and natural gas accounted for just over 50% and coal supplied 22% of the energy used worldwide. About 17% of the world's electricity energy was produced by 441 nuclear plants, which of course do not generate Greenhouse Gas from the fuel combustion.

At the same time the United Nations Population Division assessed that the Earth's population would grow from the current estimate of nearly 6 billion people to 8 billion people in 2025, and to around 9.4 billion by the year 2050 [80]. This represents a considerable increase in the potential land, water and air pollution as well as the Greenhouse Gas emissions. Notwithstanding these trends, under the Kyoto agreement United States must reduce Greenhouse Gas emissions by 7%, and the European countries agreed to 8% reduction from 1990 levels by the year 2012. Japan agreed to a 6% cut, and Australia was given a special dispensation to actually increase the 1990 levels by 8%. Developing, and less developed countries were not included in the agreement. The impact of Germany's decision to close all Nuclear Plants, on the ability to comply with the Kyoto Protocol, is difficult to evaluate.

To achieve these objectives, it is essential to implement the concept of sustainable development. Sustainability is the ability to *maintain a desired condition over time*. It is recognised, however, that the concept of *a desired condition* can vary significantly, and that it is not only a function of ethnic, historical and economical characteristics for a common group of people, but also of the existing at the time environmental, political and possibly also cataclysmic, etc., factors. *Sustainable Development is a tool for achieving sustainability, and not a desired goal.* It should help humans to achieve and maintain healthy and productive life style, by integrating environmental management into all forms of the development processes. Utilisation of wastes in the cement and concrete industries is, of course, an essential part of such development [81].

REFERENCES

1. SAMARIN, A. Encapsulation of Solid Wastes from Industrial By-products, Keynote address, Intern. Conf. on Environmental Management, Geo-Water and Engineering Aspects, Wollongong, Australia, Eds. Chowdhury and Sivakumar, A A Balkema, Roterdam, 1993, pp 63-78.

2. HANSEN, T C. Recycling of Demolition Concrete and Masonry, RILEM Report No.6, (Technical Committee 37-DR6), E&FN Spon, 1992.

3. SAMARIN, A. Hydraulic Cements - New Types and Raw Materials and Radically New Manufacturing Methods, Leader Paper, Intern. Conf. Concrete in the Service of Mankind, Appropriate Concrete Technology, Dundee, Eds. Dhir and McCarthy, E&FN Spon, 1996, pp 265-279.

4. TERREL, R L, *et al.* Recycling of Existing Bituminous Materials for Rehabilitation of Pavements in Developing Countries, Intern. Conf. Materials of Construction in Developing Countries, Bangkok, Thailand, Eds. Pama, *et al,* 1978, Vol II, pp 567-581.

5. SCHUHBAUER, A. On the Properties and Application of Semi-Rigid Pavement Material, Residential Workshop, Materials and Methods for Low Cost Road, Rail, and Reclamation Works, Leura, Australia, Eds. Lee, *et al* W H Sellen, 1976, pp 405-437.

6. WILK, W, AND TSOHOS, G. Recycling of Concrete Pavements, Intern. Conf. Concrete 2000, Economic and Durable Construction Through Excellence, Dundee, Eds. Dhir and Jones, E&FN Spon, 1993, pp 1201-1209.

7. SAEKI, N, AND SHIMURA, K. Recycled Concrete Aggregate as a Road Base Material in Cold Regions, Intern. Conf. Concrete in the Service of Mankind, Concrete for Environmental Enhancement and Protection, Dundee, Eds. Dhir and Dyer, E&FN Spon, 1996, pp 151-156.

8. SAMARIN, A. Advanced Concrete Technologies - Opportunities for the Next Decade, Concrete Inst. of Aust. 15-th Biennial Conference, Concrete 91, Sydney, pp 103-111.

9. FRONDISTOU-YANNAS, S A. Recycled Aggregate as New Aggregate, Progress in Concrete Technology, CANMET, Ed. Malhotra, 1980, pp 639-684.

10. ELAZHARI, S A, *et al.* Recycling of Concrete at an Early Age, Intern. Conf. Concrete 2000, Economic and Durable Construction Through Excellence, Dundee, Eds. Dhir and Jones, E&FN Spon, 1993, pp 401-410.

11. SPEARE, P R S, AND BEN-OTHMAN, B. Recycling Concrete Coarse Aggregates and Their Influence on Durability, Intern. Conf. Concrete 2000, Economic and Durable Construction Through Excellence, Dundee, Eds. Dhir and Jones, E&FN Spon, 1993, pp 419-432.

12. HENDRIKS, Ch, F. Recycling and Reuse as a Basis for Sustainable Development in Construction Industry, Intern. Conf. Concrete on the Service of Mankind, Concrete for Environmental Enhancement and Protection, Dundee, Eds. Dhir and Dyer, E&FN Spon, 1996, pp 43-54.

13. COLLINS, R J. Increasing the Use of Recycled Aggregate in Construction, Intern. Conf. Concrete in the Service of Mankind, Concrete for Environmental Enhancement and Protection, Dundee, Eds. Dhir and Dyer, E&FN Spon, 1996, pp 73-80.

14. AJDUKIEWICZ, A B AND KLISZCZEWICZ, A T. Properties of Concrete with Rubble Aggregate from Demolition of RC/PC Structures, Intern. Conf. Concrete in the Service of Mankind, Concrete for Environmental Enhancement and Protection, Dundee, Eds. Dhir and Dyer, E&FN Spon, 1996, pp 115-120.

15. de VRIES, P. Concrete Re-Cycled: Crushed Concrete as Aggregate, Intern. Conf. Concrete in the Service of Mankind, Concrete for Environmental Enhancement and Protection, Dundee, Eds. Dhir and Dyer, E&FN Spon, 1996, pp 121-130.

16. di NIRO, G, *et al.* Recycled Aggregate Concrete (RAC): Properties of Aggregate and RC Beams Made from RAC, Intern. Conf. Concrete in the Service of Mankind, Concrete for Environmental Enhancement and Protection, Dundee, Eds. Dhir and Dyer, E&FN Spon, 1996, pp141-149.

17. UCHIKAWA, H AND HANEHARA, S. Recycled Concrete Waste, Intern. Conf. Concrete in the Service of Mankind, Concrete for Environment Enhancement and Protection, Dundee, Eds. Dhir and Dyer, 1996, pp 163-172.

18. OZKUL, M H. Properties of Slag Aggregate Concrete, Intern. Conf. Concrete in the Service of Mankind, Concrete for Environmental Enhancement and Protection, Dundee, Eds. Dhir and Dyer, 1996, pp 543-552.

19. MATHER, B. Mineral Aggregate for Concrete - Research Needed, Progress in Concrete Technology, CANMET, Ed. Malhotra, 1980, pp 57-110.

20. MONTGOMERY, D G AND NANG, G. Strength Properties of Concrete Containing Instant-Chilled Steel Slag Aggregate, Intern. Conf. Concrete for the Nineties, The Use of Fly Ash, Slag, Silica Fume and Other Siliceous Materials in Concrete, Leura, Australia, Eds. Butler and Hinczak, 1990, pp Mont. 1 to 14.

21. SUSTERSIC, J, et al. High Performance Concrete Using High Strength Artificial Aggregate, Intern. Conf. Concrete in the Service of Mankind, Concrete for Environment Enhancement and Protection, Dundee, Eds. Dhir and Dyer, E&FN Spon, 1996, pp569-578.

22. PINTO, T de P AND AGOPYAN, V. Recycling Construction Wastes as Raw Material for Low-Cost Construction in Brazil, Intern. Conf. Concrete in the Service of Mankind, Concrete for Environmental Enhancement and Protection, Dundee, Eds. Dhir and Dyer, E&FN Spon, 1996, pp 109-144.

23. De PAUW, C, et al. Reuse of Construction and Demolition Waste as Aggregate in Concrete, Technical and Environmental Aspects, Intern. Conf. Concrete in the Service of Mankind, Concrete for Environmental Enhancement and Protection, Dundee, Eds. Dhir and Dyer, E&FN Spon, 1996, pp 131-140.

24. TREVORROW, A. The Recycling of Construction Waste - a Planning Nightmare? Intern. Conf. Concrete in the Service of Mankind, Concrete for Environmental Enhancement and Protection, Dundee, Eds. Dhir and Dyer, E&FN Spon, 1996, pp 157-162.

25. BROWN, B V. Alternative and Marginal Aggregate Sources, Intern. Conf. Concrete in the Service of Mankind, Conctrete for Environmental Enhancement and Protection, Dundee, Eds. Dhir and Dyer, E&FN Spon, 1996, pp 471-484.

26. KIBRIYA, T AND SPEARE, P R S. The Use of Crushed Brick Coarse Aggregate in Concrete, Intern. Conf. Concrete in the Service of Mankind, Concrete for Environmental Enhancement and Protection, Dundee, Eds. Dhir and Dyer, E&FN Spon, 1996, pp 495-503.

27. MANSUR, M A. Crushed Bricks as Coarse Aggregate for Concrete, Intern. Conf. Concrete in the Service of Mankind, Concrete for Environmental Enhancement and Protection, Dundee, Eds. Dhir and Dyer, E&FN Spon, 1996, pp 505-514.

28. SAMARIN, A. Use of Fine Crushed Bottle Glass Sand and Partial Cement Replacement in Concrete, Intern. Conf. Materials of Construction for Developing Countries, Bangkok, Thailand, Eds. Pama *et al*, 1978, Vol I, pp 369-92.

29. RITCHIE, A G B AND TINGARI, M B. The Use of Waste Polystyrene Chips as a Lightweight Aggregate for Concrete, First Australian Conference on Engineering Materials, Uni. of New South Wales, Eds. Morgan and Welsh, 1974, pp 157-175.

30. SRI RAVINDRARAJAH, R AND TUCK, A J. Lightweight Concrete with Expanded Polystyrene Beads, Uni. Of Technology, Sydney, Civil Eng. Monograph No. C.E. 93/1 M.E., 1993.

31. AZIZ, M A, *et al.* Lightweight Concrete Using Cork Granules, Intern. Conf. Materials of Construction for Developing Countries, Bangkok, Thailand, Eds. Pama *et al*, 1978, Vol I, pp 181-189.

32. PARAMASIVAM, P AND LOKE, YO. Study of Sawdust Concrete, Intern. Conf. Materials of Construction for Developing Countries, Bangkok, Thailand, Eds. Pama *et al*, 1978, Vol I, pp 169-179.

33. ROSTAMI, H, *et al.* Use of Recycled Rubber Tyers in Concrete, Intern. Conf. Concrete 2000, Economic and Durable Construction Through Excellence, Dundee, Eds. Dhir and Jones, E&FN Spon, 1993, pp 391-399.

34. CONSUL, A M D AND MARTINEZ, HG. Alternative Building materials and Components for Concrete Construction in Cantiago de Cuba, Intern. Conf. Concrete in the Service of Mankind, Concrete for Environmental Enhancement and Protection, Dundee, Eds. Dhir and Dyer, E&FN Spon, 1996, pp 599-608.

35. HARRISON, AM, *et al.* The Use of Alternative Aggregates in Concrete, Intern. Conf. Concrete in the Service of Mankind, Concrete for Environmental Enhancement and Protection, Dundee, Eds. Dhir and Dyer, E&FN Spon, 1996, pp 543-552.

36. PIASECKI, J, *et al.* Concrete Waste Materials, Intern. Conf. Concrete 2000, Economic and Durable Construction Through Excellence, Dundee, Eds. Dhir and Jones, E&FN Spon, 1993, pp411-418.

37. LIAUTAUD, G. State of the Art Review of Practice and Experience on Low Cost Roads in Tropical Areas, Residential Workshop, Materials and Methods for Low Cost Road, Rail, and Reclamation Works, Leura, Australia, Eds. Lee *et al,* W.H. Sellen, 1976, pp 439-537.

38. BIJEN, J. Waste Materials and Alternative Products: Pro's and Con's, Intern. Conf. Concrete in the Service of Mankind, Concrete for Environment Enhancement and Protection, Dundee, Eds. Dhir and Dyer, E&FN Spon, 1996, pp 587-598.

39. MADEJ, J, *et al.* Investigation of Individual By-Products Considered for Use as Concrete Aggregate, Intern. Conf. Concrete in the Service of Mankind, Concrete for Environmental Enhancement and Protection, Dundee, Eds. Dhir and Dyer, E&FN Spon, 1996, pp 99-108.

40. SAMARIN, A. Total Fly Ash Management: from Concept to Commercial Reality, The Australian Coal Review, Issue 4, November, 1997 pp 34-37.

41. BIJEN, J M J M. Manufacturing Processes of Artificial Lightweight Aggregate from Fly Ash, The International Journal of Cement Composites and Lightweight Concrete, Vol 8, No.3, August 1986, pp 191-199

42. HARLEY, P. Neutralysis Technology Goes OS, Australian Environment Review, Vol 8, No.8, August 1993, p 3.

43. SAMARIN, A. Effect of Type, Compatibility and Proportions of Mix Constituents on the Durability of Reinforced Concrete, Japan-Australia Workshop on Concrete Durability, CSIRO, Highett, Victoria, Australia, November 28-30, 1988.

44. SAMARIN, A. Effect of Type, Compatibility and Proportions of Mix Constituents on the Corrosion of Reinforcement in Concrete, 2-nd International Seminar, Durability of Concrete - Aspects of Admixtures and Industrial By-Products, Gothenburg, Sweden, Chalmers Univ., Document D9, June 1989, pp 179-204.

45. CEMBUREAU, European Annual Review of Cement Industry and Market Data, No. 18, Brussels, Belgium, July, 1997.

46. SAMARIN, A. Encapsulation of Hazardous Heavy Metal Wastes in High Performance Concrete, Australian Academy of Technological Sciences and Engineering Journal Focus, No. 87, May - June 1995, pp 11-14.

47. AUSTRALIAN ACADEMY OF TECHNOLOGICAL SCIENCES AND ENGINEERING, Urban Air Pollution in Australia, Melbourne, Australia, 1997, Chapter 4, Large Stationary Sources, pp 157- 168.

48. GLUCHOWSKI, W D, *et al.* Slag-alkaline Cements and Concrete: Structures, properties, Technological and Economical Aspects of Use, Silicate Industries, No. 48, 1983, pp 197-200.

49. KRIVENKO, P V, *et al.* Slag-alkaline Polymer Cement Concretes, Intern. Conf. Concrete in the Service of Mankind, Appropriate Concrete Technology, Dundee, Eds. Dhir and McCarthy, E&FN Spon, 1996, pp 309-314.

50. RUNOVA, R F. Binders from Dispersion of Calcium Hydrosilicates, Intern. Conf., Concrete 2000, Economic and Durable Construction Through Excellence, Dundee, Eds. Dhir and Jones, E&FN Spon, 1993, Vol. I, pp 533-540

51. KIRILISHIN, V P. Constructive Chemically Resistant Concrete, Intern. Conf. Concrete 2000, Economic and Durable Construction Through Excellence, Dundee, Eds. Dhir and Jones, E&FN Spon, 1993, Vol II, pp 1013-1022.

52. SINGH, M AND VERMA, C L. Properties and Applications of New Cementitious Materials, Intern. Conf., Concrete 2000, Economic and Durable Construction Through Excellence, Dundee, Eds. Dhir and Jones, E&FN Spon, 1993, Vol II, pp 1313-1324.

53. DAVIDOVITS, J. Mineral Polymers and Methods of Making Them, U. S. Patent No. 434 9386, 1994.

54. DAVIDOVITS, J, et al. Geopolymeric Concretes for Environment Protection, Concrete International, Design and Construction, July 1990, Vol. 12, No. 7, pp 30-40.

55. DAVIDOVITS, J. Properties of Geopolymer Cements, Geopolymer Institute, Saint-Quentin, France, 1994.

56. van JAARSVELD, J G S AND van DEVENTER, J S J. Geopolymerisation: a Novel Technique for Treating Mining and Other Industrial Waste Materials, 2-nd Intern. Conf. On Environmental Management, Univ. of Wollongong, Australia, Eds. Sivakumar and Chowdhury, Elsevier, 1998, Vol II., pp 847-854.

57. DAVIDOVITS, J. Global Warming Impact on the Cement and Aggregates Industries, World Resources Review, Vol 6, No.2, 1993, pp 263-278.

58. MITIAKIN, P L AND ROSENTAL, O M. Refractory Materials Based on Silica-Water-Suspension Binders, Academy of Sciences of the USSR, Nauka Publishers, Novosibirsk, 1987, (in the Russian language).

59. DATYE, K R, et al. Engineering Application of Bamboo, Intern, Conf. Materials of Construction in Developing Countries, Bangkok, Thailand, Eds. Pama et al, 1978, Vol. I., pp 3-20.

60. GHAVAMI, K, et al. Structural Behaviour of Bamboo Reinforced Concrete Beams and Slabs, Intern. Conf. Concrete in the Service of Mankind, Appropriate Concrete Technology, Dundee, Eds. Dhir and McCarthy, E&FN Spon, 1996, pp 473-482.

61. FANG, H Y AND FAY, S M. Mechanism of Bamboo - Water - Cement Interaction, Intern. Conf. Materials of Construction in Developing Countries, Bangkok, Thailand, Eds. Pama et al, 1978, Vol. I., pp 37-48.

62. SARJA, A E. Tailoring the Properties of Concrete Structures with Appropriate Non-Ferrous Reinforcement, Intern. Conf. Concrete in the Service of Mankind, Appropriate Concrete Technology, Dundee, Eds. Dhir and McCarthy, E&FN Spon, 1996, pp 376-390.

63. SAMARIN, A. Concrete Shrinkage - Causes and Effects to be Considered in a Structural Design, Concrete in Australia, Conc. Inst. of Aust., Vol. 22, No. 1., April 1996, pp 18-23.

64. PAPAYANNI, I, *et al.* Low Cost Fibre Reinforced Cement Products by Using Inexpensive Additives, Intern. Conf. Concrete in the Service of Mankind, Appropriate Concrete Technology, Dundee, Eds. Dhir and McCarthy, E&FN Spon, 1996, pp 431-441.

65. JORILLO, P Jr AND SHIMIZU, G. Durability of Coir Fibre-Reinforced Concrete, Fourth CNMET/ACI Intern. Conf. on Durability of Concrete, Sydney, Australia, Rd. Malhotra, Supplementary Papers, 1997, pp 357-375.

66. RACINES, P G AND PAMA, P R. A Study of Bagasse Fibre-Cement Composite as Low Cost Construction Material, Intern. Conf. Materials of Construction for Developing Countries, Bangkok, Thailand, Eds. Pama *et al,* 1978, Vol. I., pp 191-206.

67. SINGH, S M. Coconut Husk - a Versatile Building Material, Intern. Conf. Materials of Construction for Developing Countries, Bangkok, Thailand, Eds. Pama *et al,* 1978, Vol. I., pp 207-219.

68. WALKUS, B R. Concrete Composites with Steel Micro-Reinforcement Subjected to Uniaxial Tension, Intern. Conf. Materials of Construction for Developing Countries, Bangkok, Thailand, Eds. Pama *et al,* 1978, Vol. I., pp 357-368.

69. TERREL, R L AND RIMSRITONG, S. Wood Lignins Used as Extenders for Asphalt in Bituminous Pavements, Intern. Conf. Materials of Construction for Developing Countries, Bangkok, Thailand, Eds. Pama *et al,* 1978, Vol. II., pp 583-596.

70. SAMARIN, A. Discussion on Paper by McDonald: Temporary and Engineering Roads and Pavements, Residential Workshop, Materials and Methods for Low Cost Road, Rail, and Reclamation Works, Leura, Australia, Eds. Lee *et al,* W.H. Shellen, 1976, pp 811-812.

71. KAWAKAMI, M, *et al.* Some Physical Properties of Resin Concrete as Construction Material, Intern. Conf. Materials of Construction for Developing Countries, Bangkok, Thailand, Eds. Pama *et al,* 1978, Vol. I., pp 275-283.

72. LOOV, R E, *et al.* The Use of Sulphur as a Binder for Soils and Waste Materials, Intern. Conf. Materials of Construction for Developing Countries, Bangkok, Thailand, Eds. Pama *et al,* 1978, Vol. I., pp 285-297.

73. MALHOTRA, V M. Sulphur Concrete and Sulphur Infiltrated Concrete - Properties, Applications and Limitations, Progress in Concrete Technology, CANMET, Ottawa, Canada, 1980, pp 583-637.

74. McLAREN, K G AND SAMARIN, A. Concrete - Polymer Composite Materials Produced by Radiation Polymerisation Techniques, First Aust. Conf. on Engineering Materials, Eds. Morgan and Welch, Univ. of New South Wales, Sydney, Australia, 1974, pp 137-155.

75. van EIJK, R J AND BROUWERS, H J H. Optimising the Portland Cement Matrix in Regard to Leaching Resistance, Second Intern. Conf. on Environmental Management, Univ. of Wollongong, Australia, Eds. Sivakumar and Chowdhury, Elsevier, 1998, Vol. II, pp 839-846.

76. DEJA, J AND MALOLEPSZY, J. Resistance of Alkali-Activated Slag Mortars to Chloride Solution, Third Intern. Conf. Fly Ash, Silica Fume, Slag, and Natural Pozzolans in Concrete, Trondheim, Norway, Ed. Malhotra, ACI SP-114, 1989, pp 1547-1553.

77. SAMARIN, A. Theory and Practice of Durable Concretes for Encapsulation of Hazardous Wastes, Invited Paper, Fourth CANMET/ACI Intern. Conf. on Durability of Concrete, Ed. Malhotra, Sydeny, Australia, Supplementary Papers, pp 833-855.

78. DAVIDOVITS, J. Geopolymers: Inorganic Polymeric New Materials, Journal of Thermal Analysis, No. 37, 1991, pp 1633-1656.

79. MACKEY, T. Energy: the Next Fifty Years, Australian Energy News, Issue 9, September 1998, pp 31.

80. ENCYCLOPAEDIA BRITANNICA, Yearbook of Science and Future, 1999, pp 308-311.

81. SAMARIN, A. The Anthropogenic Paradox - is Sustainable Future Possible? Keynote Paper, Second Intern. Conf. Environmental Management, University of Wollongong, Australia, Eds. Sivakumar and Chowdhury, Elsevier, 1998, pp 33-46.

THEME ONE:
MAXIMISING USE

Keynote Paper

USE OF INDUSTRIAL BY-PRODUCTS IN CEMENT-BASED MATERIALS

T R Naik

R N Kraus

University of Wisconsin-Milwaukee

United States of America

ABSTRACT. Nearly 4.2 billion tonnes of non-hazardous by-products are generated from agricultural, domestic, industrial, and mineral sources. Large amounts of wastes generated from industrial and domestic sources are currently landfilled due to non-availability of economically attractive use options. Landfilling is undesirable because it causes not only huge financial burdens to producers of by-products, but also makes them responsible for unknown future environmental liabilities. Additionally, due to shrinking landfill space and increased environmental restrictions, cost of landfilling is escalating. To address these problems it has become essential to find cost-effective solutions to waste disposal problems. Recycling not only saves on huge disposal costs, but also conserves natural resources, and in some cases provides technical and economic benefits. This paper describes various by-product materials generated from industrial operations as well as recycling of these materials. The by-product materials include coal combustion by-products, and wood ash. For each by-product material, production, properties, and potential applications in manufacture of emerging materials are briefly addressed. Some applications include use in Controlled Low Strength Materials, concrete and cast concrete products, blended cement, lightweight aggregates, and in metal composites.

Keywords: Bottom ash, Coal combustion by-products, Concrete, Controlled low strength materials, Fly ash, Recycling, Wood ash

Dr Tarun R Naik is Director of the UWM Center for By-Products Utilization, and Associate Professor of Civil Engineering, at the University of Wisconsin-Milwaukee. Dr. Naik's contribution in teaching and research has been well recognized nationally and internationally. He has taught courses, given seminars, held workshops, and/or presented invited lectures worldwide

Rudolph N Kraus is Assistant director of the UWM Center for By-Products Utilization at the University of Wisconsin-Milwaukee.

INTRODUCTION

Nearly 4.2 billion tonnes of non-hazardous by-products are generated from agricultural, domestic, industrial, and mineral sources. The amounts of by-product materials generated are 2,100 million tonnes from agricultural sources, 200 million tonnes from domestic sources, 400 million tonnes from industrial sources, and 1,800 million tonnes from mineral sources [5]. Legislation has been passed by federal lawmakers to encourage beneficial use of by-product materials generated from various sources. The Federal Resource Recovery Act of 1970 was the first law which encouraged recycling, resource recovery, and energy recovery of by-product materials. These laws were later replaced by the Federal Resource Conservation and Recovery Act of 1976 (RCRA). This law requires selection of appropriate disposal of solid residue in order to avoid any injury to human health as well as the environment. Further amendments to the RCRA were made in 1980 and 1984, with a greater emphasis on suitable disposal of waste materials and resource recovery. The law further encourages recycling of waste materials. Large amounts of wastes generated from industrial and domestic sources are currently landfilled due to non-availability of economically attractive use options. Landfilling is undesirable because it causes not only huge financial burdens to producers of by-products, but also makes them responsible for unknown future environmental liabilities. Additionally, due to shrinking landfill space and increased environmental restrictions, cost of landfilling is escalating. To address these problems it has become essential to find cost-effective solutions to waste disposal problems. Recycling of by-product materials generated from various sources must provide innovative solutions to the above problems.

This paper briefly describes various by-product materials generated from industrial operations and post-consumer wastes, as well as recycling of these materials. The by-product materials include coal combustion by-products, and wood ash. For each by-product material, production, properties, and potential applications in manufacture of emerging materials and the environmental impact are briefly addressed. Additionally, future recycling and research needs are discussed.

COAL-COMBUSTION BY-PRODUCTS

Coal-fired power plants derive energy by burning coal in their furnaces. These power plants generally use either pulverized coal-fired furnaces or cyclone furnaces [15]. The cyclone furnaces burn relatively coarse coal particles, less than 13 mm, at very high temperature. The pulverized coal-fired furnaces use fine coal particles with particle size passing No. 200 sieve. During the process of combustion in pulverized coal-fired furnaces, the volatile matters and carbon burn off and the coal impurities fuse and remain in suspension. These fused substances solidify when flue gas reaches low temperature zones to form predominantly spherical particles called fly ash. The remaining matter which agglomerate and settle down at the bottom of the furnace are called bottom ash. The pulverized coal-fired furnaces employ either a dry bottom or wet bottom to collect bottom ash. Amount of bottom ash can range from 20 to 25% of total coal combustion by-products for dry bottom collection system. Fly ash constitutes a major component (75 - 80%) of by-product material in pulverized coal-fired power plants. The combustion of coal in cyclone furnaces occurs by continuous swirling in a high intensity heat zone [15]. This causes fusing of fly ash particles into a glassy slag, called boiler slag, which drop to the bottom of the furnace.

The boiler slag constitutes the major component of the cyclone boiler by-product (70 to 85%). The remaining combustion by-products exit along with the flue gases. Clean coal ash is defined as the ash derived from plants involving the use of SO_x and NO_x control technologies.

The production of fly ash was estimated at about 50 million tonnes in 1995, according to ACAA (1996). Fly ash is a heterogeneous mixture of particles varying in shape, size, and chemical composition. The particle types may include carbon from unburnt coal, fire-polished sand, thin-walled hollow spheres and fragments from their fracture, magnetic iron containing spherical particles, glassy particles, etc. Fly ash is predominantly composed of spherical particles which can be less than 1 μm to more than 1 mm in size. The nitrogen adsorption surface area of fly ash varies in the range of 300 to 500 m^2/kg. The density of fly ash normally varies between 1.6 and 2.8 g/cm^3. Major mineralogical component of fly ash is a silica-aluminate glass containing Fe_2O_3, CaO, and MgO. It also contains certain other oxide minerals. According to ASTM C 618, Class F fly ashes contain less than 10% total CaO, whereas Class C fly ashes normally show total CaO content greater than 10% according to ASTM C 618 [3]. Class C fly ashes can even show cementitious behavior in the presence of water. The properties of coal ashes are presented in Table 1. The particle size distribution of fly ash and bottom ash is presented in Figure 1.

According to ACAA [1], in 1996, the total amount of bottom ash and boiler slag produced was about 15 million tonnes and 2.3 million tonnes, respectively. Bottom ash and slag are generally non-spherical and are composed of particles ranging from 2 μm to 20 mm. Bottom ash particles are rounded in shape but can be also angular. They have porous structures. Boiler slag is composed of angular particles with a glassy appearance. The size distribution of bottom ash and boiler slag is also shown in Figure 1. Specific gravity for bottom ash and slag varies between 2.2 and 2.8. Their bulk densities ranges from 737 to 1586 kg/mt^3 [15].

Wet scrubbers or flue gas desulfurization (FGD) systems are most commonly used to control power plant SO_2 emissions and they produce wet by-products. Approximately 22 million tonnes of FGD materials were generated in 1996 [1]. The residue from such systems consists of a mixture of calcium sulfite and sulphate, $CaCO_3$, and fly ash in water. The fly ash amount in FGD material varies from less than 10% to 50% depending upon whether or not fly ash was collected prior to the FGD system. Particle size distribution for FGD sludge is shown in Figure 2. Recent increased concern over SO_2 emissions from power plants has resulted in development of several advanced SO_2 control systems that produce dry by-products. Therefore, these new processes avoid the complexity and operating problems encountered when handling large volumes of liquid or semi-liquid wastes produced in the case of wet FGD systems. In addition, no dewatering is needed prior to utilization or landfilling. However, these processes require costlier sorbent materials. The advanced systems include Atmospheric Fluidized Bed Combustion (AFBC), Lime Spray Drying, Sorbent Furnace Addition, Sodium Injection, and other clean coal technologies such as integrated coal classification combined cycle process (IGCC), etc. The solid waste products generated by these processes have some physical and chemical properties significantly different from those for conventional coal ashes.

Table 1 Typical chemical and physical properties of fly ash from different power plants [2]

FLY ASH SOURCE	CHEMICAL PROPERTIES, PERCENT								PHYSICAL PROPERTIES		
	LOI	CaO	SiO_2	Al_2O_3	Fe_2O_3	MgO	Na_2O	K_2O	No. 325 sieve retention, %	Blaine fineness m^2/kg	Specific gravity
Less than 10 percent CaO (Class F)											
FA-4	1.0	6.7	58.5	19.9	5.6	1.7	1.5	1.3	17	379	2.31
FA-5	0.9	0.7	60.1	27.8	3.8	1.0	0.3	2.8	18	262	2.18
FA-7	1.8	1.7	56.0	25.7	8.3	1.1	0.3	2.8	22	282	2.28
FA-8	2.6	2.4	49.0	21.8	17.9	1.0	0.4	2.7	20	282	2.45
FA-13	4.2	1.7	45.0	19.6	23.9	0.9	0.4	2.3	24	236	2.45
FA-14	3.0	1.9	47.7	29.5	9.7	0.7	0.3	1.9	28	287	2.30
FA-15	2.5	1.3	52.7	28.6	5.8	1.0	0.3	2.4	17	351	2.38
FA-16	4.0	1.6	50.6	27.6	8.2	1.0	0.4	2.5	4	508	2.49
FA-17	0.4	7.5	49.8	21.6	7.0	1.7	2.8	0.7	24	316	2.27
FA-18	4.3	2.2	43.6	26.0	16.6	0.9	0.3	1.9	17	337	2.24
No. 3	7.2	3.2	64.4	24.7	3.9	1.5	-	-	2	-	-
D-Precip	3.9	1.0	52.9	30.1	7.3	1.1	0.4	2.9	8	643	2.33
D-Mech	6.4	1.0	54.9	27.6	10.4	0.9	0.3	2.4	30	333	2.15
More than 10 percent CaO (Class C)											
FA-1	0.9	25.5	36.3	17.7	6.7	4.6	1.6	0.6	15	417	2.65
FA-2(a)	1.9	15.5	38.8	13.4	22.5	1.5	0.5	1.9	16	355	2.74
FA-9(a)	0.5	11.6	50.5	17.7	6.6	3.4	3.5	1.2	11	315	2.44
FA-10(a)	0.5	28.2	35.9	17.1	5.6	5.1	1.8	0.5	16	390	2.70
FA-11(a)	0.4	16.9	51.4	16.9	5.8	3.5	0.6	0.8	21	288	2.52
A	0.4	17.3	35.7	20.3	5.8	4.3	6.5	0.8	11	418	2.67
F	0.7	24.9	23.1	13.3	9.6	7.5	7.3	0.6	12	324	2.86
G	0.6	11.7	48.9	21.3	3.7	2.7	6.4	0.9	38	318	2.31
I	0.3	29.0	31.1	17.0	5.6	3.8	3.2	0.4	15	604	2.74

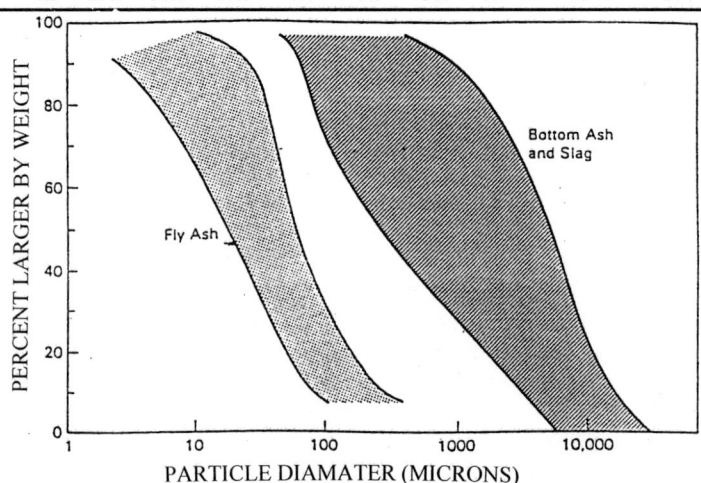

Figure 1 Particle size distribution for fly ash and bottom ash [32]

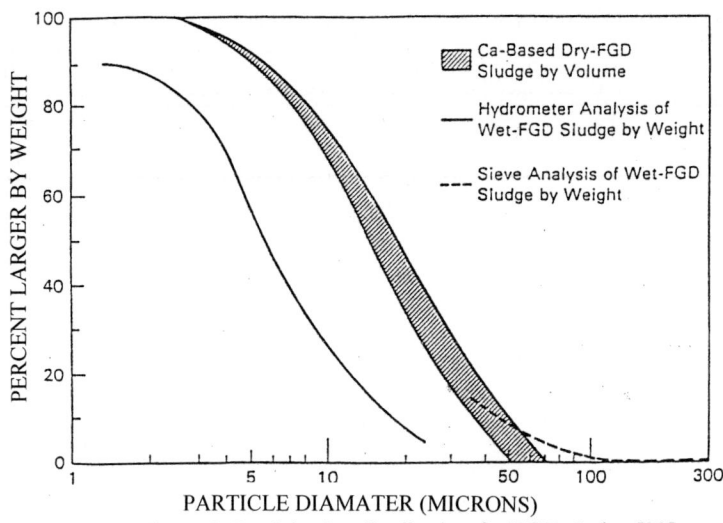
Figure 2 Particle size distribution for FGD sludge [32]

The AFBC process produces coal ash, sulfur reaction products, and calcined limestone reaction products. The sulfur reaction products are primarily composed of calcium sulfate and sulfite, and calcium oxide. The calcined limestone reaction forms primarily calcium sulfate. Chemical composition of the AFBC residues is given in Table 2. The chemical composition of the AFBC fly ash is similar to that of Class C fly ash except SO_3 and SiO_2 contents. AFBC SO_3 content is higher and SiO_2 content is lower relative to the conventional Class C fly ash.

The spray dryer by-products (Table 2) consist of primarily spherical fly ash particles coated with calcium sulfite/sulphate, fine crystals of calcium sulfite/sulphate, and unreacted sorbent composed of mainly $Ca(OH)_2$ and a minor fraction of calcium carbonate. The spray dryer by-products are higher in concentrations of calcium, sulfur, and hydroxide, and lower in concentrations of silicon, aluminum, iron, etc. compared to the conventional Class C fly ash.

The Lime Furnace Injection (LFI) by-products (Table 2) are made up of primarily coal ash, calcium sulfite and sulfate, and unreacted lime. By-products generated by LFI contain 40 to 70% fly ash, 15 to 30% free lime, and 10 to 35% calcium sulfate by weight.

The calcium injection process produces by-products (Table 2) similar to that of LFI and calcium spray dryer because of similarities in sorbents and injection methods used. The sodium injection process differs from the calcium injection in regards to type of sorbent used. This process uses a sodium-based sorbent such as sodium bicarbonate, soda ash, trona, or nahcalite [12]. By-products generated by this process include fly ash particles coated and intermixed with sodium sulfite/sulfate, and unreacted sorbent. The IGCC process produces by-products similar to the SO_2 control processes.

Table 2 Clean coal by-products chemical composition in percent by weight (a) [12]

SAMPLE NO.	Al_2O_3	CaO	Fe_2O_3	MgO	K_2O	SiO_2	Na_2O	SO_3
AFBC:								
TVO3 (bed)	2.72	45.07	4.77	0.62	0.31	3.17	0.27	6.50(b)
TVO4 (char)	7.29	30.79	13.20	0.48	0.78	7.97	0.05	20.00
TVO5 (ash)	15.04	22.64	18.88	0.51	1.93	15.26	0.34	17.25
SFO6 (comp.)	6.12	39.13	17.11	0.54	0.72	6.04	0.29	12.00
Spray Dryer:								
ARO7	25.20	21.73	3.26	0.84	1.69	21.17	3.29	17.50
STO7	12.60	31.22	10.92	2.93	1.45	15.60	1.76	12.00
LRO7	21.20	26.88	6.11	2.33	0.74	17.72	2.08	12.25
HSO5	24.90	20.02	6.51	2.62	0.75	21.30	1.81	10.25
APO7	24.90	17.67	3.11	0.65	1.35	25.72	2.05	18.25
NVO4	15.00	21.32	4.83	1.53	0.60	20.42	6.58	14.00
RSO5	19.00	28.50	15.34	2.85	0.42	15.96	2.12	13.75
AVO6	18.00	19.03	9.23	4.62	1.46	24.52	9.17	11.50
Lime Furnace Injection:								
SRO7 (lime)	16.40	28.83	14.20	2.50	2.84	17.72	1.77	12.50
SRO9 (limestone)	17.20	29.15	16.48	0.82	2.96	19.33	1.64	11.25
OLO3 (limestone)	17.80	36.13	13.17	0.63	1.11	15.75	0.48	6.25
OLO4 (limestone)	17.10	40.00	11.91	0.70	1.08	16.18	0.51	5.50
OLO8 (limestone)	29.80	16.80	16.86	0.67	2.12	27.86	1.02	3.50
Calcium Injection:								
AHO6	9.07	40.57	2.17	0.56	0.82	10.27	0.59	NA
AA1O-01	31.37	15.39	8.86	1.13	3.37	29.95	1.24	NA
AA1O-02	31.37	13.99	8.86	1.13	3.37	27.81	1.27	NA
Sodium Injection:								
NXO4	28.90	4.54	2.50	1.16	0.77	25.18	24.78	12.00
NBO4	30.50	4.40	6.60	0.70	1.45	33.94	12.89	7.75

(a) All elements expressed as their oxides, but may occur in other forms
(b) SO_3 content of the uncrushed sample; the crushed sample had a SO_3 content of 23.9%

From the above description, it is evident that most SO_2 control processes generate a by-product similar to the conventional fly ash. But due to sorbent addition, fly ash is modified to a significant extent. The modified fly ash contains fly ash particles coated with sorbent and sorbent reaction products, and smaller non-fly ash particles composed of reacted and unreacted sorbents. The solid waste products generated by these processes exhibit some physical and chemical properties significantly different than those for conventional coal ashes [12, 13].

APPLICATIONS OF COAL-COMBUSTION BY-PRODUCTS

The most widely accepted use of fly ash is in making concrete. However, in keeping with the primary emphasis of this paper, only emerging materials using fly ash is discussed.

Fly Ash

With a view to save a significant amount of energy and cost in cement manufacturing, fly ash can be utilized as a major component of blended cements, exceeding 50% of total blended cement mixture [22]. Fly ash can be used as either a raw material in the production of the cement clinker, interground with the clinker, or blended with the finished cement.

Fly ash can be used in manufacture of Controlled Low Strength Materials (CLSM) as a replacement of regular concrete sand up to 100% [16, 17, 20, 25]. Flowable slurry made with fly ash is suitable for base support and backfilling of foundations, bridge abutments, buildings, retaining walls, utility trenches, etc.; for filling abandoned tunnels, sewers, and other underground facilities; and as embankments, grouts, etc.

Both sintered (fired) and unfired (cold bonded) processing methods can be used to manufacture lightweight aggregates using fly ash [6, 10]. For manufacture of lightweight aggregate, first fly ash is pelletized. Thereafter, it is sintered in a rotary kiln, shaft kiln, or travelling gate at temperature 1000 to 1200°C.

Naik et al. [23] developed mixture proportions for paving roadway concrete using large amounts of fly ash. These mixtures were composed of 50% Class C fly ash and 40% Class F fly ash as a replacement of portland cement. Test results revealed that high volumes of Class C and Class F fly ashes can be used to produce high-quality concrete pavements with excellent performance.

Fly ash with and without silica fume can be used in manufacture of high-performance concrete [19]. High-performance concrete mixtures containing up to 30-40% fly ash can be proportioned to attain both high-strength and high-durability related properties. Past studies [22, 21] have substantiated that concrete containing large amounts (more than 50%) of either Class C or Class F fly ash can be proportioned to meet strength and durability requirements for structural applications.

Recent studies by Naik and Ramme [26] have substantiated that superplasticized Class C fly ash concrete with low water-to-cementitious materials ratio can be proportioned to meet the very early-age strength as well as other requirements for precast/prestressed concrete products. The maximum cement replacement with the fly ash was reported to be 30% for such high-early strength concrete application.

Fly ash can be used in large amounts as a fine filler material as well as a pozzolan in roller compacted concrete [31]. In manufacture of autoclaved cellular concrete, fly ash can be used as a replacement of 30 to 100% of silica sand [27]. Cenospheres derived from fly ash are an ideal filler material for manufacture of polymer matrix composites [11, 28].

The inclusion of fly ash improves mechanical properties, elastic modulus, permeability, and reduces thermal conductivity and expansion. In fresh concrete it reduces bleeding and heat of hydration. Fly ash has been used as a particulate filler in cast metal matrix composites [30]. Inclusion of the fly ash is reported to increase hardness, abrasion resistance, stiffness, and decrease density of aluminum matrix [29].

Bottom Ash/Boiler Slag

Extensive studies by Naik and his associates [24, 33] and elsewhere have revealed that bottom ash can be used as lightweight aggregates. Large size bottom ash can be used as coarse aggregate and small size bottom ash can be used as fine aggregate sand. Naik and his associates demonstrated feasibility of using bottom ash in manufacture of masonry products as a partial replacement of coarse as well as fine aggregates [33]. Bottom ash used in CLSM slurry can enhance insulating ability of the fill. The same is true for boiler slag. The most popular use of coal boiler slag is in architectural concrete as aggregates.

Clean Coal Ash

Relatively little work has been done concerning the utilization of clean coal ash. Stabilized FGD sludge can be used in construction of stabilized road base. It can be used as a raw material for production of cement. FGD can be used as gypsum for manufacture of wallboards. The advanced SO_2 by-products have a potential for use in structural fill, mineral filler in asphalt, synthetic aggregates, concrete, mineral wool, ceramic products, masonry products, etc. [12, 13]. Naik et al. [18] carried out an extensive laboratory investigation to characterize clean coal ash by-products in order to establish their applications in cement-based materials. Based on laboratory investigations, they reported that significant amounts of clean coal ash by-products can be used in concrete as well as masonry products. Naik et al. [17] have also established mixture proportions and production technologies for clean coal ash in CLSM as a replacement of sand and/or conventional fly ash.

WOOD ASH

Wood ash is the residue generated due to combustion of wood and wood products (chips, saw dust, bark, etc.). Wood ash is composed of both inorganic and organic compounds. Ash content yield decreases with increasing combustion temperature [7]. Density of wood ash decreases with increasing carbon content. Typically, wood ash contains carbon in the range of 5-30% [4]. The major elements of wood ash include calcium (7-33%), potassium (3-4%), magnesium (1-2%), manganese (0.3-1.3%), phosphorus (0.3-1.4%), and sodium (0.2-0.5%). The chemical and physical properties depend upon the type of wood, combustion temperature, etc. [4, 14]. An elemental metal, and other analyses for various types of wood are shown in Table 3.

Table 3 Elemental metals and other analyses for ash from wood (mg/kg)

ELEMENTAL METAL	REGULATORY LIMITS (U.S.EPA)	NORMAL WOOD FUEL	PARTICLE/ PLYWOOD	CREOSOTE-TREATED	PENTACH-LORO-PHENOL TREATED	CONSTRUCTION DEMOLITION WOOD	CCA-TREATED
Aluminum	N/A	4000 - 4500	4400 - 4800	3600 - 5000	3600 - 4200	4900 - 5800	3900 - 4500
Arsenic	41/75	42 - 53	22.5 - 26.9	51 - 64	24.3 - 27.7	78 - 98	8570 - 9390
Barium	N/A	220 - 300	280 - 400	200 - 280	220 - 270	480 - 590	220 - 280
Cadmium	39/85	5.5 - 6.1	7.3 - 7.9	5.1 - 5.7	8.7 - 10.1	7.1 - 8.1	10.7 - 11.7
Chromium	1200/3000	12 - 14	12 - 15	14 - 17	19 - 23	34 - 39	1710 - 1850
Copper	1500/4300	41 - 46	50 - 59	49 - 52	52 - 61	71 - 93	2610 - 2820
Iron	N/A	5900 - 6100	3700 - 4300	14900 - 17100	5000 - 5100	6900 - 7400	6400 - 6900
Lead	300/840	29 - 35	73 - 78	47 - 50	198 - 235	920 - 1010	58 - 73
Manganese	N/A	2440 - 2750	2430 - 2740	2040 - 2140	2020 - 2230	2030 - 2230	2610 - 2720
Mercury	17-57	0.05 - 0.08	0.06 - 0.10	0.12 - 0.14	0.09 - 0.16	0.36 - 0.52	0.03 - 0.32
Molybdenum	18/75	5.6 - 6.7	7.6 - 8.2	4.0 - 5.4	4.8 - 6.1	6.9 - 8.0	8.6 - 11.4
Nickel	420/420	6 - 8	6 - 7	8 - 10	9 - 9	7 - 10	6 - 8
Selenium	36/100	0.53 - 064	0.55 - 0.64	0.74 - 0.81	0.55 - 0.65	0.84 - 0.97	1.18 - 1.45
Silver	N/A	0.2 - 0.4	0.3 - 0.4	0.4 - 0.4	0.1 - 0.2	0.1 - 0.1	0.7 - 0.8
Zinc	2800/7500	380 - 420	530 - 610	450 - 510	540 - 590	1420 - 1520	520 - 620
pH	—	11.31 - 11.67	10.64 - 10.85	10.69 - 11.09	10.18 - 10.39	10.76 - 11.12	10.68 - 10.84
Alkalinity (%)	—	12.0 - 13.2	13.4 - 14.6	10.2 - 11.6	9.1 - 11.3	11.7 - 12.5	11.1 - 12.2

APPLICATION OF WOOD ASH

The majority of wood ash is either landfilled or land applied. In Europe, wood ash is used as a feedstock for cement production and roadbase material [9].

Wood ash can be used in manufacture of low-strength concrete and controlled low-strength materials [8]. It can also be used as an admixture in concrete through proper mixture proportioning for encapsulating heavy metals and other pollutants present in the ash. However, this technology is yet to be developed.

SUMMARY AND CONCLUSIONS

Large volumes of by-product materials generated from industrial and post-consumer sources are landfilled. The amount of waste generation is increasing, while landfill space is decreasing. Additionally, due to stricter environment regulations, it is difficult to obtain approval for developing new disposal facilities. Thus, cost of disposal is escalating. Recycling not only saves on huge disposal costs, but also conserves natural resources, and in some cases it provides technical and economic benefits. Various uses of by-products generated from industrial and post-consumer sources exist. Uses of these by-products in emerging materials is as follows:

1. Fly ash can be used in manufacturing of Controlled Low Strength Materials as a replacement of regular concrete sand up to 100%.

2. Fly ash can be used as a major component of blended cement, exceeding 50% of total blended cement mixture.

3. Fly ash can be used in manufacture of lightweight aggregates.

4. Significant amounts of fly ash can be used in the manufacturing of high-performance concrete (HPC) in the range of 15 to 35% depending upon type of fly ash.

5. More than 50% of cement can be replaced with fly ash in the manufacturing of superplasticized structural-grade concrete.

6. Fly ash can be used as a cement replacement up to 30% in manufacture of precast/prestressed concrete products.

7. Fly ash can be used as a fine filler as well as a pozzolan in roller compacted concrete.

8. Fly ash can be used as a replacement of 30 to 100% silica sand in the manufacturing of autoclaved cellular concrete.

9. Fly ash can be used as a filler in polymer matrix composites as well as metal matrix composites.

10. Bottom ash/boiler slag can be used as both fine and coarse lightweight aggregates.

11. Clean-coal ash can be used as a raw material in production of cement. FGD can be used as a gypsum for the manufacturing of wall boards. It can also be used in concrete as well as cement-based masonry products.

12. Wood ash can be used in the manufacturing of CLSM.

REFERENCES

1. ACAA. (1996). "1996 Coal combustion products (CCP) - production and use." Alexandria, VA, 2 pages.

2. ACI COMMITTEE 226. (1987). "Use of fly ash in concrete." *ACI Materials Journal.* 381- 409.

3. ASTM. (1995). "Standard specification for coal ash and raw or calcined natural pozzolan for use as a mineral admixture in portland concrete." *ASTM C 618, Annual Book of ASTM Standards*, Section 4, construction, 04.02, Concrete and Aggregates, 304-306.

4. CAMPBELL, A.G. (1990). "Wood ash disposal and recycling source book." *TAAPI Journal* (September) 141-145.

5. COLLINS, R.J., AND CIESIELSKI, C.K. (1994). "Recycling and use of waste products and by-products in highway construction." *Synthesis of Highway Practice, Transportation Research Board, National Research Council, National Academy Press,* Washington, D.C.

6. COURTS, G. D. (1991). "The aggregate of the future is here today." *Proceedings of the Ninth International Ash Use Symposium, ACAA, EPRI Report No. GS-7162*, Palo Alto, CA, (Jan), 1, 21-1 - 21-10.

7. ETIÉGNI, AND CAMPBELL, A.G. (1991). "Physical and chemical characteristics of wood ash." *Bioresource Technology: Biomass, Bioenergy, Biowastes, Conversion Technologies, Biotransformations, Production Technologies,* Elsevier Science Publishers Ltd, England, 37, 173-178.

8. FEHRS, J.E., (1996). "Ash from the combustion of treated wood: characterization and management options." *Presented at the National Bioash Utilization Conference*, Portland, ME, (April), 20 pages.

9. GREENE, T.W., (1988). "Wood ash disposal and recycling source book." *Prepared for the Coalition of Northeast Governors by OMNI environmental Sources*, Beaverton, Oregon.

10. HAY, P. D., AND DUNSTAN, E. R. (1991). "Lightweight aggregate production and use in Florida." *Proceedings of the Ninth International Ash Use Symposium*, ACAA, *EPRI Report No. GS-7162*, Palo Alto, CA, (Jan), 1, 22-1 - 22-10.

11. HEMMINGS, R., AND BERRY, E. (1986). "Evaluation of plastic filler applications for leached Fly ash." *EPRI Report No. CS-4765*, Palo Alto, CA, (Sep).

12. ICF NORTHWEST. (1988). "Advanced SO_2 control by-product utilization laboratory evaluation." *EPRI Report No. CS-6044*, Palo Alto, CA, (Sep).

13. ICF TECHNOLOGY, INCORPORATED. (1988). "Laboratory characterization of advanced SO$_2$ control by-products: spray dryer wastes." *EPRI Report No. CS 5782*, Palo Alto, CA (May).

14. MISRA, M.K., RAGLAND, K.W., AND BAKER, A.J., (1992) "Wood ash composition as a function of furnace temperature." *Biomass & Bioenergy*, 4(2), 103-116.

15. MURARKA, I.P. (1987). "Solid waste disposal and reuse in the United States." *CRC Press, Inc.,* Boca Raton, FL, 1, 187 pages.

16. NAIK, T.R., AND SINGH, S.S. (1997). "Flowable slurry containing foundry sands." *Journal of Materials in Civil Engineering,* 9 (2), 93-102.

17. NAIK, T.R., KRAUS, R.N., STURZL, R.F., AND RAMME, B.W. (1997). "Design and testing of controlled low-strength materials (CLSM) using clean coal ash." To be published in ASTM STP 1331, A.K. Howard and J.L. Hitch, Eds.

18. NAIK, T.R., BANERJEE, D.O., KRAUS, R., SINGH, S.S., AND RAMME, B.W. (1997). "Characterization and application of Class F fly ash coal and clean-coal ash for cement-based materials." Proceedings of the Twelfth International Symposium on Management of Use of Coal Combustion by-Products (CCBS), Vol. 1.

19. NAIK, T.R., SINGH, S.S., OLSON, JR., W.A., AND BEFFEL, J.C. (1997). "Temperature effects on strength and durability of high-performance concrete." *Proceedings of the PCI/FHWA International Symposium on High Performance Concrete*, New Orleans, Louisiana, (Oct).

20. NAIK, T.R., AND SINGH, S.S. (1997). "Permeability of flowable slurry materials containing foundry sand and fly ash." *Journal of Geotechnical and Geoenvironmental Engineering,* ASCE, 123(5), 446-452.

21. NAIK, T.R., SINGH, S.S., AND MOHAMMAD, M. (1995). "Properties of high performance concrete incorporating large amounts of high-lime fly ash." *International Journal of Construction and Building Materials*, 9(6), 195 - 204.

22. NAIK, T.R., AND SINGH, S.S. (1995). "Use of high-calcium cement-based construction materials." *Proceedings of the fifth CANMET/ACI International conference on Fly Ash, Silica fume, slag and Natural Pozzolans in Concrete*, Milwaukee, 1-44.

23. NAIK, TARUN R., RAMME, B.W., AND TEWS, J.H. (1994). "Use of high volumes of class C and class F fly ash in concrete." *ASTM Cement, Concrete & Aggregates*, (16)1, 12-20

24. NAIK, T.R., WEI, L., AND SINGH, S.S. (1992). "Low-cost ash-derived construction materials: state-of-the-art assessment." *EPRI Report No. TR-100563*, Palo Alto, CA, (April).

25. NAIK, T. R., AND RAMME, B. W., AND KOLBECK, H.J. (1990). "Filling abandon underground facilities with CLSM slurry." *ACI Concrete International*, 12(7), 1 - 7.

26. NAIK, T.R., AND RAMME, B.W. (1990). "High early strength fly ash concrete for precast/prestressed products." *PCI Journal*, (Nov-Dec), 72-78.

27. PYTLIK, E.C., AND SAXENA, J. (1991). "Fly ash based autoclaved cellular concrete: The building material of the 21st century." *Proceedings of the Ninth International Ash Use Symposium, ACAA, EPRI Report No. GS-7162,* Palo Alto, CA, (Jan), 1, 25-1 - 25-12.

28. QUANTTRONI, G., LEVITA, G., AND MARONETTI, A. (1993). "Fly ashes as modifier for low-cost polymeric materials." *Proceedings of the Tenth International Ash Use Symposium,* ACAA, *EPRI Report No. TR-101774,* Orlando, Florida, (Jan), 2, 75-1 - 75-9.

29. ROHATGI, P.K. (1994)."Low-cost, fly ash-containing aluminum-matrix composites." *Journal of Materials Science,* (November), 55-59.

30. ROHATGI, P.K., KESHAVARAM, B.N., HUANG, P., GUO, R., GOLDEN, D.M., REINHARDT, S., AND ODOR, D. (1993). "Microstructure and properties of cast aluminum fly ash particle composites." *Proceedings of the Tenth International Ash Use Symposium,* ACAA, *EPRI Report No. TR-101744,* Palo Alto, CA, (Jan), 2, 76-1 - 76-22.

31. SCHRADER, E.K. (1994). "Roller compacted concrete for dams: the state of the art." *Advances in Concrete Technology, Second Edition,* V.M. Malhotra, Ed., Ottawa, Ontario, Canada, 371-417.

32. SUMMERS, K.V., RUPP, G.L., AND GHERINI, S.A. (1983). "Physical-chemical characteristics of utility solid waste." *EPRI Report No. EA-3236,* Palo Alto, CA.

33. WEI, L.H., (1992). "Utilization of cal combustion by-products for masonry construction." *EPRI Report No. TR-100707*, Palo Alto, CA, 1-1 - B-12

STUDIES ON THE EFFECTIVE USE OF CONCRETE SLUDGE – SLUDGE WATER AND PULVERIZED DRY SLUDGE

Y Sato   C Kiyohara   L Chia-Ming
Oita University
Y Takeda
Nishi-Nippon Institute of Technology
S Taguchi
Fukuoka Ready-Mixed Concrete Industry Association
T Yakushiji
Oita Ready-Mixed Concrete Industry Association
Japan

ABSTRACT. In this paper, the potential of the effective use of sludge water and "pulverized dry sludge (PDS)" was investigated. The so-called pulverized dry sludge is a very fine powder with a Blaine specific surface area of about 5,000 cm^2/g made from concrete sludge by a process of drying and crushing. Tests on the properties of fresh and hardened concrete incorporating with sludge water, and containing PDS were performed. It was found that concrete sludge has no harmful effects on hardened concrete, although the dosage has sometimes to be limited to about 10% of the cement content because high content of concrete sludge significantly increases the demand of mixing water

Keywords: Ready-mixed concrete, Waste water, Returned and surplus concrete, Concrete sludge, Landfill, Sludge water, Sludge cake, Pulverized dry sludge.

Dr Yoshiaki Sato is a Professor in Director of Concrete Technology of the Department of Human Welfare Engineering, Oita University, Japan. His research interests include the creep and shrinkage cracking of concrete, fiber reinforced concrete, and the repair of concrete.

Miss Chizuru Kiyohara is a Research Associate of the Department of Architecture Engineering, Oita University, Japan. She received her Master's degree from Kyushu University, Japan. Her research is in the field of high strength and high performance concrete.

Mr Liu Chia-Ming is a Graduate Student (Master's Course) of the Department of Architectural Engineering, Oita University, Japan.

Mr Yoshitsugu Takeda is an Associate Professor of the Department of Architectural Engineering, Nishi-Nippon Institute of Technology, Japan.

Mr Shigehisa Taguchi is a Technical Manager of Fukuoka Ready-Mixed Concrete Industry Association, Japan.

Mr Teruo Yakushiji is a Technical Manager of Oita Ready-Mixed Concrete Industry Association, Japan.

INTRODUCTION

The disposal of waste water from concrete mixers, returned and surplus concrete is becoming a greater concern because of the environmental disruption [1,2]. Waste (or Recycled) water is divided into two groups as shown in Figure 1; one is a relatively clean water and called clarified water, and the other grey (or sludge) water containing cement solids and other fine particles. The clarified water can be used as mixing water in new batches, replacing a part of the city water. Regulation JIS A 5308 (Japanese Industrial Standard, "Ready-Mixed Concrete") states that the amount of the solid content in gray water should contain less than 3% of cement content by weight, when used as mixing water. It is said that all the sludge water with 3% of sludge solid content ratio can be used up as mixing water if the constant amount of ready-mixed concrete is produced every day. However, since the ready-mixed concrete is an order-made product, fine particles in unused sludge water settle out. The settled material is usually dehydrated, and this is called sludge cake. Most of the sludge cake is discharged to the controlled landfills, because the settled material falls under the category of dirty mud according to the law on the disposal of waste materials in Japan. The total disposal cost of sludge cake is very high. It is necessary to investigate strategies for managing the effective use of concrete sludge. In response to this need, Japan Concrete Institute (JCI) organized a research committee on the effective use of concrete sludge in 1994, and a report on research activities was published in 1996 [3].

Since it is very difficult to treat the sludge cake as concrete material because of it's high water content (50-60% by weight), the process of drying and crushing was introduced to make pulverized dry sludge (PDS) from the sludge cake. This research project was designed to investigate the potential of the effective use of sludge water and PDS.

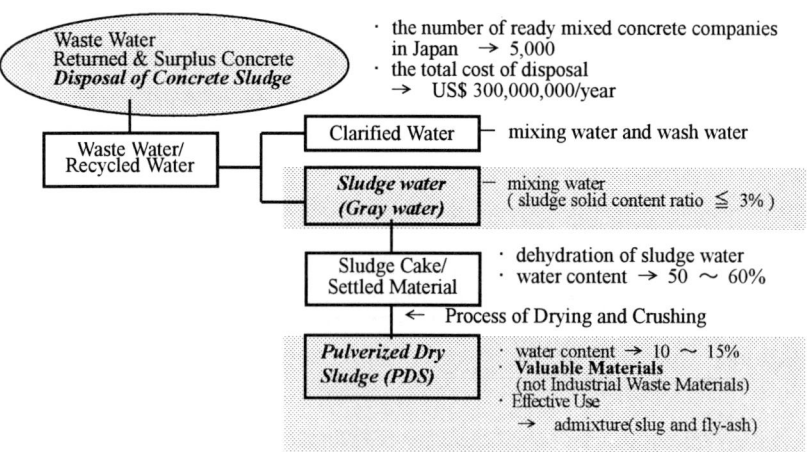

Figure 1 Effective use of concrete sludge

OUTLINE OF EXPERIMENT

Experimental Program

Table 1 shows the outline of experiment. The experiment consists of two series. In series I, the effects of the sludge solid content in mixing water on the workability, bleeding, compressive strength, freeze and thaw resistance, and carbonation of concrete was investigated. In series II, tests on the properties of concrete containing PDS were performed to investigate the limitation of the dosage of PDS.

Materials

Materials used in this experiment are commercially available and currently used in the production of ready-mixed concrete in Oita Prefecture, Japan. Physical properties of materials are shown in Table 2. PDS was obtained by a process in which the sludge cake was dried to some extent, and then crushed by a forced air (cyclone system) [4].

Table 1 Outline of experiment

W/C %	SLUMP cm	SERIES I SLUDGE WATER	SERIES II PDS	CONTENTS OF EXPERIMENT				
				Bleeding	Compressive Strength	Drying Shrinkage	Freezing & Thawing	Carbonation
50		0 %	0 %	○	○	○	○	○
55		3 %	10 %	□	○	○	□	□
60	8, 18	5 %	20 %	□	○	□	□	□
65		10 %	30 %	○	○	○	○	○
70		0 %	0 %	□	○	○	□	□

○: measured, □: not measured

Table 2 Physical properties of materials used

Cement	Ordinary Portland cement Specific gravity : 3.16, Specific surface area : 3,190 cm^2/g
Fine Aggregate	Blended sand (sea sand + pit sand) Maximum size : 2.5 mm, Absorption : 2.88%, Fineness modulus : 2.36 Specific gravity under saturated surface-dry = 2.55
Coarse Aggregate	Crushed stone Maximum size : 20mm, Absorption : 0.62%, Fineness modulus : 6.58 Specific gravity under saturated surface-dry = 2.66
Chemical Admixture	Air-entrained water reducing agent
Concrete Sludge	Series I= Sludge water: Waste water with the age of 1 day SeriesII= PDS: Specific gravity : 2.17, Specific surface area : 4,800cm^2/g

Mix Proportions

The mix proportions of concrete with water cement ratio, W/C (or water binder ratio, W/B) of 0.5, 0.55, 0.60, 0.65 and 0.70, and slumps of 8 and 18 cm, respectively, were selected as reference concretes. In each reference concrete except in the case of W/C of 0.70, concrete incorporating with sludge water (series I) and containing PDS (series II) was produced.

In series I, mixing water with the sludge solid content ratio of 3, 5, and 10% by weight was used, and the amount of sludge solid was regarded as an additive to cement content. In series II, a part of cement content was replaced by 10, 20, and 30% of PDS by weight. Mix proportions were decided according to the guideline for the utilization of recycled water issued in 1979 by National Ready-Mixed Concrete Industrial Association in Japan [5]. Detailed mix proportions are shown in Table 3 and Table 4.

TEST RESULTS AND DISCUSSION

Fresh Concrete

The water content increased as the content of concrete sludge increased, in order to obtain the same value of slump of a reference concrete. For both series of concrete with the slump of 18 cm, the increase in the water content was remarkable compared to the others. In these mix proportions, the dosage of air entrained water reducing agent was calculated with respect to the cement content.

In series I, the troweling workability of all sludge water concretes was found to be better than that of reference concretes, and all PDS concretes except for 30% of PDS content were also better than reference concretes. Concrete sludge tends to improve the cohesiveness and workability of concrete.

Compressive Strength

Figure 2 (a) shows the relationship between the sludge solid content and compressive strength at the age of 28 days and 1 year (series I). Since all concretes incorporating with sludge water are designed to have the same W/C of reference concretes, it is found that compressive strength is not affected by the sludge solid content, and sludge water concretes show almost the same value of reference concretes.

Figure 2 (b) shows the test results of series II on the relationship between PDS content and compressive strength. For concretes designed to have the constant slump of 8 and 18 cm, there is a similar trend that the strength decreases as the content of PDS increases, since the water binder ratio (W/B) increases. This is due to the fact that the actual (effective) water cement ratio becomes high since dried sludge is used as a part of cement. However, the 1-year strength of concrete with PDS content of 10% showed almost the same value as that of a reference concrete. This fact indicates that PDS has a slight hydraulicity.

Table 3 Mix proportions of concrete containing concrete sludge- Series I

MIX NO.	W/C %	SLUMP cm	SL %	S/A %	UNIT WEIGHT, kg/m³						SLUMP cm	AIR %	UNIT MASS kg/m³	BLEE-DING cm³/cm²
					W	C	S	G	SL	A				
						Series I - Sludge Water								
1			0		165	330	719	1077	-	0.825	8.0	4.4	2.32	0.12
2			3		172	344	649	1059	10.3	0.860	9.0	4.5	2.31	[]
3		8	5	41.0	174	348	684	1053	17.4	0.870	8.5	4.7	2.30	0.07
4			10		176	352	660	1048	35.2	0.880	6.0	3.7	2.31	0.07
5	50		0		189	378	694	998	-	0.945	18.0	4.9	2.28	0.20
6			3		203	406	658	963	12.2	1.015	19.0	3.6	2.29	-
7		18	5	42.0	206	412	643	955	20.6	1.030	19.0	4.2	2.27	0.14
8			10		210	420	609	944	42.0	1.050	18.0	6.0	2.23	0.10
9			0		164	298	750	1075	-	0.745	10.0	6.0	2.28	-
10			3		170	309	730	1059	9.27	0.772	8.5	4.7	2.30	-
11		8	5	42.0	172	313	717	1056	15.6	0.782	9.0	5.0	2.29	-
12			10		174	316	696	1051	31.6	0.790	7.0	4.5	2.29	-
13	55		0		188	342	724	100	-	0.855	18.5	4.7	2.28	-
14			3		196	356	699	979	10.7	0.890	19.0	4.5	2.28	-
15		18	5	43.0	198	360	686	976	18.0	0.900	19.0	5.3	2.26	-
16			10		201	365	660	968	36.5	0.912	18.5	4.7	2.26	-
17			0		163	272	778	1069	-	0.680	7.0	4.9	2.32	-
18			3		169	282	758	1056	8.46	0.705	9.5	4.2	2.30	-
19		8	5	43.0	171	285	748	1053	14.2	0.712	9.0	4.4	2.29	-
20			10		173	288	727	1048	28.8	0.720	8.0	3.7	2.30	-
21	60		0		187	312	753	998	-	0.780	19.5	4.7	2.27	-
22			3		194	323	732	982	9.69	0.808	19.5	4.7	2.27	-
23		18	5	44.0	196	327	721	976	16.4	0.818	19.0	4.1	2.27	-
24			10		199	332	694	971	33.2	0.830	17.0	5.2	2.25	-
25			0		163	251	804	1061	-	0.628	6.5	4.6	2.30	0.17
26			3		169	260	786	1048	7.80	0.650	7.5	4.8	2.29	-
27		8	5	44.0	171	263	773	1045	13.2	0.658	8.0	4.9	2.28	0.13
28			10		173	266	755	1040	26.6	0.665	7.5	4.4	2.30	0.09
29	65		0		187	288	781	990	-	0.720	19.5	4.8	2.26	0.29
30			3		194	298	758	976	8.94	0.745	19.0	4.8	2.26	-
31		18	5	45.0	196	302	745	971	15.1	0.755	20.0	4.9	2.26	0.25
32			10		199	306	724	963	30.6	0.765	19.0	4.2	2.26	0.17
33	70	8	0	45.0	163	233	827	1051	-	0.582	6.0	4.5	2.29	-
34		18	0	46.0	187	267	806	982	-	0.668	18.0	5.2	2.26	-

SL: Concrete Sludge, S/a: Sand Percentage, W: Water, C: Cement, S: Sand, G: Gravel, A: Air Entrained Water Reducing Agent

Table 4 Mix proportions of concrete containing concrete sludge- Series II

MIX NO.	W/C %	SLUMP cm	SL %	S/A %	UNIT WEIGHT, kg/m³					SLUMP cm	AIR %	UNIT MASS kg/m³	BLEE-DING cm³/cm²
					W	C	S	G	SL	A			
Series II - PDS													
1	50	8	0	41.0	165	330	719	1077	-	0.825	8.0	4.5	2.293	0.06
2			10		170	306	704	1051	38.8	0.765	7.5	4.2	2.283	0.05
3			20		176	282	686	1024	80.3	0.705	8.0	4.0	2.272	-
4			30		185	259	660	990	126.5	0.648	6.5	3.6	2.255	0.03
5		18	0	42.0	189	378	694	998	-	0.945	19.5	4.5	2.272	0.11
6			10		195	351	676	968	44.5	0.878	18.0	4.8	2.233	0.10
7			20		209	334	640	920	95.3	0.835	17.0	4.5	2.217	-
8			30		220	308	612	878	150.5	0.770	17.0	4.0	2.186	0.08
9	55	8	0	42.0	164	298	750	1075	-	0.745	9.0	5.5	2.255	-
10			10		169	276	735	1053	35.0	0.690	8.0	4.1	2.285	-
11			20		175	254	717	1029	72.5	0.635	7.0	3.8	2.259	-
12			30		184	235	691	995	114.6	0.588	8.0	4.1	2.243	-
13		18	0	43.0	188	342	724	100	-	0.855	18.5	4.3	2.269	-
14			10		194	318	707	971	40.2	0.795	18.0	4.3	2.249	-
15			20		200	292	686	947	83.2	0.730	18.5	4.4	2.232	-
16			30		211	269	658	907	131.3	0.672	17.5	4.5	2.199	-
17	60	8	0	43.0	163	272	778	1069	-	0.680	8.0	4.9	2.273	-
18			10		168	252	763	1048	31.9	0.630	8.0	4.5	2.272	-
19			20		174	232	748	1027	66.1	0.580	7.5	4.4	2.259	-
20			30		183	213	724	995	104.3	0.532	8.5	4.3	2.230	-
21		18	0	44.0	187	312	753	998	-	0.780	18.5	5.1	2.239	-
22			10		193	290	735	974	36.7	0.725	18.0	4.6	2.243	-
23			20		200	266	717	947	75.9	0.665	19.5	4.5	2.229	-
24			30		210	245	689	912	119.7	0.612	18.5	4.9	2.188	-
25	65	8	0	44.0	163	251	804	1061	-	0.628	8.0	5.4	2.250	0.7
26			10		168	232	788	1043	29.4	0.580	8.5	4.4	2.257	0.10
27			20		174	214	771	1021	61.1	0.535	9.0	4.6	2.236	-
28			30		183	197	750	992	96.4	0.492	10.0	5.0	2.210	0.08
29		18	0	45.0	187	288	781	990	-	0.720	19.5	4.6	2.245	0.16
30			10		193	267	763	968	33.9	0.668	18.0	5.2	2.216	0.14
31			20		200	246	742	944	70.2	0.615	19.5	4.1	2.225	-
32			30		210	226	717	910	110.5	0.565	19.5	4.3	2.192	0.12
33	70	8	0	45.0	163	233	827	1051	-	0.582	7.0	4.7	2.262	-
34		18	0	46.0	187	267	806	982	-	0.668	18.0	5.2	2.225	-

SL: Concrete Sludge, S/a: Sand Percentage, W: Water, C: Cement, S: Sand, G: Gravel, A: Air Entrained Water Reducing Agent

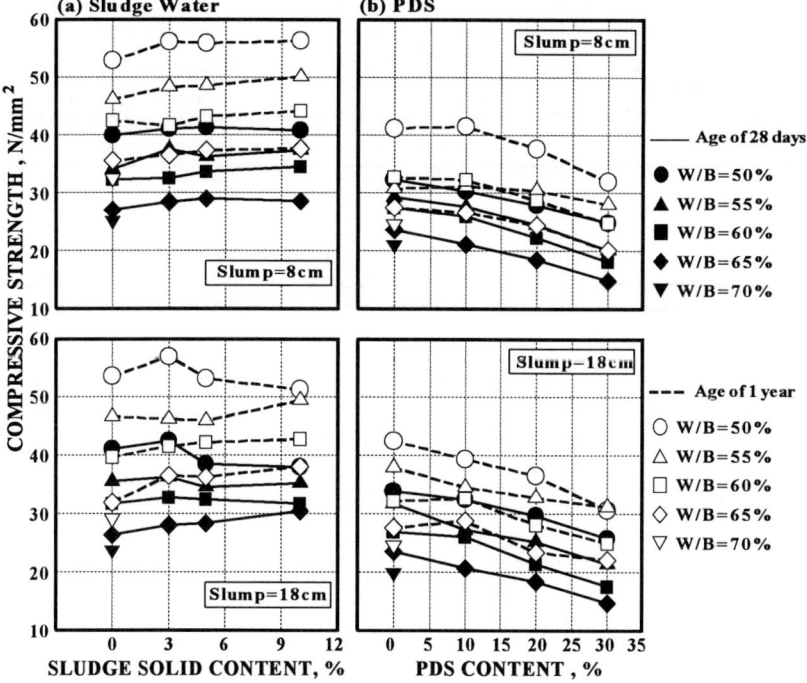

Figure 2 Compressive strength

Drying Shrinkage Strain

Figure 3 shows the results of drying shrinkage, in which the relationship between the content of concrete sludge and drying shrinkage strain at the drying time of 1 year is expressed for each W/C (or W/B). The drying condition was 20±0.5°C and 60±5% R.H. As can be seen from Figure 3 (a), it is found that drying shrinkage is not affected by the sludge solid content, and sludge water concretes show almost the same value of reference concretes. It is evident from Figure 3 (b) that drying shrinkage increases as the content of PDS increases regardless of the slump and W/B, and drying shrinkage of sludge concrete is approximately 10 to 30% higher than that of a reference concrete. However, if the content of concrete sludge is limited to 10%, the increase in drying shrinkage is fairly small.

Freeze and Thaw Resistance

Figure 4 shows the test results of series I and II on the relative dynamic modulus of elasticity, f (f = fn^2 / fo^2×100%, fn: fundamental transverse frequency after n cycles of freezing and thawing (Hz), fo: fundamental transverse frequency at 0 cycles of freezing and thawing (Hz)). The freezing and thawing tests of concrete with a slump of 8 cm and 18 cm were carried out at the different research laboratory, respectively.

In the case of concrete with a slump of 18 cm, the effects of concrete sludge content were not recognized, and the relative dynamic modulus of elasticity was almost same, regardless of W/C (or W/B).

In the case of concrete with a slump of 8 cm, concrete having sludge solid content of 10% and PDS content of 30% showed the decrease in the relative dynamic modulus of elasticity, and particularly concrete with W/B of 0.5 showed a remarkable decrease.

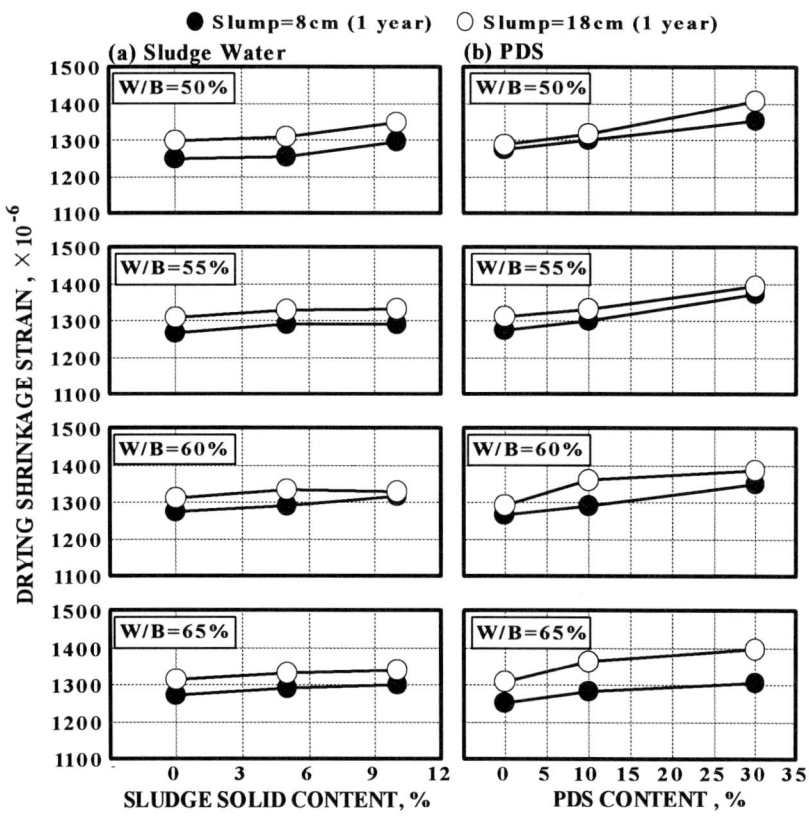

Figure 3 Drying shrinkage strain

Carbonation

Figure 5 shows the results of both series on the accelerated carbonation test. The accelerated carbonation test was carried out at the different research laboratory, as well as the rapid freezing and thawing test. From these figures, concrete with concrete sludge content of less than 10 % shows the similar tendency of the reference concrete, regardless of W/C (or W/B) and slump. The carbonated thickness of concrete with PDS content of 30% after 26 weeks was about 1.3 to 1.7 times of that of the reference concrete.

Effective Use of Concrete Sludge 45

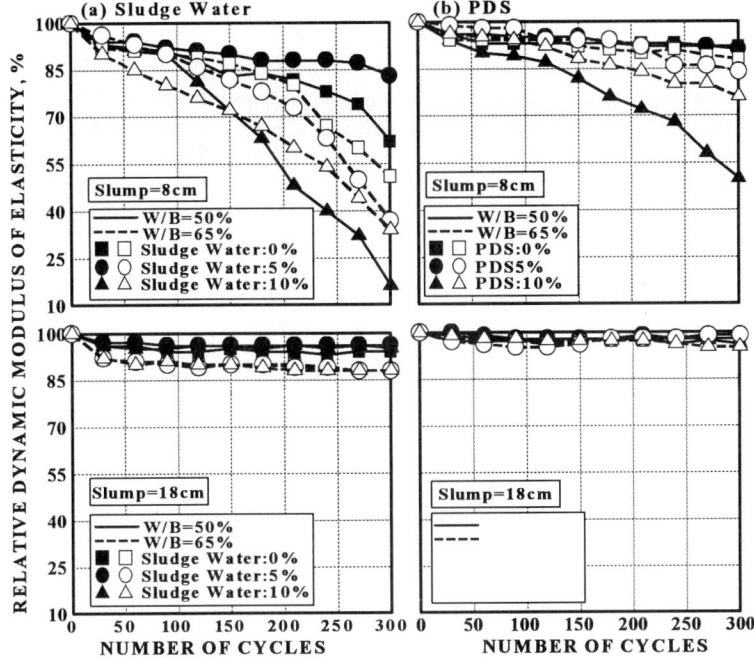

Figure 4 Relative dynamic modulus of elasticity (freeze and thaw resistance)

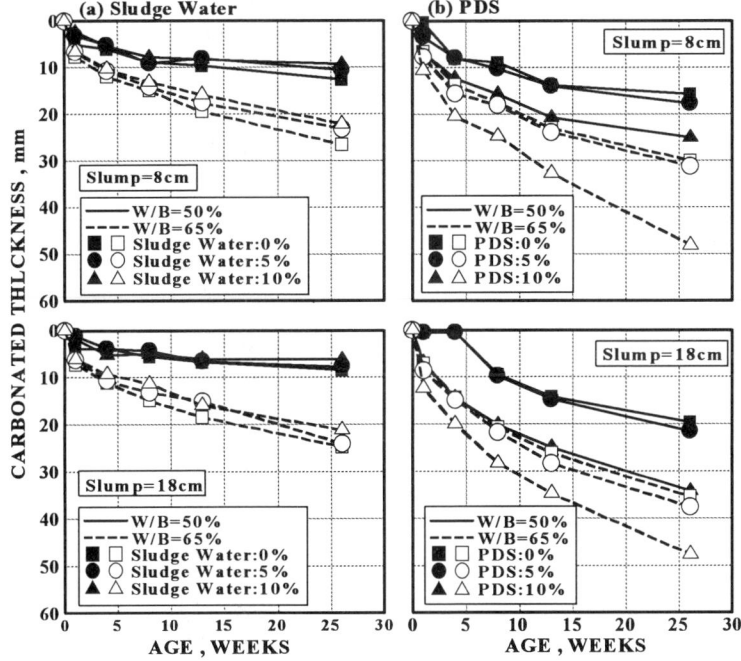

Figure 5 Depth of carbonation

CONCLUSIONS

Regarding the effective use of concrete sludge, that is, both cases of the use of sludge water as a mixing water and that of PDS as replacement materials for cement, concrete with various kinds of mix proportions were produced, and the properties of fresh and hardened concretes were investigated. It was found that concrete sludge has no harmful effects on the properties of concrete, although the dosage has sometimes to be limited to about 10% of the cement content because high content of concrete sludge significantly increases the demand of mixing water.

ACKNOWLEDGEMENTS

The experiments were performed in the laboratory of Kyushu Tokuyama Ready-Mixed Concrete Company, Oita Factory. A part of tests was performed in the Research and Development Center of Tokuyama Co., Ltd., NMB Co., Ltd., and Takemoto-yushi Co., Ltd. PDS sample was produced by Mr. C. Suemoto of Environmental Technology Development Co., Ltd. The authors wish to express the acknowledgements to their help. The financial assistance provided by Oita Ready-Mixed Concrete Industry Association is gratefully acknowledged.

REFERENCES

1. BORGER, J, CARRDSGUILLO, R L and FOWELER, D W Use of Recycled Wash Water and Returned Plastic Concrete in the Production of Fresh Concrete, J. of Advanced Based Materials, 1994, pp.267-274

2. SOUWERBREN, C. Re-use of Wash Water in the Concrete Industrial Production Circle Without Waste. European Standard for Mixing Water and the Re-use of Wash Water, Proc., of Int. Conf., Concrete In The Service of Mankind, Concrete for Environment and Protection, Edited by R. K. Dhir & T. D. Dyer, 1996, pp.183-197

3. JAPAN CONCRETE INSTITUTE(JCI); Report of JCI committee on Effective Use of Concrete Sludge, JCI-C40, May, 1996, pp.1-139

4. SUEMOTO, C. Technical Report, Environmental Technology Development Co., Ltd., 1996, pp.1-31

5. NATIONAL READY-MIXED CONCRETE INDUSTRY ASSOCIATION; Guideline for the Utilization of Recycled Water, 1979

EXPERIMENTAL BASIC ASPECTS FOR REUSING SEWAGE SLUDGE ASH (SSA) IN CONCRETE PRODUCTION

J Monzó

J Payá

M V Borrachero

Polytechnical University of Valencia

Spain

ABSTRACT. This study deals with the effects of Sewage Sludge Ash (SSA) on basic properties of mortars. Chemical and mineralogical analysis of SSA showed a high sulphate concentration, probably generated during the oxidation of organic compounds and inorganic sulphates present in sewage sludge. Setting determination, soundness, pozzolanic activity, workability and mechanical strength were studied. An increase of initial and final setting times occurred with SSA percentage increase. This behaviour can be attributed to the increase of water/cement+SSA ratio with SSA proportion increase. No unsoundness due to the high sulphate percentage in pastes containing up to 60% of original and ground SSA was detected. Thermogravimetric analysis shows a fixation of $Ca(OH)_2$ by SSA accompanied by the formation of hydrates, suggesting a rapid pozzolanic reaction of SSA. A decrease of workability with SSA percentage for all water volumes studied was observed. This workability could be improved using a superplasticizer, grinding SSA or adding Fly Ash (FA). Mechanical strength of mortars containing SSA were in the majority of cases similar to control mortar strengths, suggesting an active role of SSA.

Keywords: Sewage sludge ash (SSA), Fly ash (FA), Portland cement (PC), Waste management, Workability, Mechanical strength, Soundness, Pozzolanic activity, Setting.

Dr José Monzó is a Lecturer in Building Materials Chemistry, Polytechnical University of Valencia, Spain and belongs to GIQUIMA (Research group of building materials chemistry). His main research interests include waste reutilization in construction. He belongs to Construction Engineering Department.

Dr Jordi Payá is a Lecturer in Building Materials Chemistry, Polytechnical University of Valencia, Spain. He is Director of GIQUIMA (Research group of building materials chemistry). His main research interests include waste reutilization in construction. He belongs to Construction Engineering Department.

Dr María V Borrachero is Associate Professor in Building Materials Chemistry, Polytechnical University of Valencia, Spain and belongs to GIQUIMA (Research group of building materials chemistry). His main research interests include waste reutilization in construction. He belongs to Construction Engineering Department.

INTRODUCTION

Waste is an inevitable consequence of human activities. The main problem is that these wastes can pollute the environment (1). From the introduction of the Urban Wastewater Treatment Directive which will close the sewage sludge sea disposal option by 1998 and the increase of wastewater volume treatment, significant amounts of sewage sludge are being and will be generated.

The properties of sewage sludge depend on the source of the wastewater (urban or industrial). For example, the concentration of heavy metals in sewage sludge, that have their origin in industrial activities, is determining for reusing of this solid residue in agriculture. The excess of sewage sludge production and, in some cases, their undesirable chemical properties have lead to scientist to study other ways of disposing of sewage sludge. The most common sewage sludge disposal alternative is to incinerate, this method permits an important reduction of volume. The sewage sludge ash (SSA) obtained can be placed in controlled landfills or used in construction to improve some properties of building materials. Many incinerator residues like rice husk ash (2,3) and municipal solid waste ash (4-6) have been successfully used in construction. SSA has been used in mortars (7), concrete mixtures (8,9), to manufacture bricks (10), as fine aggregate in mortars (11) and in asphaltic paving mixes (12). The aim of this work is to study several aspects of the incorporation of SSA in mortar mixtures.

EXPERIMENTAL DETAILS

Materials, Apparatus and Procedures

Portland cement (PC) used for mortar preparation was conforming to the specifications of ASTM type I, ASTM C-150 (13). Sikanol-M was used as a plasticizer. Fine aggregate was natural sand with 2.94 fineness modulus. SSA was collected from the sewage treatment plant of Pinedo (Valencia, Spain). Fly ash (FA) was collected from the precipitators of thermoelectric power plant of Andorra (Spain). Table 1 shows SSA, FA and PC chemical data. From these data can be emphasized the high concentration of sulphur in SSA (12.4% expressed as SO_3 content). Figure 1 shows SEM micrographs of SSA, these micrographs make clear the irregular morphology of particles and the presence of crystalline aggregates (Figures 1a, 1b and 1c). Plate 1d shows a wide range of diameters of particles. Shape of particles and granulometric distribution will have a decisive influence in workability of mortars.

Thermogravimetric analysis (TGA850 Themogravimetric Measuring Module, Mettler-Toledo): curves were obtained with 10°C/min heating rate, using 70 µL alumina crucibles and heating in a nitrogen atmosphere (75 mL/min purging flow gas). Scanning electron microscopy (JEOL JSM-6300): equipped with energy dispersive X-ray microanalysis capability, operated at 20 kV and using an ultra thin window. X-ray diffraction (PW1710 Based Diffractometer): operating with Cu-Kα radiation at 40 kV and 20 mA, step size 2θ=0.020° and 4 seconds per step. Preparation of mortars, setting determination, consistency of volume and mechanical strength measurements were performed according to ASTM C-305 (14), ASTM C-191 (15) and UNE 80-102 (16). Workability of mortars was determined using a flow table (17), measuring the spreading of mortar cones after 15 drops. Flow table spreading (FTS) is given as a mean of maximum and minimum diameters of the spread mortar cone. Mortars were put in a mold with internal dimensions of (40x40x160) mm for obtaining specimens, which were stored in a moisture room (20±1°C) for 24 hours. Afterwards, the specimens were demoulded and cured by immersion in 40±1°C or 20±1°C water. Samples of SSA were ground using a laboratory ball-mill, Gabrielli Mill-2, for ten minutes: 300g of SSA were introduced into the bottle mill containing 98 balls of alumina (2 cm diameter). Samples obtained was named GSSA.

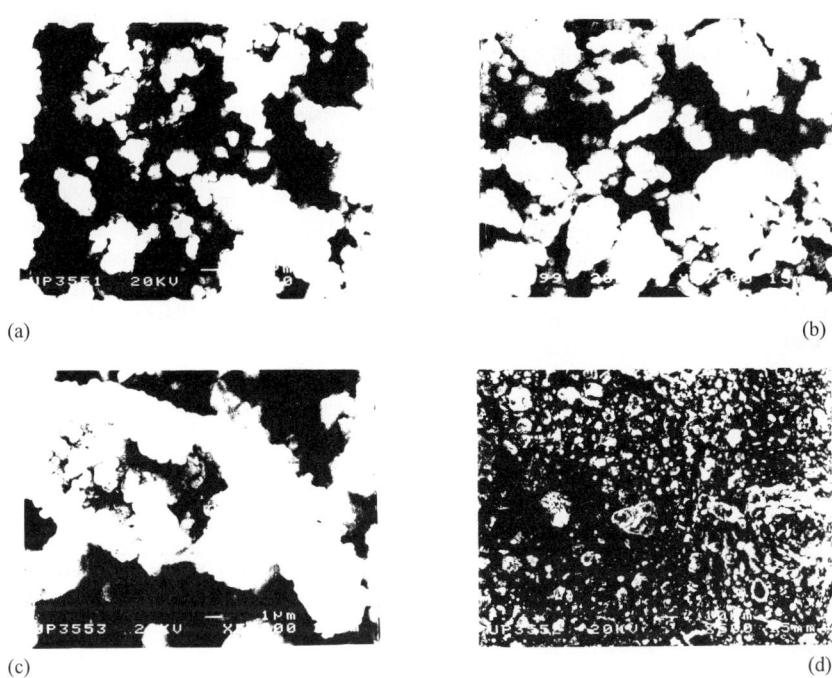

Figure 1 SEM Micrographs of SSA particles

Table 1 Chemical Composition of Sewage Sludge Ash (SSA),
Fly ash (FA) and Portland cement (PC)

%	SSA	FA	PC
Moisture	0.5	--	--
Loss on ignition	5.1	3.0	--
Insoluble Residue	16.1	1.0	--
SO_3	12.4	3.5	3.3
Fe_2O_3	7.4	2.9	4.0
SiO_2	20.8	20.2	20.9
CaO	31.3	62.9	64.6
MgO	2.6	1.1	1.2
Al_2O_3	14.9	4.9	4.9
P_2O_5	6.7	--	--

RESULTS AND DISCUSSION

Physical Properties of Portland Cement-SSA Blends and Pozzolanic Reactivity of SSA

The composition of SSA (see Table 1) indicates the possibility of chemical interaction between SSA and cement in the hydration processes (setting and curing), particularly internal sulphate attack, due to the high percentage of sulphur in SSA (12.4% in SO_3). On the other hand, the X-ray diffraction pattern of SSA shows the presence of several crystalline products such as quartz [SiO_2], magnetite [Fe_3O_4], calcite [$CaCO_3$], hidroxilapatite [$Ca_5(PO_4)_3(OH)$], calcium phosphate hydrate [$Ca_3(PO_4)_2.xH_2O$] and anhydrous calcium sulphate [$\gamma\text{-}CaSO_4$]. This last compound, also known as insoluble anhydrite, was in high proportion, suggesting that mostly sulphur in SSA would be as this chemical form. It may be taken into account that in the combustion process of sewage sludge, oxidation of organic sulphur containing compounds and reaction with lime take place. Calcium sulphate from these reactions together with inorganic calcium sulphates present in the sewage sludge, due to high temperature (600-800°C) into the combustion chamber, are converted to the most stable form in these conditions, which is $\gamma\text{-}CaSO_4$. This compound could be considered as an inert material in the hydration of cement. However, test on setting and constancy of volume in cement-SSA pastes have been carried out.

The setting of cement-SSA pastes was investigated using a standard Vicat needle. Initial (T_i) and final (T_f) setting times for several cement-SSA pastes (from 0 to 60% of cement replaced by SSA) were determined. Obtained values are represented in Figure 2. Two observations can be emphasised. Firstly, an increase of both, initial and final setting times occurred when SSA replacing percentage increased too. This behaviour can be attributed to the increase of water /cement+SSA ratio when the proportion of SSA increases (see Figure 2). Due to the high adsorption of water in the surface of SSA particles (see Workability section), high proportion of water is liable to cause a delay in setting times. In second place, the interval between the partial loss of fluidity (T_i) and the acquisition of sufficient firmness to resist certain pressure (T_f) is not apparently affected for different replacing percentages (from 29 to 57 minutes).

Expansive processes in cement pastes after setting could be taking place due to slow hydration of certain of the constituents. In this case, the presence of sulphates in SSA may achieve the reaction between calcium sulphates and calcium aluminate hydrates, yielding calcium sulphoaluminates (particularly ettringite), which are concerned in considering the soundness of cements. The expansive reaction will be due, in this case to internal factors. Because unsoundness in cements is a slow process in many cases, accelerated test using Le Chatelier apparatus was carried out (16). The value of the expansion is obtained by measuring the increase of the distance between the indicator needles, after a thermal treatment of the paste immersed in boiling water for 3 hours. This distance must not exceed 10 mm when tested. Portland cement/SSA blends were tested for constancy of volume. Prepared blended cements containing SSA were in the range 15% to 60% of SSA in weight. In all studied cases, the increase in the distance between the indicator needles were less than 0.2 mm, suggesting that the high SO_3 content in SSA did not produce unsoundness in cement.

In order to increase the reactivity of SSA in cement pastes and, consequently, to make possible higher reactivity of the ash towards hydrated cement compounds, ground SSA(GSSA) was used for preparing blended cements containing GSSA in 15% to 60% range.

Again, no expansions of cement pastes were observed.

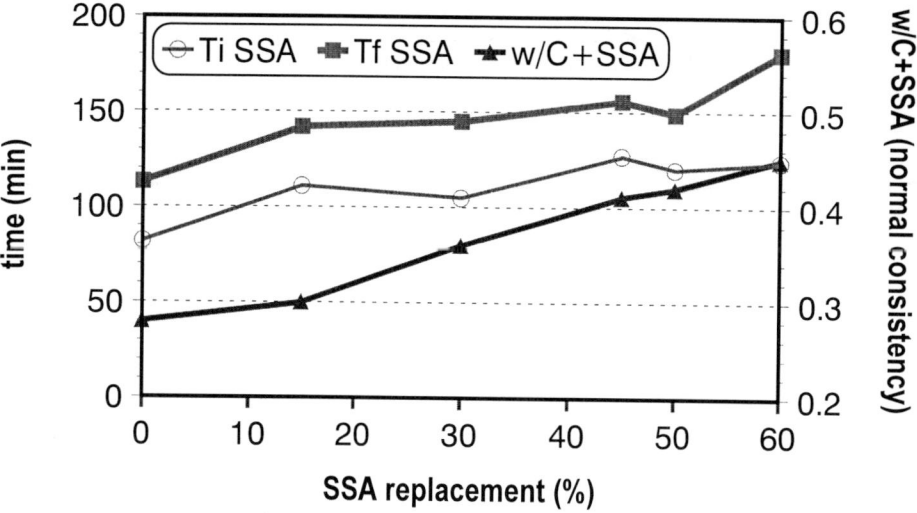

Figure 2 Initial and final setting times in mortars containing SSA

Thus, unsoundness due to the excessive sulphate percentage in SSA or GSSA was not detected by this test. Probably the insoluble γ-CaSO$_4$ present in the ashes did not react easily with calcium aluminate hydrates and, consequently ettringite was not formed.

Additionally, unsoundness may be attributed to the presence of free lime (CaO) in relatively high contents (greater than 2%). Used SSA contains 31.30% of CaO as the total calcium content. When SSA is mixed with water no formation of calcium hydroxide by hydration of free lime was confirmed by thermogravimetric analysis. Thus, calcium in SSA would be chemically combined with carbon as carbonates, with sulphur as sulphates, with phosphorous as phosphates and possibly with silicon as silicates.

From pozzolanic reactivity point of view, the SSA reactivity was tested preparing SSA/hydrated lime mixes, and their evolution in water at 40°C was studied by thermogravimetric analysis. Surprisingly, the calcium hydroxide content (CH) of the pastes decreased from 29% to 16% in the first three days curing period at, suggesting a rapid pozzolanic reaction for SSA. The fixation of CH with time is accompanied of the formation of hydrates (see Figure 3). The water of these hydrates, H, was measured as the weight loss produced in the 110-400°C range.

The pozzolanic activity of SSA could be attributed to the presence of reactive silica and/or alumina in SSA. There are two sources of these reactives in SSA. On the one hand, the presence of clays or other related minerals in sewage sludge, which, after heat treatment in the combustion process, evolved to pozzolanic materials like metakaolin. fluidized bed, and it

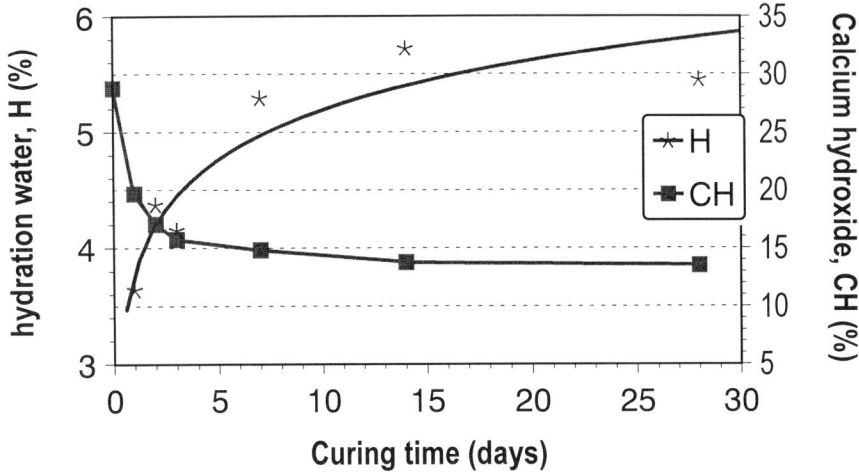

Figure 3   Thermogravimetric data for SSA/hydrated lime mixes cured at 40°C

On the other hand, it may be taken into account that the combustion takes place in finely divided quartz was lost together ash due to mechanical friction. In these conditions, part of the total silica and/or alumina is potentially reactive towards hydrated lime.

Workability of Mortars Containing SSA

Flow Table Spread (FTS) of mortars containing SSA has been studied as explained in experimental section. Figure 4 shows FTS values versus percentage of SSA in mortars containing 200 and 225 mL water volumes respectively (0.44 and 0.5 water/cement+SSA ratios). Mortars were prepared replacing growing amounts of cement by ash. In this experiment a decrease of FTS with SSA percentage for both water volumes studied has been observed. Very low mortars workability mortars was obtained for high replacement percentages (30%). This behavior can be explained taking into account the irregular morphology of SSA particles observed in SEM micrographs (Figure 1). This irregular morphology does not permit the "lubricant effect" as it is produced by coal fly ashes which present spherical shape. Additionally, high water adsorption on SSA particles surface would occur.

Influence of SSA in Strength of Mortars

Mortars containing SSA were prepared replacing 15 or 30% of cement content in control mortar (1:3 cement/sand ratio) and cured at 40°C. Specimens were tested for flexural (Rf) and compressive (Rc) strength for a given curing ages (3,7,14 and 28 days curing period). The results are plotted in Figure 5.

Figure 5a shows the results obtained for compressive strength. It can be observed that the replacement of cement by SSA does not produce a significant reduction of Rc compared to control mortar, except for 3 days curing time. In this case a decrease of Rc with SSA percentage is observed.

This decrease observed at short curing time may be produced by a delay in pozzolanic reaction. However, for longer curing times mortars containing SSA reached similar Rc than control mortar proving that pozzolanic activity studies on SSA /hydrated lime pastes.

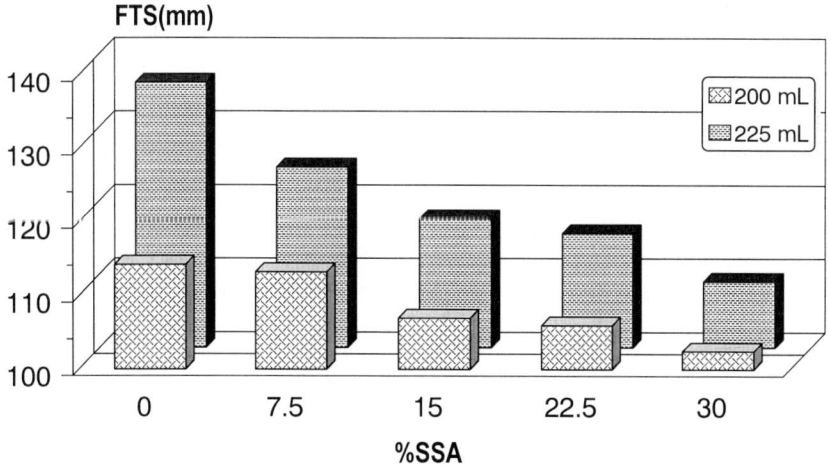

Figure 4 FTS values versus %SSA for mortars with 200 and 225 mL of water

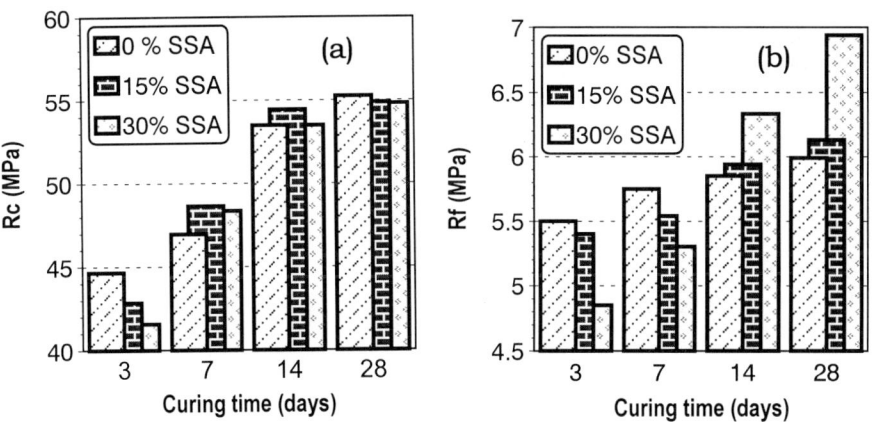

Figure 5 Strength of mortars cured at 40°C containing 15 and 30 % of SSA (1 % of superplasticizer was used) a) Compressive strength and b) flexural strength.

Figure 5b plots the values obtained for flexural strength. In this case, a reduction of Rf with SSA percentage for short curing times (3 and 7 days) occurs. However, Rf values for SSA mortars reached again control mortar.

Mechanical strengths of mortars containing SSA cured at 40°C were in the majority of the cases similar to control mortar strengths, suggesting an active role of SSA in cement paste strength development. Moreover, SSA reactivity at early age (3 days) is higher than found for FA because SSA-containing mortars cured at 40°C yielded similar or greater compressive strength values than control mortar, but at 28 days curing time, apparently, mortars containing FA reached higher Rc values than SSA/cement mortar ones (18,19). Similar behavior was found for flexural strength, but, in this case, better results for FA mortars at 28 days curing time were found.

Second set of experiments consist of studying long term behavior of mortars containing SSA. These mortars were prepared replacing 15 and 30% of cement by SSA and cured at 20°C from 3 days to two years. The results obtained are shown in Figure 6.

Figure 6 plots relative compressive strength (Rc_i/Rc_o) versus curing time, being Rc_i and Rc_o compressive strength of mortars containing SSA and control mortar respectively. From the analysis of the results, it can be established that the increase of curing time does not imply an increase of Rc_i/Rc_o except for short curing times (3 to 28 days) and 30% SSA. Although there are not a defined trend it could be conclude that SSA behaves as an active material which produce most of it activity at 28 days curing time in mortars containing 30% of SSA and no significant differences in mortars containing 15% of SSA were observed.

Most of Rc_i/Rc_o values for 15% of SSA were near to the unit (no dependence of curing time was observed) and two clear trends in mortars containing 30% were observed. First one at early curing times (3 to 28 days) and second one at medium and long term curing times (2 to 24 months) (see Figure 6). On the other hand, it is very important to not that no drop in mechanical properties were found for longest curing period, suggesting that the possible damage interaction between SSA constituents and cement hydration products did not take place and/or did not affect mechanical properties of SSA mortars.

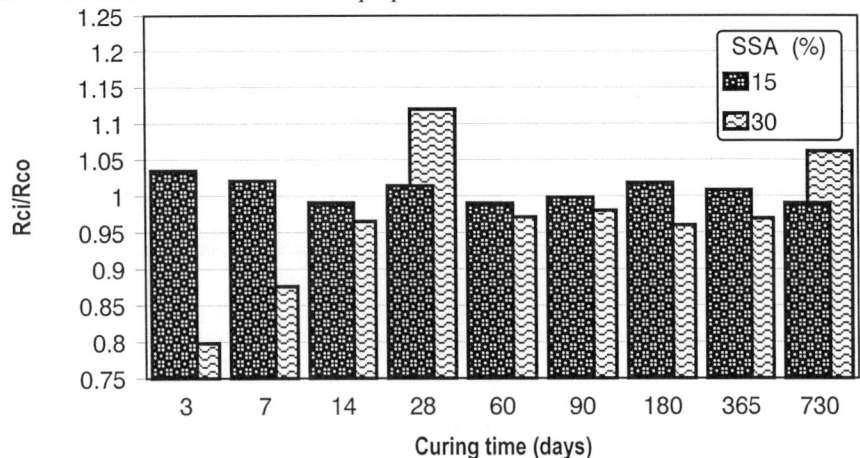

Figure 6 Rc_i/Rc_o of mortars cured at 20°C containing 15 and 30% of SSA
(1 % of superplasticizer was used)

CONCLUSIONS

1. No unsoundness due to the high sulphate percentage in sewage sludge ash for pastes containing up to 60% of SSA or GSSA were detected.

2. SSA showed pozzolanic activity. This mainly could be attributed to the presence of reactive silica, produced in the combustion process which take place in a finely divided quartz fluidized bed.

3. A decrease of workability with SSA percentage was observed. Probably due to irregular shape of particles and high water absorption on particle surfaces.

4. The replacement of cement by SSA up to 30% does not produce an important strength reduction at 24 months curing time compared to control mortar. SSA behaves as an active material which produces most of its activity after long term curing times.

ACKNOWLEDGEMENTS

The authors would like to express their appreciation to the following students, who have collaborated in the present study: Ms Amparo Bellver, Ms María José Blanquer, Mr José Ramón Martínez-Alcántara and Mr Angel Córcoles.

REFERENCES

1. MANAHAN, S,E. Environmental Chemistry, Lewis Pub, 6th, 1994.

2. MEHTA, P, K. Properties of Blended Cements: Cements Made from Rice Husk Ash. J Amer. Conc. Inst., 1977, Vol. 74, pp 440-442.

3. MEHTA, P, K, AND PIRTZ, D. Use of Rice Husk to Reduce Temperature in High-Strength Mass Concrete. J. Amer. Conc. Inst., 1978, Vol. 75, pp 60-63.

4. GRESS, D, L, ZHANG, X, TARR, S, PAZIENZA, I, AND EIGHMY, T, T. Municipal Solid Waste Combustion Ash as an Aggregate Substitute in Asphaltic Concrete. Proceedings of the International Conference on Environmental Implications of Construction with Waste Materials (WASCON 91), Maastricht, The Netherlands, November 10-14, 1991, pp161-175.

5. HAMERNICK, J, D, AND FRANTZ, G, C. Strength of Concrete Containing Municipal Solid Waste Fly Ash. ACI Mat. J., Sep-Oct ,1991, Vol. 88, No. 5.

6. ALI, M, T, AND CHANG, W, F. Strength Properties of Cement-Stabilized Municipal Solid Waste Incinerator Ash Masonry Bricks. ACI Mat. J., May-Jun, 1993, Vol. 91, No. 3.

7. MONZÓ, J, PAYÁ, J, BORRACHERO, M, V, AND CÓRCOLES, A. Use of Sewage Sludge Ash (SSA)-Cement Admixtures in Mortars, Cem. Concr. Res., 1996, Vol. 26, No. 9, pp 1389-98.

8. TAY, J, H. Sludge Ash as Filler for Portland Cement Concrete. J. Envirom. Eng. Div. ASCE, 1987, Vol. 113, pp 345-51.

9. TAY, J, H, AND SHOW, K,Y. Clay Blended Sludge as Lightweight Aggregate Concrete Material. J. Envirom. Eng. Div. ASCE, 1991, Vol. 117, pp 834-44.

10. ALLEMAN, J, E, AND BERMAN, N, A. Constructive Sludge Management: Biobrick. J. Envirom. Eng. Div. ASCE, 1984, Vol. 110, pp 301-11.

11. BHATTY, J, I, AND REID, J, K. Compressive Strength of Municipal Sludge Ash Mortars. ACI Mater., 1989, Vol. 86, pp 394-400.

12. AL SAYED, M, H, MADANY, I, M, AND BUALI, A, R, M. Use of Sewage Sludge Ash in Asphaltic Paving Mixes in Hot regions. Constr. Build. Mater., 1995, Vol. 9,No. 1, pp 19-23.

13. ASTM C-150: Standard Specification of Portland Cement.

14. ASTM C-305: Standard Method for Mechanical Mixing of Hydraulic Cement Pastes and Mortars of Plastic Consistency.

15. ASTM C-191: Standard Test Method for Time of Setting of Hydraulic Cement by Vicat Needle.

16. UNE 80-102: Determinación de tiempos de fraguado y de la estabilidad de volumen.

17. ASTM C-230: Standard Specification for Flow Table for Use in Tests of Hydraulic Cement.

18. MONZO, J, PAYA, J, AND PERIS-MORA, E. A preliminary study of fly ash granulometric influence on mortar strength. Cem. Concr. Res. 1994, Vol. 24, No. 4, pp 791-796.

19. PAYA, J, MONZO, J, PERIS-MORA, E, BORRACHERO, M, V, TERCERO, R, AND PINILLOS, C. Early-strength development of portland cement mortars containing air classified fly ashes. Cem. Concr. Res. 1995, Vol. 25, No. 2, pp 449-456.

THE POZZOLANIC NATURE OF PONDED FLY ASH

T L Robl
J G Groppo
A Hobbs
University of Kentucky
United States of America

ABSTRACT. Fly ash stored in ponds and landfills represent a major reserve of pozzolan for cement and concrete applications. This resource may become important in the future as pollution control devices and burn conditions modify the nature of fly ash. Stored ash is often heterogeneous and frequently has a high loss on ignition (LOI). A good pozzolan requires small particles and low LOI. Several ponds were cored analyzed for particle size, LOI and ASTM strength index (SI). Some samples were cut at 200 mesh and subjected to froth flotation for carbon reduction. All of the fly ash was ASTM Class F. The ponded ash was found to have undergone a high degree of particle sorting, with coarse ash typically found near slurry feed points and fine ash in more distal portions of the ponds. The ash was also highly stratified which reflected a complex sedimentary regime. In general, it was found that the >75 mµ material had higher LOI values and showed a greater degree of variability of LOI than the finer fractions. Ponded or landfilled storage did not appear to affect the pozzolanic nature of the fly ash. Samples stored for as long as 25 years were compared with recent fly ash of equivalent particle size and LOI from the same source and no difference was found in SI.

Keywords: Landfilled and ponded fly ash, Strength index, Particle size distribution, Effect of ponding, Hydraulic classification, Froth flotation, LOI reduction.

Dr Thomas L Robl is an Associate Director of the University of Kentucky Center for Applied Energy Research (CAER) and is in charge of the Waste Management Research program. He is interested in the utilization of coal by-products in cement, concrete and polymers.

Dr John G Groppo is a Senior Mining Engineer at the CAER. His primary interest is minerals processing technology, particularly dewatering, beneficiation and froth flotation. He has developed technology for the recovery of fine waste coal and for the beneficiation of coal combustion ash via hydraulic classification and froth flotation.

Ms Allie Hobbs is a Research Assistant at the CAER. She holds a BA in Civil Engineering and her interests are in geotechnical engineering. She is currently working on the nature of ponded fly ash and the use of FGD material as mine grouts.

INTRODUCTION

There are more than 1 billion tons of fly ash and bottom ash stored in ponds and landfills in the United States. This material represents a substantial reserve of pozzolanic materials if it can be recovered. Changes in burn conditions dictated by modern pollution control technology has resulted in a decline of fly ash quality. Ash from new low NO_x emitting burners typically has higher loss on ignition (LOI) and more unfused quartz [1]. Uncertainties in the quality of pozzolan supply makes the recovery of stored fly ash attractive.

Stored fly ash has not been extensively used as a pozzolan. Problems with heterogenity make its recovery difficult. And there is a general perception that storing and "aging" ash in landfills or ponds somehow affects its pozzolanic activity in a negative way.

The objective of this research was to investigate how pond storage and time affect the physical and chemical nature of utility ash. To date, six ponds and five landfills have been investigated. All of the fly ash was ASTM Class F with low calcium content as typically produced from eastern U.S. bituminous coal.

Quality Criteria for Coal Combustion Fly Ash as a Cement Pozzolan

The major criteria for qualifying fly ash for cement pozzolan is defined by the American Society of Testing and Materials (ASTM) under the C-618 specification [2]. Included are: loss on ignition (LOI), which must be less than 6%; fineness, which requires that no more than 34% of the ash may be retained by a 325 mesh (45 μm) screen; moisture, which must be less than 3%; and, strength index (SI), which requires that a cube of mortar mixed with 20% fly ash achieve 75% of the compressive strength, after either 7 days or 28 days, of a cube made without fly ash.

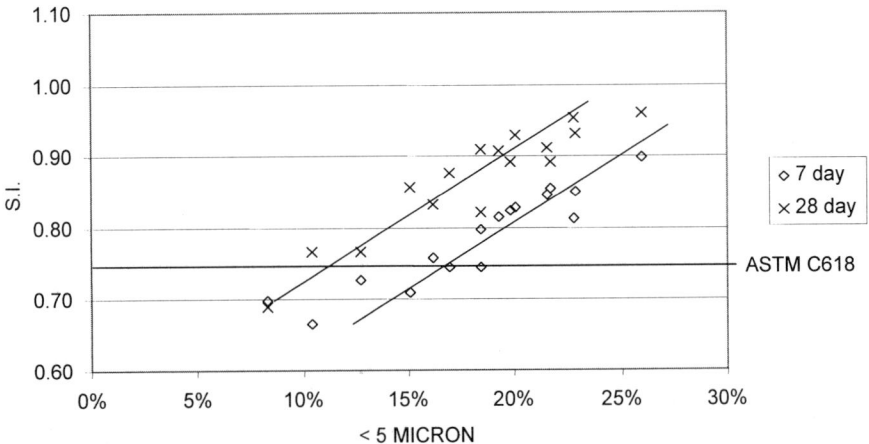

Figure 1 Plot of percent of sample less than 5 micron in diameter versus ASTM Strength Index value determined at 7 and 28 days

Loss on ignition in a Class F ash is primarily a measure of the unburned carbon in the fly ash. The carbon has comparatively high surface area, anywhere from 20 to 70 m^2/g, and is found to adsorb the air entrainment reagent which reduces the concrete's ability to resist freeze thaw reactions. The ASTM C-618 LOI specification allows a maximum of 6%.

Strength index (SI) and fineness are to some extent related. Class F fly ash generally does not begin to react in the cement until the paste has reached a pH of ~13. Then the glassy portion of the ash goes through partial solution-precipitation reactions that also involve the $Ca(OH)_2$ present in the cement paste to form cementitious calcium silicates. The fineness requirement insures that the fly ash has a high surface area and, therefore, good chemical reactivity. This reactivity is an important, but not the sole, factor in determining if the ash will meet strength index criteria.

The purpose of the moisture limit is to insure that the material can be handled by pneumatic methods.

THE CHARACTERISTICS OF STORED ASH

Three ponds have been investigated in detail and will be reported herein. They will be referred to as ponds A, W and G. They are all fed from coal burning utilities which consume Eastern U.S. bituminous coal.

Analytic Approach and Sample Recovery

Our research to date has focused on ponds or landfills with materials that do not meet C-618 criteria. The mingling of the bottom ash with the fly ash results in materials which did not often meet the fineness criteria. Also the materials had high LOI's which often exceeded 6%.

The recovery of a pozzolanic material requires that the materials be classified by size and that sufficient carbon be removed to reduce LOI's to acceptable levels. An extensive assessment of classification technologies was made as part of this work. Screens, hydrocyclones and hydroclassifiers were investigated. It was concluded that hydroclassfication was the most cost effective and efficient. Studies were conducted with the Lewis Econosizer^{TM} , a co-current hydraulic classifier, and ash could effectively be cut at 200 mesh (75 μm) with little displacement of materials [3]. Thus 200 mesh was used as the cutoff in this study for the upper size for the material to be recovered as a pozzolan.

There are several commercial methods for separating carbon from ash. However since the materials of this study had high moisture, froth flotation technology was used [4,5]. All the materials of the study were found to be amenable to this approach. The samples used to determine ASTM strength index were subjected to flotation to reduce their total LOI to a 2.5% to 3.0% range.

The reconnaissance of a large number of ash samples for pozzolanic properties is problematic. Strength index is time consuming, relatively imprecise and expensive. It is not a good screening tool. Particle size distribution, as determined by laser diffraction, and LOI can be measured more readily and quickly. The laser diffraction data can be correlated to strength index. The laser diffractometer used in the study was a Granulometer model 1064 which effectively measures particle diameters from 500 μm to 0.1 μm. Combined with sieve information, a full range of particle sizes is obtained.

The Granulometer data was also used to calculate the area/volume ratio for the <200 mesh (<75 μm) size fractions of the samples. This value was found to correlate well with the SI data. It was determined that the very fine material (<5 mμ in diameter) was responsible for much of the surface area of the fly ash and, in turn, was also relatable to the SI results (Figure 1). In general, samples which had calculated area/volume ratios of less than ~2.0 $\mu m^2/\mu m^3$, containing less than ~18% of <5 mμ diameter fly ash, produced SI's which were only marginal, or failed to pass criteria at 7 days. Pozzolan index values of less than ~1.5 with ~15% or less <5 μm particles could also be expected to fail or only marginally pass the SI criteria at 28 days. The data appear to be linearly related for samples with up to about 30% <5 μm particles and area/volume ratio's of about $3\mu m^2/\mu m^3$. For higher values the strength index appear to reach a limiting value of about 90% -95% at 7 days and 100%-110% at 28 days for the Class F ashes investigated to date. Loss on ignition was determined at 900 °C for 30 minutes in accordance to the specifications of ASTM Procedures [6]. Carbon was determined with a LECO analyzer.

The characterization of the materials in the ponds began with the W pond. A series of cores were collected using simple plastic tubes that were forced into the sediment as part of a reconnaissance effort. The total core depth ranged from 30 to 55 cm. These materials were divided into appropriate intervals and analyzed for size. These cores proved difficult to recover. The core barrel was difficult to insert into the coarse ash and difficult to withdraw from the fine ash. These cores did provide a good general indication of the particle size trends in pond materials. Additional cores were drilled later using heavy core drilling equipment typical of that used in the recovery of soil for engineering foundation work. These cores penetrated to ~3 m in depth. There were 14 short cores which were subdivided into 31 samples and 5 longer cores which were divided into 18 samples in the W pond.

The 15 cores from the A pond were up to 6 m in length with 3-4 m being typical. The cores were divided into 34 samples of ~1.5 meters in length. Three grab samples were also collected. There were four cores drilled in the G pond of ~4.5 m in length which were divided into 24 samples on a ~0.7 m basis.

Sample collection from the ponds proved difficult and at times hazardous. The areas of the pond which contained fine materials could quickly turn unstable and heavy equipment or even personnel can quickly flounder. At one point in the investigation a truck was almost lost in the waterlogged fines. It was found that the materials, particularly the fines were best sampled from the banks of the ponds. It is recommended that extreme caution be used in sampling this material.

Description of the Ponds

The ash is fed to the ponds in the form of a ~10% ash-water slurry. The materials sent to all of the ponds includes both bottom ash and fly ash, but no pyrites. The slurry for pond A and W flows into a secondary catch basin for final clarification. This basin contains a significant quantity of very fine fly ash. Pond G flows into a drainfield which is seeded with hydrophilic plants for filtration. This water is then recycled to the power plant in a closed loop.

The total pond areas and estimated ash content for the study ponds is presented in Table 1. Data for the A and W clarification reservoirs are included. The recovery of ash from the filtration field for Pond G was not considered feasible and this material was ignored in our assessment.

Pozzolanic Nature of Ponded Fly Ash 61

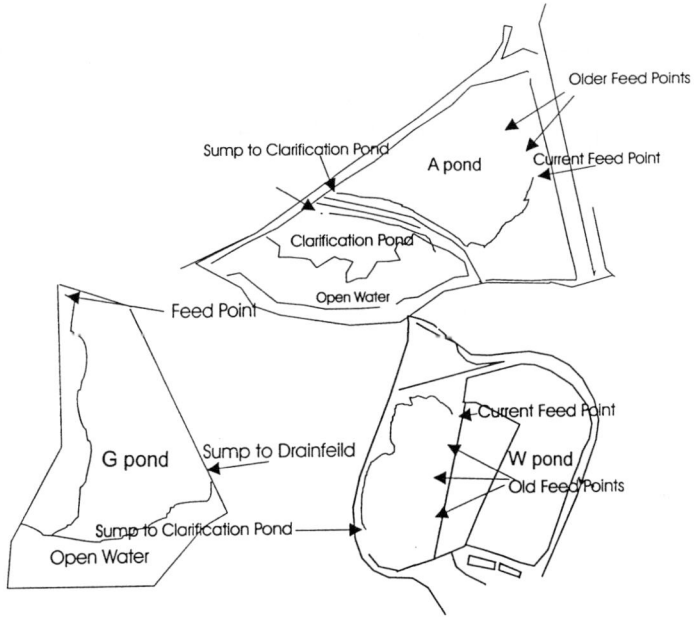

Figure 2 Maps of study ponds with ash slurry streams indicated

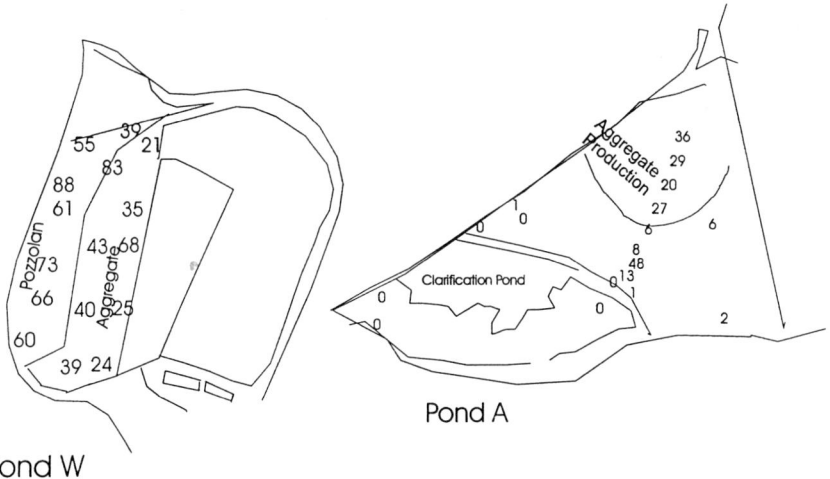

Figure 3 relative percent of <200 mesh material in near surface samples collected at the W pond and percent >16 mesh (1mm) in the near surface samples of A pond. Plot illustrates areas best suited to either pozzolan or aggregate. In A pond aggregate can be co produced with pozzolan

Table 1 Data for study ponds

POND	AREA (m²)	ASH (Mt)	POZZOLAN (Mt)
Pond A (primary)	360,500	2,060,000	1,130,000
Pond A (clarification)	258,000	655,000	470,000
Pond W (primary)	254,000	2,010,000	645,000
Pond W (clarification)	300,000	580,000	430,000
Pond G (primary)	235,000	635,000	280,000

The ash slurry feed is on the eastern side of pond A (Figure 2) and it flows to the western corner of the pond. From there it flows under an embankment into another pond for final clarification. Historically the slurry for Pond A has been fed from the eastern part of the pond. The slurry for the W pond was originally fed from the southeast corner of the W pond and, as the pond was filled with material, the feed point was moved from south to north. The slurry for the W pond exits on the western side and flows into a final clarification pond located to the south (not shown). The G pond is fed from the north and the slurry moves north to south. The pond drains through a weir located in the southeast. The G pond is drained and the material landfilled on a periodic basis.

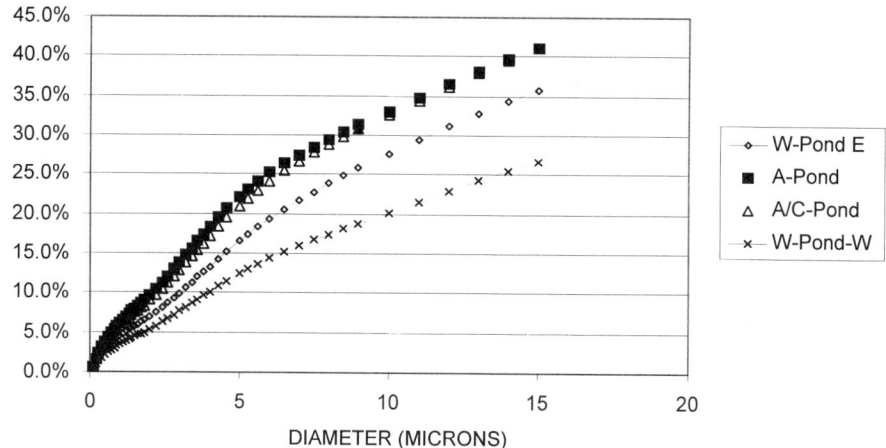

Figure 4 Average distribution of <15 micron particles in <200 mesh fraction of samples collected from the A pond, the A clarification pond and the east and west sides of the W pond. Note how the fine materials have been washed out from The east side of the W pond

The flow regime of the slurry is reflected in the ash particle size distribution. The coarsest ash is deposited near where the slurry is introduced into the ponds and the finer ash is generally deposited more distant. This is illustrated in the plot of the relative abundance of <200 mesh and <16 mesh ash in the samples (Figure 3).

The ponds do show a contrast in the degree of material sorting. For example, the W Pond had much coarser materials in the south and eastern side of pond. This size distribution even affected the <200 mesh size distribution. Material sorting also affected the A pond but not to the same extent. Samples collected from the A clarification pond did not contain any >16 mesh material but the <200 mesh material had approximately the same size distribution as the primary pond. Thus, a pozzolan could be produced via hydraulic classification at 200 mesh almost anywhere in the A pond but the materials in the eastern portion of the W pond were too coarse to produce a good pozzolan when cut at 200 mesh (Figure 4).

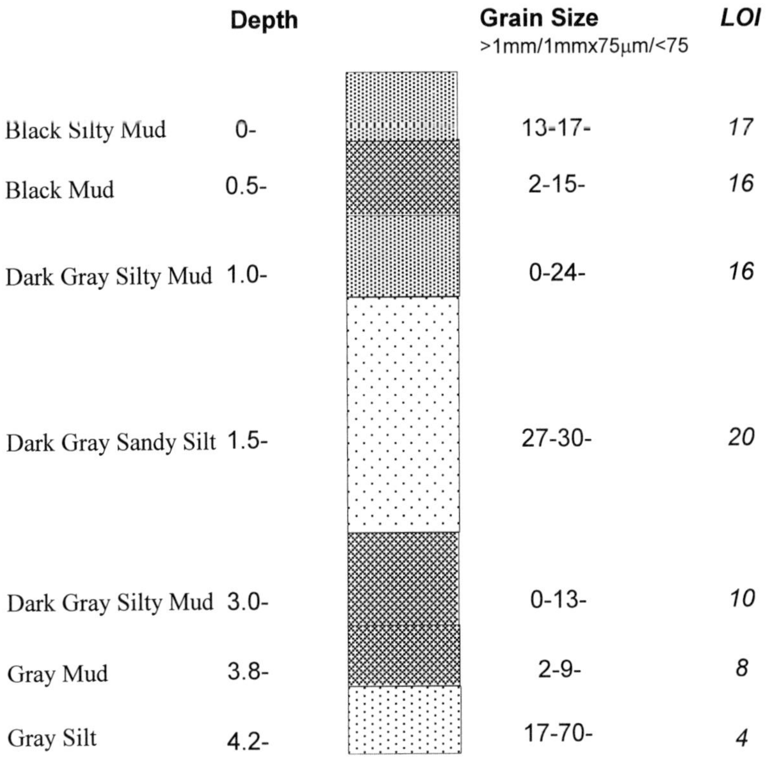

	Depth	Grain Size >1mm/1mmx75µm/<75	LOI
Black Silty Mud	0-	13-17-	17
Black Mud	0.5-	2-15-	16
Dark Gray Silty Mud	1.0-	0-24-	16
Dark Gray Sandy Silt	1.5-	27-30-	20
Dark Gray Silty Mud	3.0-	0-13-	10
Gray Mud	3.8-	2-9-	8
Gray Silt	4.2-	17-70-	4

Figure 5 Plot of changes in ash bed characteristics in core 1 from the G pond, depth is in meters, LOI in percent, particles size corresponds to Sieve sizes 16 and 200 mesh

The ponds themselves are not homogenous from top to bottom, but instead reflect a complex sedimentary regime. This is illustrated in Figure 5, which presents data for Core 1 from pond G. The ash at this location is found to consist of beds which range from fine mud with particles which are almost entirely smaller than 75 µm micron in diameter to silty beds with substantial components of sand and even gravel sized material. The LOI distribution in this core also shows highly variable distribution of LOI, with values as low as 4% near the base of the core and 17% at the top. This is not unexpected as the sedimentary regime must change significantly at any location due to changes in the depth of water in the pond over time and changes in where the ash-slurry entered the pond.

LOI and Particle Size

The distribution of LOI has a direct bearing on the material's potential use and has processing implications. The LOI, as determined for the ashes of the W pond samples, is due almost entirely to carbon and a strong linear correlation (r =0.998) between carbon and LOI is found. This is not surprising as the volatile matter in these samples is very low and they do not contain minerals that carbonate or hydrate easily.

Table 2 Average LOI and size distribution data for the core samples from Pond A (35 samples) and Pond G (24 samples), all data is in weight percent

MESH	WHOLE	>16	16x30	30x50	50x100	100x200	<200
Pond A							
Wt %		12.4	3.3	6.1	9.2	14.3	54.8
LOI	7.0	Nd	10.5	25.0	16.9	8.3	4.7
% of LOI		Nd	3.3	17.2	20.5	16.7	42.3
Pond G							
Wt%		11.0	2.9	6.5	10.9	13.4	55.3
LOI	14.4	Nd	11.9	22.8	24.8	22.3	9.9
% of LOI		Nd	2.7	11.4	20.9	23.0	42.1

Typically the >16 mesh sized materials consisted of fused or scoriaceous bottom ash and had very low LOI. Thus, LOI was not routinely measured on this size fraction. In the W pond ash, the LOI values for the ash recovered from the pozzolan portion of the pond on the whole samples averaged 12.9%. The carbon in the <200 mesh fraction the ash was only 9.1%, and that of the >200 mesh fraction was found to be much higher in carbon, averaging 22.4% C.

The distribution of LOI in the samples from the A and G pond was investigated extensively. It was found that the fractions of fly and bottom ash in the 600 μm x 150 μm (30 x 100 mesh) size have the highest LOI's (Table 2). The average LOI for 300 μm x 150μm (50 x 100 mesh) fraction for the G pond was 25% and samples as high as 40% were found. The size of the carbon is generally coarser than that of the silicate portion of the fly ash. For example, on average 55% of the A and G pond samples passed 200 mesh while 42% of the LOI was found in this ash fraction. Like the W pond samples, the <200 mesh fractions were found to have much lower LOI values, by almost one third, than the whole sample.

The Effect of Storage and Time Upon Pozzolan Strength Index

The amount of time the samples have been stored in a landfill or pond was somewhat difficult to determine. However, in one case, a landfill sample could be located and precisely dated. This sample was originally stored in 1974. A current sample was collected at the same power plant as the stored sample and both were submitted for strength index testing. No significant difference was found in the size distribution or in the measured strengths at 7, 28 or 56 days (Figure 6). Strength index testing was also conducted samples collected from the A pond.

Two samples from the upper portions of the cores and 2 from the lowest intervals of the cores were run. Based upon the history of the pond, the lower samples were deposited approximately 20 to 25 years ago and the upper samples were deposited in the last five years. No appreciable difference was found in the strength index data for the samples. The upper samples had SI's of 77% and 81% at 7 days and 82% and 86% at 28 days and the lower samples had values of 81% and 77% at 7 days and 93% and 86% at 28 days.

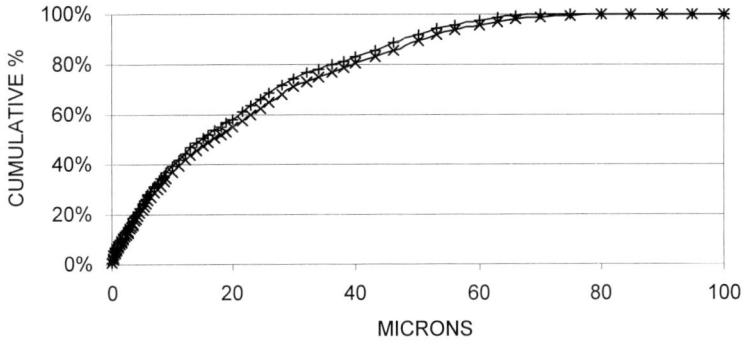

Figure 6 Plot of particle size distribution for ash samples stored in landfill since 1974 (upper) and fresh samples from 1997 (lower curve), SI was 83% and 84% for landfill and 83% and 87% for fresh at 7 & 28 days

CONCLUSIONS

The ash stored in the ponds investigated in this study are complex and potentially useful resources. The nature of the material varies considerably both among and within the ponds. An important cause for this is the nature of how it is deposited, by a water-ash slurry, which is a powerful mechanism for sorting the ash by size. This was of particular importance for the W pond ash which was so coarse that in the southern and western portion of the pond, even the <200 mesh fraction of the ash is too coarse to make a good pozzolan.

Conversely, the coarse nature of the W pond ash combined with the sorting induced by the ponds give rise to the opportunity to produce quality lightweight aggregate. Significant levels of production of aggregate have already been achieved from Pond A. Pozzolan production and aggregate production require substantially different feedstocks and should use materials from the opposite ends of the ponds.

There are also considerable differences among the ponds. The W pond was found to be much coarser in average particle size than the ash in A pond or G pond. However, all of the ash must be classified at a minimum of 200 mesh to improve its consistency and produce a premium pozzolan. Also, the ponded ash investigated must also be processed to remove carbon and lower LOI. It is estimated that about 51% of the total ash in the ponds investigated can be recovered for pozzolan.

The sulfur concentration in the samples was found to be very low. It averaged 0.07% on an as-received basis for the W pond samples. The average C/S weight ratio was 181, which is more than double typical values for high quality Appalachian steam coal. Thus any recycle-fuel derived by separating and concentrating the carbon would be environmentally friendly, as least with respect to sulfur emissions.

The data collected to date does not indicate that aging in a pond or a landfill changes the pozzolan properties of the materials to any significant extent. Yasuda et al., [7] did find that samples collected from ponds in Japan were partially flocked. This has also been noted in some samples from our research. However the flocks are not found to be highly stable and would not impact the overall surface area of the ash or its usefulness as a pozzolan. More important than any changes due to aging in the pond is the overall effect of particle size sorting. Clearly, some areas of the ponds are enriched or depleted in fine materials which greatly affects their suitability to produce a quality pozzolan. The fine materials also proved more difficult to sample, which may be at least partly responsible for the concept that reclaimed ponded materials have diminished pozzolan properties.

REFERENCES

1. TYSON, S S. 1997, Unintended Effects of NO_x Emission Control Strategies on Unburned Carbon and CCP Marketability, Proceedings, Third Annual Conference on Unburned Carbon on Utility Fly Ash, U.S. Dept. of Energy, F.E.T.C., Pittsburgh, PA, pp 3-5.

2. AMERICAN SOCIETY FOR TESTING AND MATERIALS. 1990, Standard Specification for Fly Ash and Raw or Calcined Natural Pozzolan for use as a Mineral Admixture in Portland Cement Concrete, C-618a, Annual Book of ASTM Standards, Vol. 4.02, pp 298-300.

3. LEWIS, W M. 1998, Simplified Tailings Recovery Utilizing a Lewis Econsizer, 1998 SME Annual Meeting and Exhibit, Orlando, FL, Preprint, pp 98-128.

4. GROPPO, J G, ROBL, T L, GRAHAM, U M and MCCORMICK, C J. 1996, Selective Beneficiation for High Loss-of-Ignition Fly Ash: *Mining Engineering*, **48**, No. 6, SME, pp 51-53.

5. GROPPO, J G, ROBL, T L, GRAHAM, U M, MCCORMICK, C J, EDENS, T F and MEDINA, S S, 1997, Pilot Test Results of APT's Fast Float^{TM} Technology for the Reclamation of Stored Fly Ash, Proceedings of the 1997 International Ash Utilization Symposium, Univ. of Ky, CAER, The Journal Fuel and the U.S. Dept. of Energy, Lexington, KY pp 637-642.

6. AMERICAN SOCIETY FOR TESTING AND MATERIALS, 1990, Standard Test Methods for Sampling and Testing Fly Ash or Natural Pozzolans for use as a Mineral Admixture in Portland-Cement Concrete, C-311, Annual Book of ASTM Standards, Vol. 4.02, pp 186-190.

7. YASUDA, M, NIIMURA, T, IIZAWA, M, and SHIMADA, Y, 1991, Possibility of Utilizing Wet Stored Ash as Material for Concrete, Proceedings, 1991 American Coal Ash Association Symposium, Paper 18-1, 15 pp.

ADMIXTURE COMPATIBILITY AND MIXER REQUIREMENTS FOR CONDITIONED PFA CONCRETE

M J McCarthy

P A J Tittle

R K Dhir

University of Dundee

United Kingdom

ABSTRACT. The paper describes the work of a study carried out to resolve practical issues associated with the use of conditioned (moistened) pulverized-fuel ash (PFA) as a component of cement in concrete. This included the influence of conditioned PFA on the action of chemical admixtures, with both water-reducers and air-entrainers considered. In addition, work was carried out to examine the influence of concrete mixer type (forced action, rotary drum and ready mix) on the uniformity (by strength variation) of conditioned PFA concrete. The study demonstrated that the effects of conditioned PFA on plain and admixture concrete in terms of workability and strength properties were similar and hence admixture efficiency was unaffected. Furthermore, air contents at a fixed admixture dosage were also comparable in dry and conditioned PFA concrete. Similar results in terms of concrete uniformity were obtained between different concrete mixers, indicating that no special provisions are required in the production of conditioned PFA concrete.

Keywords Conditioned pulverized-fly ash, Air-entraining admixtures, Super-plasticizing admixtures, Concrete mixer requirements, Uniformity.

Dr M J McCarthy is a Lecturer in Concrete Technology, University of Dundee, Scotland, UK. His main research interests include developing technology for extending PFA use in concrete construction and most aspects of concrete durability, in particular carbonation and chloride induced corrosion. He is actively involved in technology transfer of his research activities.

Mr P A J Tittle is a Research/Teaching Assistant in Concrete Technology, University of Dundee. His research interests include extending the use of industrial by-products as cement components in concrete, and examining the role of mix limitation specifications in ensuring concrete durability.

Professor R K Dhir is Director of the Concrete Technology Unit, University of Dundee. He specialises in the use of cements and the durability and protection of concrete, with particular reference to carbonation and chloride ingress. Professor Dhir has published widely and serves on many Technical Committees. He is past chairman of the Concrete Society, Scotland.

INTRODUCTION

After production at coal-fired power stations and collection from precipitators, pulverized-fuel ash (PFA) is normally treated one of the following ways, (i) stored dry in bags or silos, (ii) moistened under control and stockpiled (conditioned), or (iii) mixed to produce a slurry and pumped to a lagoon. According to most national standards, it is only the material kept dry that is permitted as a cement component in concrete. This limitation is principally in place to prevent possible deterioration through contact with water and to ensure consistent handling properties.

Recent work carried out by the authors [1,2] has examined the effects of moisture addition to the properties of conditioned PFA. This has demonstrated that agglomeration of the material, both as a result of physical and chemical processes occurs. The degree of this effect is mainly influenced by the lime content of the PFA and the storage period and is less affected by other material properties or conditioning variables such as fineness or moisture content.

Studies of the influences of conditioned PFA as a component of cement in concrete indicate that despite the changes in PFA characteristics noted, these have a relatively minor impact on concrete performance, providing account is taken of the water present in the PFA. Indeed, while slight workability loss may occur, equivalent hardened properties to those of dry PFA are generally achievable to 28 days if certain requirements, mainly relating to storage time, (should be used between 3 and 6 months after conditioning depending on lime content) are met [3].

The work, described above, has demonstrated feasibility for use of conditioned PFA as a cement component in concrete. There are, however, changes in PFA characteristics, not least the agglomeration that occurs, which could conceivably influence concrete production and hence before wider use at a practical level is achieved, these need to be addressed. The aims of the work described in this paper were to examine two of these issues, namely (i) compatibility of conditioned PFA with chemical admixtures and (ii) its influence on mixer requirements.

TEST PROGRAMME

Admixture Compatibility

A recent survey [4] indicates that the most widely used admixtures in UK practice are (i) air-entraining and (ii) superplasticizing admixtures and the work of the study, therefore, considered both of these. For the tests carried out, trial mixing with dry PFA concrete was made to establish suitable admixture dosage levels and these were then used in conditioned PFA concretes to test the influence of moisture addition.

Mixer Type

The work was concerned with the uniformity of conditioned PFA concrete produced using different types of concrete mixer, with both forced action and rotary drum mixers examined in the laboratory. The former, which was used for the majority of the work of the study, was of 0.04 m^3 capacity and had rotating paddles and scraper blade, while the latter of similar capacity worked by free fall, with vanes in the barrel which rotated about a central axis. This work was extended by carrying out tests on concrete produced at a ready-mix plant and a pre-cast factory.

EXPERIMENTAL DETAILS

Materials

A fine and a coarse PFA were used in this study, with fineness at either end of the BS EN 450 range [5]. Table 1 gives the main properties of both materials, in dry form and after storage at 10% moisture content for up to 6 months. These data indicate coarsening of PFA and slight increases in LOI following conditioning, with the effects being progressive with storage period. The changes occurring reflect the physical and chemical phenomena highlighted earlier associated with moisture addition to PFA concrete [1,2].

Table 1 Properties of dry and conditioned PFA (10% moisture and storage for up to 6 months)

PFA STORAGE, months	PFA 1				PFA 2			
	Fineness, % ret 45 μm	LOI, %	Total lime, %	Free lime, %	Fineness, % ret 45 μm	LOI, %	Total lime, %	Free lime, %
Dry	15.1	10.9	2.8	<0.1	36.6	3.5	1.8	<0.1
0	21.5	11.2	2.5	--	37.6	3.5	2.1	--
1	30.8	11.3	2.7	--	44.7	3.6	2.2	--
6	33.2	11.2	2.7	--	54.4	4.0	2.1	--

Mains water was used for PFA conditioning and concrete mixing, curing and testing.

Table 2 gives the main properties of the BS 12 [6] Class 42.5 N Portland cement (PC) used in the concrete mixes. Natural gravel aggregate in 20-10 and 10-5 mm fractions and Zone M sand, conforming to BS 882 [7], were used in a surface dry condition. Concrete mixes at the ready-mix plant and pre-cast factory used gravel and natural sand having similar properties to those used in the laboratory at Dundee University. However, at the pre-cast factory, only coarse aggregate in the 5-10 mm fraction was used, in line with normal operating procedures.

Table 3 gives the main properties of the air-entraining and super-plasticizing admixtures used, which were selected for their compatibility with PFA and conformed to BS 5075, Part 2 [8] and BS 5075, Part 3 [9] respectively.

Table 2 Main properties of Portland cement (PC, 42.5 N)

MAIN OXIDE COMPOSITION, %									FINENESS, m²/kg
SiO_2	Al_2O_3	Fe_2O_3	CaO	MgO	SO_3	K_2O	Na_2O	LOI	
20.8	5.0	2.9	63.5	2.8	3.1	0.7	0.3	0.9	350

Table 3 Properties of chemical admixtures

ADMIXTURE	APPEARANCE	SPECIFIC GRAVITY	CHLORIDE CONTENT, %
Air-entrainer	Pale straw coloured liquid	1.01 at 20°C	Nil
Super-plasticizer	Brown liquid	1.17 at 20°C	Nil

Conditioning of PFA

PFAs 1 and 2 were conditioned to a moisture level of 10% by dry mass, through the addition of mains water and mixing in a $0.025m^3$ horizontal pan mixer for 4 minutes. The material was then stored in sealed containers at 20°C for 24 hours, 1 or 6 months, prior to testing and use in concrete mixes. Related work [1] has shown that the properties of conditioned PFA produced by this laboratory procedure are comparable to those of material treated using conventional plant and stored in stockpiles at power stations.

Mix Proportions

Admixture compatibility of conditioned PFA concrete was tested on concrete of cement content 350 kg/m^3, with a PFA level of 30% and w/c ratio of 0.50, see Table 4. The proportions for the plain dry PFA concrete mix (M1, no admixture) were developed in a parallel study [10] to give nominal 75 mm slump with the fine PFA.

These mix proportions were used for air-entrained concretes (M2), with the admixture dosage determined for dry PFA concrete by trial mixing to give a nominal 4.0% air content, and adjustment made to the sand content to account for the volume of entrained air. No compensation was made for the strength change associated with the air-entrainment.

When using super-plasticizer (M3), the water-contents of the mixes were reduced by 15 kg/m^3, compared to the plain mix (M1), with adjustment made to the cement and sand contents to maintain w/c ratio and yield, respectively. The degree of water reduction made enabled the 75 mm nominal slump to be achieved with super-plasticizer dosage levels recommended by the admixture manufacture.

Conditioned PFA concretes were produced for the above mixes, after adjusting for PFA moisture content, i.e. reduced water and increased PFA quantities added at the mixer. For mixes M2 and M3, the admixture dosages established for dry PFA were used.

Mix M1 was used to consider the significance of concrete mixer type (forced action and rotary drum) in the laboratory. For the trials at the ready-mix plant, mix M1 was used. At the pre-cast factory, the company's PC mix for edge stones was used (design strength 40.0 N/mm^2), after modification for inclusion of 30% PFA, see Mix 4 in Table 4. It was noted that, in line with normal procedures, the water was 'eyed in' to give a slump of 50-70 mm.

Table 4 Concrete mix proportions used with dry PFA

MIX REF	MIX CONSTITUENTS, kg/m²									W/C
	Total Binder	PC	PFA	Free Water	ADMIXTURES*		NATURAL AGGREGATES			
					AEA (PFA 1 & 2)	SP	Sand	5-10 mm	10-20 mm	
M1	350	245	105	175	0	0	650	405	805	0.50
M2	350	245	105	175	800 & 200	0	540	405	805	0.50
M3	320	225	95	160	0	0.6	720	405	805	0.50
M4	430	300	130	120	0	0	820	1125	0	0.28

M1 Horizontal pan, rotary drum and ready-mix mixers; 75 mm nominal slump
M2-3 Horizontal pan; *AEA dosage in ml / 100 kg binder to give 4.0% nominal air content, SP as % of total binder to give 75 mm nominal slump
M4 Pre-cast factory mixer; 50-70 mm nominal slump

Concrete Mixers

Concrete was produced in the laboratory using both forced action and rotary drum mixers following the procedure described in BS 1881: Part 125 [11].

At the ready-mix plant, PFA was discharged from a small-scale sand hopper [1] onto a conveyor belt. Mixing then followed normal operating procedures, with all materials carried to the 6.0 m³ capacity ready-mix truck. A batch size of 0.2 m³ was used, which required approximately 20 kg of conditioned PFA.

At the pre-cast factory, a Cumflow 0.2 m³ capacity mixer was used. Conditioned PFA concrete was produced following normal operating procedures, with all materials shovelled into a balance hopper, prior to discharge into the concrete mixer. A batch size of 0.1 m³ was used with approximately 15 kg of conditioned PFA included.

Test Methods

Immediately following mixing, slump tests were carried out in accordance with BS 1881: Part 102 [12] to assess workability. For air-entrained mixes (forced action mixer only), the air content was determined in accordance with Method B of BS 1881: Part 101 [13]. This involved monitoring the change in air pressure within a sealed 8 litre chamber (containing the compacted fresh concrete sample), following the introduction of a fixed volume of air at a specific pressure.

Fresh concrete was compacted into 100 mm cube moulds, which were stored under damp hessian for 24 hours, prior to demoulding and water curing at 20°C until testing. At 28 days, two specimens were tested for compressive strength in accordance with BS 1881: Part 116 [14], and the mean result reported.

For all mixer types, the within-batch uniformity of concrete mixes was determined by similarly testing 20 cubes from a batch at 28 days. Uniformity was tested in this way rather than through measurement of fresh concrete for composition, as suggested elsewhere [15,16], since it represented a simpler and more practical means of evaluation.

RESULTS AND DISCUSSION

Admixture Compatibility

Air-entrained concrete

Air contents

Figure 1 compares the performance of plain and air-entrained concretes containing PC/PFA1 and PC/PFA2 in dry form and after storage at a moisture content of 10% for up to 6 months. The concrete mixes containing dry PFAs 1 and 2 required 800 and 200 ml of admixture per 100 kg cement, respectively. These follow expected behaviour, with the high dosage requirements of PFA 1 reflecting the relatively high LOI of this material.

Using these admixture dosages for the corresponding conditioned PFA concretes, the levels of entrained air measured, Figure 1(a), were comparable to the dry PFA concretes, although there was a suggestion of slight progressive increases with conditioned PFA storage period, particularly with PFA 1.

Given the progressive changes in properties occurring, in particular agglomeration, with PFA conditioning and storage, it is probable that despite general breakdown of conditioned PFA particles in the concrete mixer, some remain and these coarser particles may represent regions where air entrapment may occur during compaction. It is additionally possible that the chemical changes in PFA with conditioning [2] may have minor influences on air bubble formation.

The data suggest that admixture dosages established for dry PFA concrete mixes can be used with conditioned PFA and the differences in air content obtained fall within acceptable specification tolerances.

Workability

The results from the tests for concrete workability are given in Figure 1(b). In the case of dry PFA concrete, the slump of air-entrained concrete was 20 to 30 mm higher than equivalent plain concrete. Perhaps surprisingly, the inclusion of a higher admixture dosage level in PFA 1 concrete, therefore, did not have any further influence on workability.

This increased workability, associated with the introduction of air-entraining admixtures, is expected behaviour, given that these admixtures are known to have a plasticizing effect, tending to promote the dispersion of PC flocs in concrete.

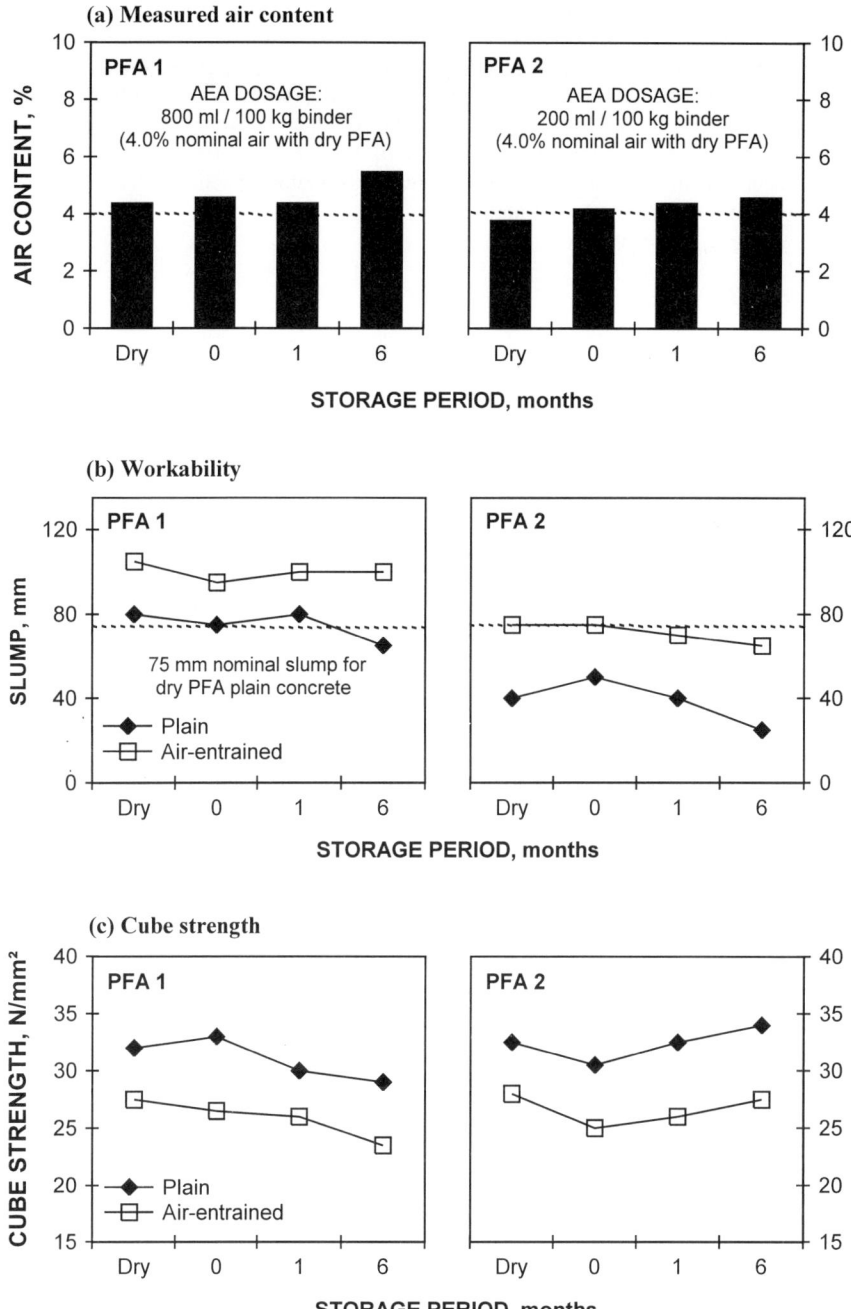

Figure 1 Performance of air-entrained dry and conditioned PFA concrete (10% PFA moisture)

The slump losses associated with conditioning in plain PFA concrete appears to reflect particle agglomeration and increased water demand associated with wet storage, which offset the dispersing and water reducing effects normally associated with PFA. It would appear that the effect of the air-entraining admixture on workability of conditioned PFA is almost constant, with near parallel lines obtained between plain and air-entrained concretes. This suggests that the admixture does not contribute to breakdown of PFA agglomerates when the material is conditioned, but simply provides for better overall dispersion of solids in the mix, working as efficiently with both dry and conditioned PFA.

Cube strength

The 28 day strength results obtained for plain and air-entrained concrete mixes are given in Figure 1(c). The effect of air-entrainment in dry PFA concrete, as expected, lead to strength reductions of typically 5.0 to 7.0 N/mm^2. These are broadly in line with expected effects, with approximately 5.0% strength loss associated with each 1.0% air entrained in the mix.

The effect of PFA conditioning and storage on 28 day cube strength was relatively minor, with differences of no more than 3.0 N/mm^2 measured. As noted for air content and workability measurements, similar strength reductions for all conditioned PFA air-entrained concretes was obtained to those containing dry PFA.

These data confirm that conditioning of PFA has little influence on the use of air-entraining admixtures in PFA concrete. Indeed, the results suggest that LOI of PFA and its consistency remains the critical parameter in controlling air-entrainment in PFA concrete. Early data [17] from work considering the freeze-thaw (scaling) resistance of conditioned PFA concrete, indicates that with air-entrainment similar performance is obtained to that with dry material.

Super-plasticized concrete

Workability

Figure 2 gives the results obtained for super-plasticized concrete mixes containing PC/PFA1 and PC/PFA2 in dry form and after conditioning to 10% moisture and storage for up to 6 months. For the dry PFA plain concrete mixes, the nominal 75 mm slump was achieved with PC/PFA1, but PC/PFA2 gave a slump of only 40 mm, reflecting the material's relative coarseness. For the super-plasticized concretes, which had water contents 15 kg/m² lower than the plain concretes, admixture dosages of 0.6% by mass of cement were required to achieve the nominal slump (75 mm) for both PFAs.

These admixture dosage requirements suggest that, the quantity of fine material in the mix, requiring dispersion, is the main controlling influence. Indeed, this appears to be independent of PFA fineness and the effect of differences in water reducing properties between PFAs of different fineness is eliminated.

The workability of super-plasticized conditioned PC/PFA1 concrete mixes, which contained the same level of admixture as equivalent dry PFA concrete, generally followed the same trend as the plain concrete, Figure 2(a), i.e. a progressive reduction with increasing storage period, compared to dry PFA concrete. While the same general trend was obtained for PC/PFA2 concrete, the super-plasticized concrete gave higher workability than that of the plain. These suggest that the admixture was more efficient with PC/PFA2 concrete and that water and admixture demand are clearly controlled by different mechanisms.

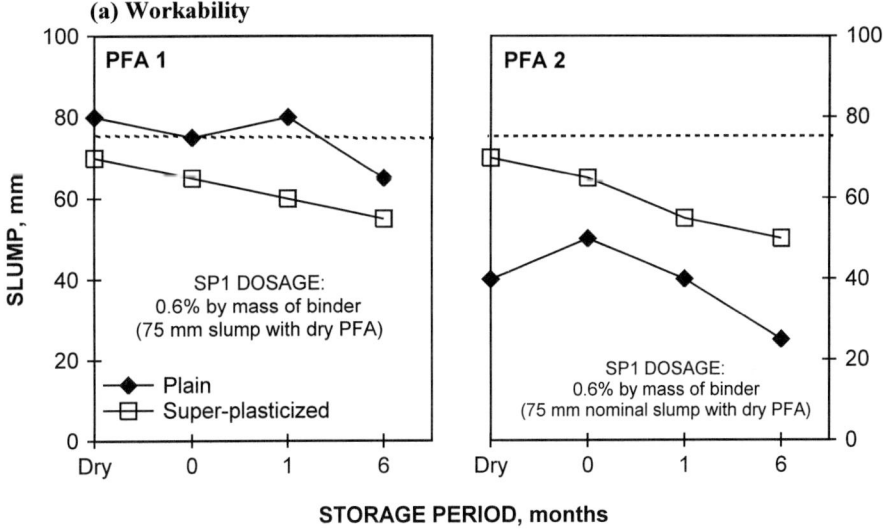

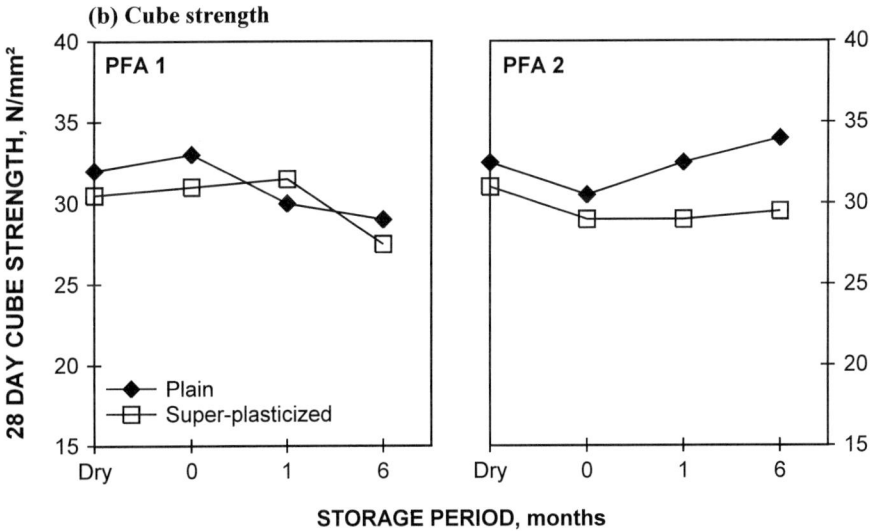

Figure 2 Performance of super-plasticized dry and conditioned PFA concrete (10% PFA moisture)

The progressive reductions in workability noted for fixed admixture dosage in PC/PFA concretes seem likely to reflect greater agglomeration of PFA particles with storage period and indicate that at a fixed concentration of admixture, a constant amount of particle dispersion in the mix occurs. The results appear to suggest that as with the air-entraining admixture, the super-plasticizing admixture leads to little breakdown of conditioned PFA agglomerates themselves and its effect is again simply to provide overall solids dispersion.

Cube strength

The 28 day strength results of super-plasticized concretes are given in Figure 2(b). For dry PFA there were essentially only minor differences between the strengths of the PC/PFA1 and PC/PFA2 concretes. The results obtained for plain and super-plasticized concretes were within 2.0 N/mm^2 for both PC/PFA1 and PC/PFA2 concrete. This is in line with expected behaviour, which for fixed w/c ratio and 15 kg/m^3 water reduction should be similar.

The data obtained for conditioned PC/PFA1 concretes containing super-plasticizing admixture were similar to those of the plain concretes, exhibiting general reductions in strength with conditioned PFA storage, but differences of no more than 2.0 N/mm^2 noted. The results of the PC/PFA2 concrete indicate that with storage time, differences between plain and super-plasticized concretes increased, with the latter exhibiting reduced strength, by approximately 4.0 to 5.0 N/mm^2 by 6 months. This is slightly surprising given that the workability data suggested greater break down of agglomerates to a finer material in the mixer, due to the dispersing effect of the admixture. It is possible that the increased dispersion associated with super-plasticized concrete leads to possible release of water held within conditioned PFA agglomerates as they breakdown, increasing the quantity of water in the mix and hence leading to strength reductions.

Further work is required to establish the compatibility of conditioned PFA and super-plasticizing admixtures in concrete mixes, although the data obtained indicate relatively minor differences and comparable performance to dry PFA concretes.

Mixer Requirements

Workability

Table 5 compares the workability (slump) data for dry and conditioned PFA concrete mixes produced in horizontal pan and rotary drum mixers. The slump measured for concrete mixes in the rotary drum mixer was typically 10 to 20 mm higher than the horizontal pan mixer. The use of conditioned PFA in concrete had the same effect on workability for both mixers, i.e. reductions of 10 to 15 mm after storage for 6 months, compared to dry PFA concrete.

Visually there appeared to be little difference between corresponding mixes produced in each of the two mixers. The higher workability obtained with the rotary drum mixer may be due to increased water availability through segregation of the mix, caused by the free-fall action of the mixer [16]. In addition, since there is no scraping action, mortar could remain at the edges of this mixer and hence produce a mix rich in coarse aggregate particles, which due to fines deficiency could also contribute to increased workability. However, since slump loss following PFA conditioning was observed consistently for both mixers, and the differences obtained between mixer types relatively minor, it would appear that the mixer action was consistent for all materials and given the relatively minor differences obtained conditioned PFA could be used in either type and at least give satisfactory workability. To overcome the material effect of conditioned PFA, adjustments to the mix proportions may then be necessary.

Table 5 Effect of mixer type on the workability of dry and conditioned PFA concrete (10% moisture)

MIXER TYPE	SLUMP, mm							
	PFA 1 Storage, months				PFA 2 Storage, months			
	Dry	0	1	6	Dry	0	1	6
Horizontal pan	80	75	80	65	40	45	40	30
Rotary drum	95	85	90	80	60	65	50	45

Uniformity of Concrete

Table 6 gives the 28 day strength statistical data obtained for the mixes described above. In general, the strength of concrete in the rotary drum mixer was between 2.0 to 4.0 N/mm^2 lower than in the horizontal pan mixer, but the same trend of slightly lower strength when using conditioned PFA, compared to dry PFA concrete, was observed for both mixers.

Similarly, the effect of PFA conditioning on within-batch uniformity was not influenced by mixer type, i.e. increases in coefficient of variation of 1.0 to 2.0%, compared to dry PFA concrete, were obtained in both cases. However, the coefficient of variation with rotary drum mixing (typically 6.0%) was generally slightly higher than that for horizontal pan mixing (typically 4.0 to 5.0%).

Within-batch uniformity was also assessed for the mixers used at the ready mix plant and pre-cast factory. Figure 3 shows that coefficients of variation of approximately 8.0 and 9.0% were obtained on 28 day strength, which are 2.0 to 3.0% higher than for comparable concrete mixes produced using the laboratory mixers, and approximately 4.0 to 5.0% higher than for the dry PFA concrete mix produced in the forced action mixer. The variability of all concrete mixes, however, remained below acceptable levels for concrete uniformity [18], suggesting suitability for use with conditioned PFA.

Testing of concrete uniformity at a fuller load in a ready-mix mixer is clearly desirable for a more complete evaluation and this is currently in progress [17].

SUMMARY OF FINDINGS

Air-entraining and super-plasticizing admixture dosage requirements established for dry PFA concrete mixes can generally be used for conditioned PFA mixes. Indeed, for the former similar air contents for a given dosage are achievable.

Table 6 Effect of mixer type on the uniformity of dry/conditioned PFA concrete (10% moisture)

MIXER TYPE AND STATISTICAL PROPERTY	CUBE STRENGTH, N/mm²							
	PFA 1 Storage, months				PFA 2 Storage, months			
	Dry	0	1	6	Dry	0	1	6
Horizontal Pan								
MEAN	32.0	33.0	30.0	29.0	30.0	27.0	28.5	28.5
SD	1.12	1.45	1.57	1.62	1.10	1.27	1.60	1.62
CV, %	3.50	4.42	5.21	5.58	3.64	4.68	5.62	5.70
95% CI*	0.52	0.68	0.73	0.76	0.51	0.59	0.75	0.76
Rotary Drum								
MEAN	29.0	29.0	28.5	26.5	27.5	26.0	26.5	25.5
SD	1.58	1.67	1.85	1.69	1.53	1.50	1.69	1.80
CV, %	5.39	5.73	6.43	6.33	5.51	5.80	6.43	7.03
95% CI*	0.74	0.78	0.87	0.79	0.72	0.70	0.79	0.84

Number of tests: 20 cubes from a single mix   * Confidence interval limits

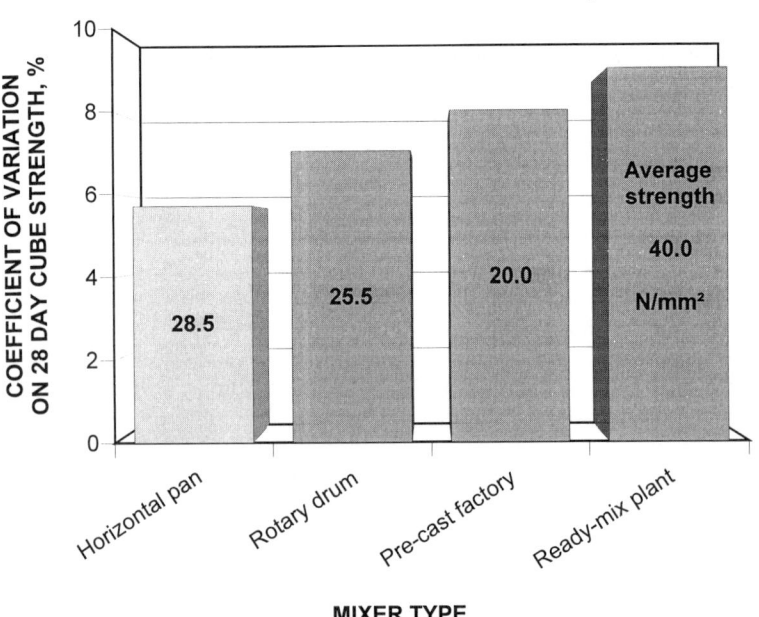

Figure 3 Within-batch uniformity of conditioned PFA produced in the laboratory and on site

The effects of conditioned PFA on fresh and hardened concrete observed for plain concrete are also observed when admixtures are included. However, they appear to work as effectively in conditioned PFA concrete as in dry PFA concrete. In order to offset possible workability loss obtained in conditioned PFA concrete, small increases in super-plasticizer may be necessary.

In general, the use of a rotary drum mixer increases slump by 10 to 20 mm and reduces 28 day strength by 2.0 to 4.0 N/mm^2, compared to that produced in a forced action mixer. This was observed for both dry and conditioned PFA concretes.

The within-batch coefficient of variation of concrete produced in a rotary drum mixer is slightly higher than for a forced action mixer, but the influence of conditioning is the same in both cases (i.e. slight reductions in uniformity). As expected, site mixers (ready-mix and pre-cast) exhibit higher within-batch coefficients of variation than laboratory mixers, but all values remain within acceptable limits for concrete production.

ACKNOWLEDGEMENTS

The Authors would like thank the Department of the Environment Transport and the Regions and industrial partners, Boral Pozzolan Ltd, BAA plc, BRMCA, National Power plc, Powergen and Scottish Power plc for their financial support and guidance of the project.

REFERENCES

1. DHIR, R K, McCARTHY, M J and TITTLE, P A J. 1998. Use of conditioned and lagoon PFA in structural concrete. Report CTU/598, University of Dundee, Scotland, 199 pp..

2. McCARTHY, M J, TITTLE, P A J and DHIR, R K. In press. Characterisation of conditioned PFA for use as a cement component in concrete. Magazine of Concrete Research.

3. McCARTHY, M J, TITTLE, P A J and DHIR, R K. Submitted. Influences of conditioned PFA as a cement component on the properties of concrete. Magazine of Concrete Research.

4. SPOONER, D C, DHIR, R K and LIMBACHIYA, M C. Targeted technology transfer in construction. Report CTU/199, University of Dundee, Scotland, 76 pp

5. BRITISH STANDARDS INSTITUTION. 1995. BS EN 450. 1995. Fly ash for use in concrete: Definitions, requirements, and quality control. London.

6. BRITISH STANDARDS INSTITUTION. 1996. BS 12: Specification for Portland Cement. London.

7. BRITISH STANDARDS INSTITUTION. 1992. BS 882: Specification for aggregates from natural sources for concrete. London.

8. BRITISH STANDARDS INSTITUTION. 1982. BS 5075, Part 2: Specification for air-entraining admixtures. London.

9. BRITISH STANDARDS INSTITUTION. 1985. BS 5075, Part 3: Specification for superplasticizing admixtures. London.

10. DHIR, R K, McCARTHY, M J and MAGEE, B J. 1998. Impact of BS EN 450 PFA on concrete construction. Construction and Building Materials. Vol 12, No 1, pp 59-74.

11. BRITISH STANDARDS INSTITUTION. 1983. BS 1881: Part 125: Testing of concrete Method for mixing and sampling fresh concrete in the laboratory. London.

12. BRITISH STANDARDS INSTITUTION. 1983. BS 1881: Part 102: Testing of concrete Method for determination of slump. London.

13. BRITISH STANDARDS INSTITUTION. 1983. BS 1881: Part 101: Testing of concrete Method for determination of air content. London.

14. BRITISH STANDARDS INSTITUTION. 1983. BS 1881: Part 116: Testing of concrete Method for determination of compressive strength of concrete cubes. London.

15. BRITISH STANDARDS INSTITUTION. 1974. BS 1305: Specification for batch-type concrete mixers. London.

16. BRITISH STANDARDS INSTITUTION. 1974. BS 3963: Testing the mix performance of concrete mixers, London.

17. DHIR, RK, McCARTHY, M J and KII, H K.1999. Conditioned PFA concrete: engineering properties and critical aspects of durability. In preparation.

18. NEVILLE, A M. 1995. Properties of concrete. Longman, England, 844 pp.

FLOWABLE ROCKDUST – CEMENT SLURRY AS BACKFILL MATERIAL

B K Baguant

University of Mauritius

Mauritius

ABSTRACT. Rockdust is an unavoidable fine waste of aggregate manufacture in Mauritius, and is produced in sufficient quantities to be an environmental nuisance. This study investigates the potential of this material to be stabilised with cement and produced as a flowable slurry for backfilling roadway trenches and other excavations. Mechanical stabilisation of the rockdust by compaction is satisfactory only if the material is prevented from becoming saturated. The characteristics of rockdust-cement slurry were determined in the flowable and hardened states for cement contents ranging from 2-20% by weight of rockdust, while maintaining flow constant. A simple method is proposed for measuring this flow. The hardened slurry possesses lower strength and CBR values when stored in a saturated condition than in air, but, it achieves in 4 days target CBR values which are adequate for backfilling purposes with cement contents of 75-100 kg/m^3 (6-8% by weight of rockdust). With higher cement contents the target CBR is reached in less time.

Keywords: Rockdust, Flowable slurry, Cement stabilisation, Bleeding, Compressive strength, California bearing ratio (CBR), Modulus of elasticity, Drying shrinkage.

Dr B K Baguant is Associate Professor in the Department of Civil Engineering, University of Mauritius. His research interests include construction materials in general and concrete technology in particular.

INTRODUCTION

Mauritius is a small island of volcanic origin and the only natural source of aggregate for use in construction is basalt rock, which occurs in the form of surface boulders, sub-surface boulders, bedrock and small hills. Thus, basalt, either quarried or in the form of boulders, is crushed to produce aggregates. The crushing process generates a significant quantity of rockdust which, if left in the crushed aggregates and crushed sand, is considered to be a harmful contaminant for concrete. In practice, these fines in the crushed aggregates are reduced to acceptable levels by washing before using the aggregates for structural concrete. Around 3 million Tonnes of rock are crushed annually in the country. The crusher run material contains about 6% of rockdust, that is, material passing 75 µm sieve. Crushed sand, if left unwashed, would contain 18-25% rockdust, which after washing is reduced to 5-8%. The local stone crushing industry disposes of some 120,000 Tonnes of rockdust annually, typically by hauling to disposal grounds located near the crushing plants, but being very fine, the dust is an environmental nuisance, especially once it dries (1).

This paper investigates the potential for using rockdust for backfill where excavatability is later required. Mechanical stabilisation by compaction of the rockdust alone can achieve, at optimum moisture contents, CBR values of 18-20%. On soaking, however, these values drop to 3-5%, which are well below the minimum of 10% soaked CBR commonly specified in Mauritius for backfills for roadway trenches. Alternatively, rockdust can be stabilised by mixing with additives, such as, lime, cements, bitumen, and polymer resins.

Results are reported here of flowable and hardened rockdust-cement slurry tested with various cement contents ranging from 2-20% by weight of rockdust, while maintaining flowability constant. In the flowable state plastic density and bleeding were measured, while in the hardened state, measurements were made of strength development from 1 to 28 days, corresponding CBR values, drying shrinkage, static modulus of elasticity, and ultrasonic pulse velocity. The advantages of using a flowable slurry are ease of filling excavations under gravity, self-compaction, less equipment and labour, minimal need for supervision if control is exercised at the batching plant, and the possibility of using the material in unsaturated or saturated soil conditions.

THE ROCK CRUSHING PROCESS

Boulders from surface sources, or rock obtained by explosive blasting in quarries, are transported by lorry to stone crushing plants. The process typically involves three, sometimes four stages of crushing. A schematic layout of the stone crushing process is shown in Figure 1. Primary jaw crushers reduce the rockfeed down to sizes of about 200 mm. This method of crushing tends to produce particle shapes and textures which are flaky, angular and rough.

Secondary crushers are fed with primary crusher material which reduce them to sizes in the range 0 - 50 mm. These crushers can also be of the jaw-type, but in most cases, are of rotary-type, such as impacters, which produce more abrasive crushing. The main advantage of the latter technique of crushing is a significant improvement in particle shapes, whereby these are more cubical and better suited for making fresh concrete workable and pumpable with lower water demands.

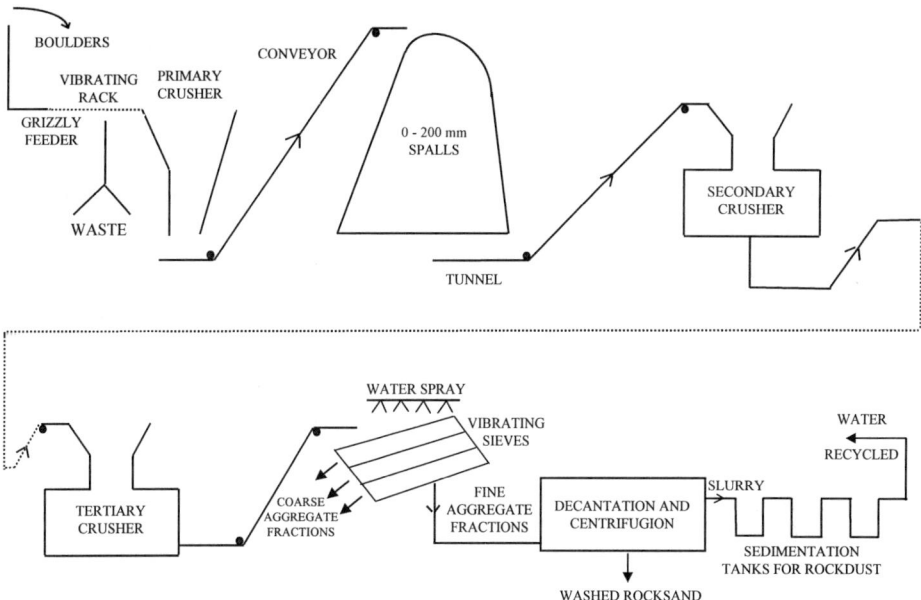

Figure 1 Schematic diagram of the stone crushing process

The tertiary crushers are invariably rotary-type crushers which reduce the output of the secondary crushers to sizes in the range 0 - 20 mm. The processes mentioned so far are typically dry crushing processes with only light water spraying to keep the dust down. The 0 - 20 mm crusher material is then fed onto a deck of vibrating sieves with simultaneous and relatively abundant water spraying so as to wash and to separate the particles into coarse fractions (20, 14 and 10mm nominal sizes) and fine aggregate (less than 5 mm).

The material passing 5mm (sometimes 4mm) is fed to decantation tanks and centrifugal suction extractors to remove excess fine material (rockdust) leaving washed rocksand. The fine particles are carried in the flow as a slurry and are allowed to settle in sedimentation tanks for disposal, while the water is recycled into the system.

PHYSICAL AND CHEMICAL CHARACTERISTICS OF ROCKDUST

Some physical properties of rockdust are shown in Table 1 and the particle size distribution is indicated in Table 2. Rockdust is mainly a coarse to medium silt, with a relatively low clay content varying between 1 and 5%. This is corroborated by the fact that it shows little or no plasticity and the liquid limit varies between 27 and 30%. The chemical composition of rockdust is very similar to that of rocksand (see Table 3), and not surprisingly so, because both products originate from the same parent rock material, that is, basalt (2).

Table 1 Some physical properties of rockdust

PROPERTY	TEST METHOD	TEST RESULT
Relative Density	BS 812 : 1975	
• Ovendry basis		2.79
• SSD basis		2.86
Water absorption	BS 812 : 1975 (Funnel Method)	2.65%
Loose Bulk Density	BS 812 : 1975	1530 kg/m³
% Voids		45%
Compacted bulk density	BS 812 : 1975	1660 kg/m³
% Voids		40%
Specific Surface	ASTM C-204-1975 (Blaine Air Permeability Method)	1655 m²/kg
Liquid Limit	BS 1377 : 1990	27.3 - 30.6%
Plastic Limit	BS 1377 : 1990	No plastic limit
Linear Shrinkage	BS 1377 : 1990	0.9%

Table 2 Particle size distribution of rockdust (M.I.T classification)

CLAY	SILT			SAND		
	F	M	C	F	M	C
0-2 microns	2-6	6-20	20-60	60-200	200-600	600-2000
1-5 %	2-5 %	17-20 %	60-64 %	11-12 %	2%	Nil

Table 3 Chemical composition

MATERIAL	ROCKDUST	ROCKSAND
Constituents	% By Weight	% By Weight
SiO_2	80.96	81.79
$Al_2O_3+Fe_2O_3$	15.72	15.31
CaO	0.10	0.08
Cl_2	0.44	0.28
SO_4	0	0
MgO	2.19	2.43
P_2O_5	0.09	0.10
L.O.I	0.59	0.67
PH	8.3	7.9

FLOW MEASUREMENT

Viscosity Method

If we were dealing with traditional concrete, the workability would have been measured by means of any of the standard methods and kept constant. For a flowable slurry it was thought appropriate to measure viscosity as the physical property related to flow, and attempt to maintain it constant. Bearing in mind that high water content seriously affects hardened properties such as strength and drying shrinkage, the aim was to minimise the total water content while ensuring adequate flowability.

Different viscometers were tried in order to assess their suitability for determining the water content-viscosity relationship with a view to choosing a target viscosity which will result in adequate flowability of the slurry. A sliding plate viscometer, a capillary viscometer, a standard tar viscometer, and a standard penetrometer were tried before finally selecting a torsion viscometer. In the latter, a steel cylinder or a paddle is suspended in the liquid by means of a wire and the wire is twisted, imparting a torque to the cylinder or paddle, which then rotates. The speed of rotation is related to the viscosity of the surrounding liquid. This instrument is widely used in sugar technology for the determination of the viscosity of molasses and is convenient for measuring high to low viscosities. The variation of viscosity in Poise with water-rockdust ratio by weight is shown in Figure 2. The slurry became stiff and non-flowable as the water-rockdust ratio reached 0.26. As expected, the viscosity decreased with increasing water content.

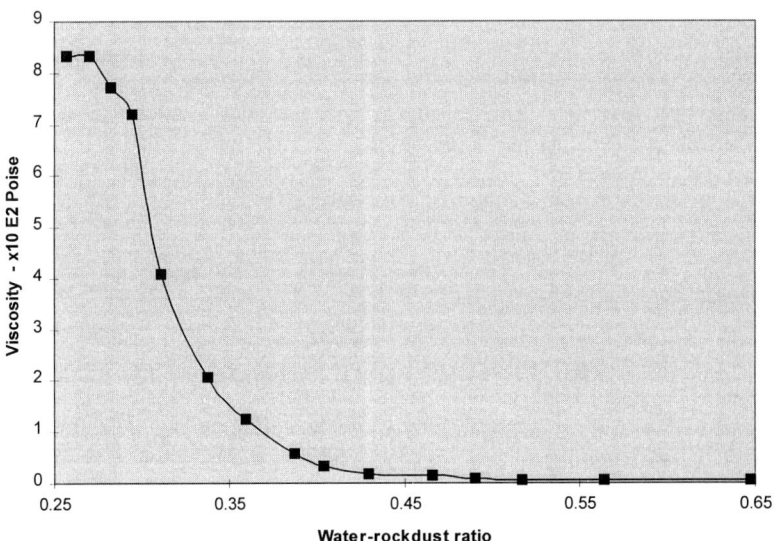

Figure 2 Variation of viscosity with water-rockdust ratio

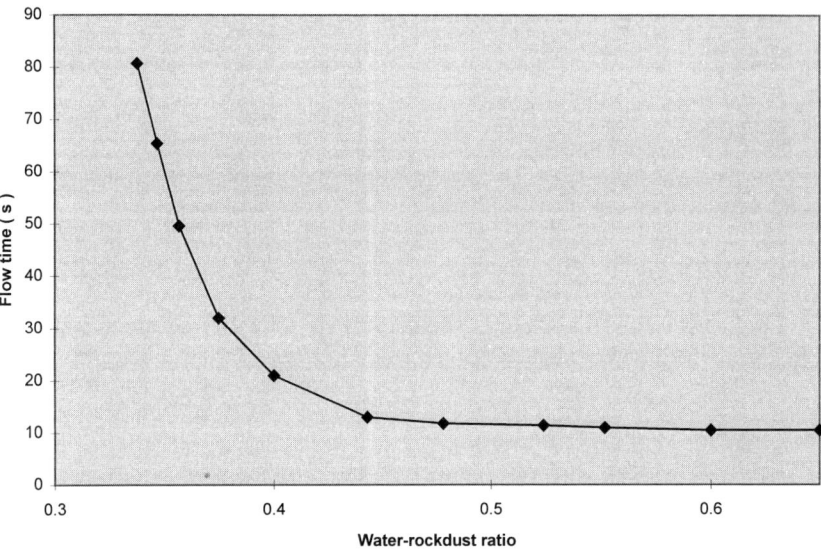

Figure 3 Variation of flow time with water-rockdust ratio

Alternative Method Using a Funnel

The principle underlying this method is similar to that of the standard tar viscometer, but is a far simpler and more straightforward test to perform. A sufficiently large volume funnel was selected so that the time taken for water filled to the brim of the funnel to flow out (the flow time) was at least 10 seconds. The funnel used had a volume of 350 mL, a rim diameter of 115 mm and a cone apex angle of 60^0. The slurry being more viscous than water had flow times greater than that for water. The variation of flow time with water-rockdust ratio is shown in Figure 3. The flow time for water was 10.3 s. The close similarity between the viscosity/water-rockdust relationship (Figure 2) and the flow time/water-rockdust relationship (Figure 3) suggests that the latter method is equally reliable. From Figure 3, a flow time of 12.5-13.0 s was chosen as the parameter to keep constant in order to maintain flow of the slurry constant. Once this was established, it was possible to measure the water-rockdust ratio as cement increased from 2-20% by weight of rockdust (2). This is illustrated in Figure 4.

PROPERTIES INVESTIGATED

The cement content of the cement-rockdust slurry varied from 25 kg/m^3 to 250 kg/m^3, and for most of these mixtures the properties investigated and the test details were as shown in Table 4. The low cement content mixtures were found to be quite weak, and therefore, the tests on hardened slurry were carried out only for mixtures containing more than 75 kg/m^3 of cement. The test methods were those applicable to concrete except that no vibration or other form of compaction was effected on the slurry as the consistency was such that it was self-compacting.

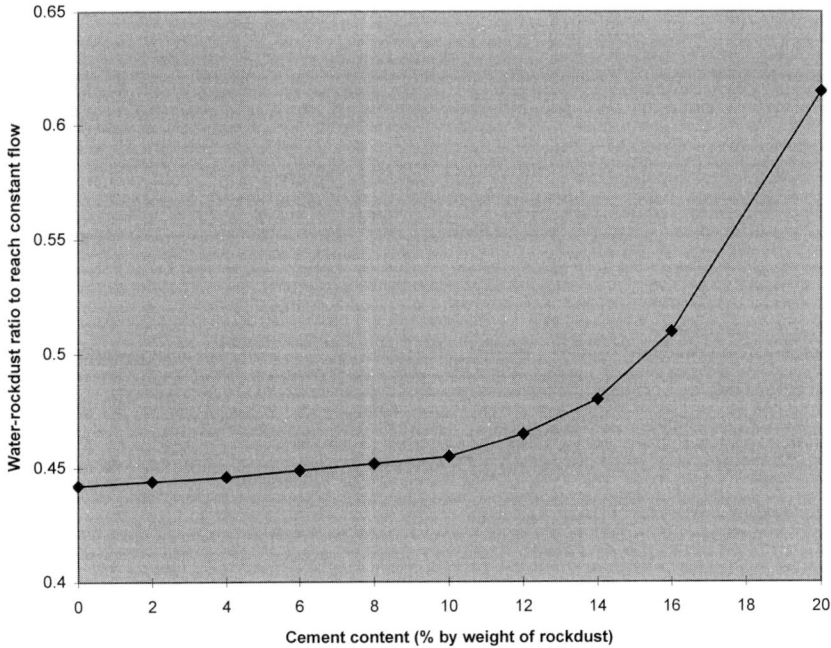

Figure 4 Change in water-rockdust ratio with cement content

Table 4 Properties investigated

PROPERTY	SPECIMENS/TEST DETAILS	PROCEDURE
• Plastic density		BS 1881: Part 107:1983
• Bleeding		ASTM C232-87
• Compressive Strength and Strength development	100 mm cubes cured in water at 23° C and in ambient air conditions and tested up to 28 days	BS 1881: Part 116:1983
• Elastic deformation	150 mm ∅ x 300 mm cylinders cured in water at 23°C and in ambient air conditions and tested at the age of 28 days for stress-strain curve and elastic modulus	BS 1881: Part 121:1983
• Drying shrinkage	75 x 75 x 300 mm prisms cured in water at 23°C and tested by oven-drying at the age of 28 days	BS 1881: Part 5:1970
• Ultrasonic Pulse Velocity	150 mm x 300 mm cylinders cured in ambient air for 28 days	BS 1881: Part 5:1970
• CBR	unsoaked and after soaking for 4 days	BS 1377 : 1990

RESULTS

Flowable Slurry

The plastic density varied within a narrow range (Table 5) as cement content increased from 25 to 250 kg/m³, but the water demand increased significantly for cement contents greater than 175 kg/m³ in order to maintain the flow constant. Bleeding was highest when no cement was present (24.7% of total water added) and decreased as cement content increased to 150 kg/m³. But for higher cement contents the bleeding started increasing again because of increased water demand to keep flow constant.

Table 5 Properties of flowable slurry

CEMENT CONTENT % of Rockdust	kg/m³	WATER DEMAND L/m³	PLASTIC DENSITY kg/m³	BLEEDING % OF TOTAL WATER
2	25	565	1865	21.3
4	50	555	1855	18.1
6	75	550	1845	14.3
8	100	540	1835	11.4
10	125	535	1825	7.7
12	150	540	1845	7.5
14	175	555	1860	11.0
16	200	570	1870	14.6
20	250	640	1890	15.2

Hardened Slurry

Both strength and CBR values of the hardened slurry increased steadily over the first 28 days (Tables 6 and 7). The cubes stored continuously in water had lower strengths than those left in ambient air conditions. Similarly, the CBR values of the soaked specimens were lower than those for the unsoaked. In the CBR tests the top face of the specimens had systematically lower values of penetration resistance than the bottom face. This was due to the high bleeding which caused a water content gradient with depth between the top and bottom faces. The excess water near the top face hindered the strength of the cement binder. Compressive strengths between 0.3 and 2.2 MPa were obtained at 28 days with cement content ranging from 75 kg/m³ to 250 kg/m³, while the CBR values varied from 45% to 180%.

The drying shrinkage of the material was very high (Table 8). This is an indication that there is a great risk for the hardened slurry to develop shrinkage cracks if it is allowed to dry to a significant extent. In trench backfilling applications the extent of drying is likely to be limited, and even if cracks do develop, it may not constitute a serious disadvantage as the loading is usually distributed over a large area. The values of static modulus of elasticity ultrasonic pulse velocity measured correspond with those of very low strength concrete, and the trend in the values indicate that on hardening the slurry was able to achieve satisfactory consolidation (2,3).

Table 6 Strength development of hardened slurry

	COMPRESSIVE STRENGTH (N/mm²)						
Age (Days)	1	2	3	4	7	14	28
	AIR CURING						
Cement Content (kg/m³)							
75	0.045	0.074	0.102	0.151	0.235	0.411	0.51
100	0.077	0.095	0.155	0.205	0.395	0.68	0.83
125	0.115	0.275	0.345	0.420	0.650	1.11	1.32
150	0.121	---	0.355	---	0.620	1.03	1.45
175	0.110	---	0.390	---	0.600	0.90	1.57
200	0.152	---	0.426	---	0.960	1.18	2.01
250	0.233	---	0.436	---	0.960	1.93	2.20
	WATER CURING						
75	----	0.046	0.065	0.100	0.155	0.26	0.31
100	----	0.083	0.125	0.155	0.300	0.50	0.629
125	----	0.240	0.340	0.401	0.59	0.91	1.05
150	----	----	0.30	----	0.48	0.85	1.18
175	----	----	0.26	----	0.30	0.735	1.25
200	----	----	0.34	----	0.702	1.02	1.58
250	----	----	0.33	----	0.87	1.29	1.86

Relationship Between Compressive Strength and CBR

The unconfined cube compressive strengths of the air-cured and water-cured hardened slurry specimens obtained at the various ages from 1 to 28 days were correlated with the CBR values obtained at corresponding ages and for corresponding test conditions. A close was found between these two properties of the material as illustrated in Figure 5.

The correlation coefficient was 0.957 and the relationship for cement-stabilised rockdust was found to be: CBR (%) = 66.1 * compressive strength (N.mm2) + 4.2

Table 7 CBR values of hardened slurry

	CBR %						
Age (Days)	1	2	3	4	7	14	28
UNSOAKED							
Cement Content (kg/m³)							
75	4	5	6	11	22	35	55
100	5	6	9	13	26	45	75
125	9	13	19	21	50	80	100
175	9	--	25	--	48	76	125
200	10	--	33	--	55	83	140
250	10	--	52	--	65	97	180
SOAKED							
75	--	5	7	10	22	35	45
100	--	6	8	12	27	45	65
125	--	12	18	24	40	55	80
175	--	--	19	--	39	62	90
200	--	--	30	--	49	76	97
250	--	--	35	--	52	77	100

Table 8 Deformation of hardened slurry

CEMENT CONTENT (kg/m³)	SECANT MODULUS* OF ELASTICITY (GPa)		ULTRASONIC PULSE VELOCITY (km/s)	DRYING SHRINKAGE (%)
	Air - Cured	Water - Cured	Air - Cured	
75	8.5	8.1	2.20	0.198
100	9.4	8.6	2.37	0.177
125	10.5	10.2	2.62	0.158
150	11.1	10.3	2.64	0.170
175	12.7	11.8	2.69	0.181
200	13.5	12.9	2.81	0.202
250	14.4	13.2	2.89	0.208

* corresponding to 50% ultimate strength

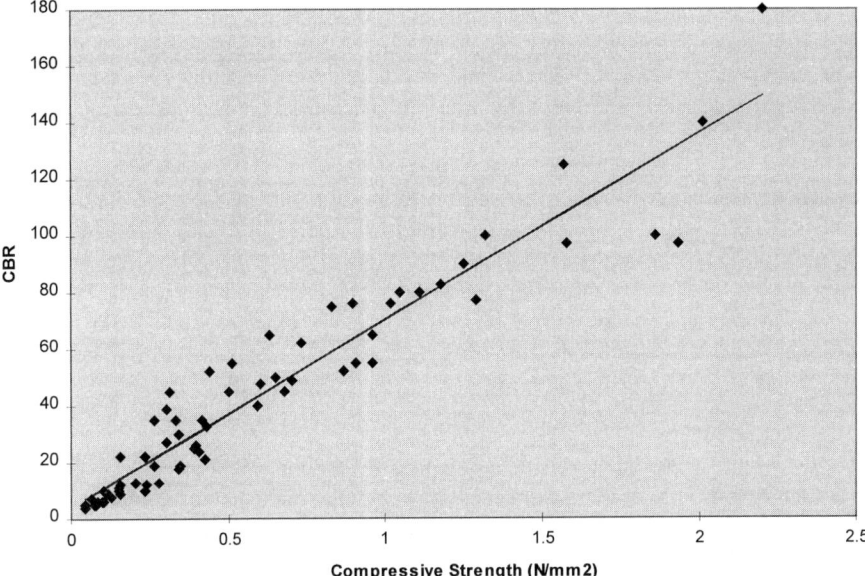

Figure 5 Relationship between CBR and compressive strength for hardened rockdust slurry

DISCUSSION

Construction requires large quantities of low-cost materials. Rockdust is an unavoidable fine waste of aggregate manufacture in Mauritius, produced in sufficient quantities to be noticeable as an environmental air pollutant and a nuisance. The ideal approach to addressing this problem would be to investigate ways in which the material can be put to constructive use. Such a potential application in the local context is as backfill material in roadway trenches and other excavations, in which the expectation would be that the material performs at least as well, if not better, than a well-graded compacted soil.

The specifications used by the Ministry of Works for backfilling of trenches before reinstatement of road surfaces in relation to the fill material are as follows : -

- the 4-day soaked CBR at 95% BS Heavy Compaction should not be less than 10%
- the plasticity index should not be more than 25%
- the liquid limit should not be more than 35%.

Stabilisation of rockdust by mechanical compaction does achieve the latter requirements as long as the material is not allowed to become saturated. On soaking, the CBR values drop to 3 to 5%, that is, well below the minimum of 10% required. This study on stabilising rockdust with cement and producing it in a flowable form was inspired from a series of research papers presented in a workshop on "Flowable Slurry Containing Fly Ash and other Mineral By-Products" in June 1995 in Milwaukee, USA (4).

It was clearly demonstrated that fly ash was being used in the United States in numerous applications including blended cement, lightweight aggregate, concrete products, bricks, blocks, paving stones, embankment fills, road construction, agriculture as soil conditioner, and as waste/sludge stabiliser. The chemical composition of rockdust is not any different from that of rocksand which has long been used in concrete. The rockdust is, thus quite compatible for mixing with cement.

The American Concrete Institute published a technical report on Controlled Low-Strength Materials (CLSM) in 1994 which was prepared by the ACI Committee 229. This report (5) gave information on applications of CLSM, their properties, mix proportioning, construction, and on quality control procedures. It defined CLSM flowable slurry as a "cementitious material that is in a flowable state at placement and has specified compressive strength of 1200 p.s.i (about 8 MPa) or less at the age of 28 days". CLSM is produced as a flowable slurry so that it flows like a liquid, is self-levelling without compacting, and hardens to a solid which can take load.

The flow/spread of the slurry can be determined by the method (6) specified by the ACI Committee 229. But, as at the time this study was started the equipment was not available, flow was assessed by measuring viscosity using a standard torsion viscometer. It was, however, demonstrated that a far simpler method, which measures the time for the slurry to flow out of a funnel of specified dimensions, can give equally reliable measurements of flow.

The results show that, apart from reasons of economy, it is beneficial to keep the cement content of the flowable rockdust slurry to a maximum of 125 to 150 kg/m^3 (6 to 8% by weight of rockdust), in order to minimise bleeding. The hardened slurry has lower strength and CBR values when it is stored in a saturated condition as compared to storage in ambient air conditions. This tends to suggest that the material behaves as a hybrid between a granular soil and hardened concrete. Nevertheless, the development of strength and the increase in CBR with age are both sustained even in the saturated condition, and a close correlation was found between these two properties of the hardened slurry. The 10 percent 4-day soaked CBR specified for backfill is achieved with a cement content of 75 to 100 kg/m^3 (6 to 8% by weight of rockdust). With cement contents of 125 kg/m^3 and above, this is achieved in 2 days, and if the specified CBR is required in 1 day, cement contents of 200 kg/m^3 or more must be used. However, the rich mixes, apart from being more costly, tend to bleed significantly because of increased water demands, and have high drying shrinkages.

CONCLUSIONS

1. Rockdust, a waste material from rock crushing for aggregate production, can be put to good use as a competent backfill for road trenches and other excavations, by blending with cement and enough water to make the mixture flowable.

2. Flow of the cement-rockdust slurry can be measured using laboratory viscometers, such as a torsion viscometer, but, a much simpler method, which measures time taken by the slurry to flow out of a specified funnel, is proposed and is shown to give equally reliable measurements to the viscometer.

3. The mechanical characteristics of hardened cement-rockdust slurry are those of a hybrid material between granular soil and hardened concrete. Thus, the hardened slurry has lower strength and CBR values when it is stored in a saturated condition compared to storage in ambient air conditions.

4. The increases in strength and CBR of the hardened slurry with age are sustained, even in the saturated condition, and a close correlation exists between these two properties.

5. A 4-day soaked CBR of 10% can be achieved in the hardened slurry with a cement content of 75-100 kg/m3 (6-8% by weight of rockdust). With higher cement contents the target CBR is reached in less time.

REFERENCES

1. DEPARTMENT OF ENVIRONMENT. Technical Report on Stone Crushing Plants. Ministry of Environment and Quality of Life, Port-Louis, Mauritius, April 1993.

2. PAHARY, S M A. Flowable Rockdust Slurry as a Backfill Material. B.Tech (Hons) Civil Engineering Final Year Project. March 1996, Faculty of Engineering, University of Mauritius.

3. PERRETTE, G. Cement-Stabilised Rockdust Slurry as Road Base and Blinding. B.Eng (Hons) Civil Engineering Final Year Project. March 1998. Faculty of Engineering, University of Mauritius.

4. PROCEEDINGS OF WORKSHOP on "Flowable Slurry Containing Fly Ash and Other Mineral By-Products" presented at the Fifth CANMET/ACI International Conference on Fly Ash, Silica Fume, Slag and Natural Pozzolans in Concrete. June 1995, Milwaukee, USA.

5. ACI COMMITTEE 229. "Controlled Low-Strength Materials (CLSM)". ACI Manual of Concrete Practice. 1995.

6. NAIK, T.R., RAMME, B.W., and KOLBECK, H.J. "Filling Abandoned Underground Facilities With CLSM Slurries". Concrete International, Vol. 12, No. 7, July 1990.

STRUCTURAL LIGHTWEIGHT CONCRETES WITH FLY ASH PELLETIZED COARSE AGGREGATE - MIX PROPORTIONING

T S Nagaraj

Indian Institute of Technology

India

T Ishikawa

Saga University

Japan

ABSTRACT. An integral part of concrete mix proportioning is the preparation of trial mixes and effect adjustments to such trials striking the balance between the requirements of placement and strength apart from satisfying durability needs. In the case of coarse aggregate characteristic strength being lower than the levels of strength of concrete desired, different methods, viz., British, ACI and country's standard code cannot be directly used. In this paper a generalized method is advanced to proportion structural lightweight concretes with fly ash pelletized coarse aggregates by considering concrete as composite of cement mortar matrix and coarse aggregate. The basic principle of the method is to assess the characteristic strength of coarse aggregate from compressive strength data of concrete failed by aggregate crushing and for that characteristic strength assess the cement mortar strength required so as to ensure the development of strength of concrete as desired. To arrive at the exact water-cement ratio of mortar matrix the generalized Abrams' law is used.

Keywords : Structural lightweight concrete, Pelletized fly ash aggregate, Two component composite concrete.

Professor T S Nagaraj is Emeritus Fellow (AICTE) in the Department of Civil Engineering, Indian Institute of Science. Bangalore, India after thirty five years of academic involvement in the same institution. He obtained his PhD (1967) and DSc. (1992) from the same Institution. His main research interests are geo and concrete materials with particular reference to Cemented soft clays, utilization of marginal materials and high performance concretes. He has over 220 publications and three books to his credit.

Professor Tatsuo Ishikawa is working in the Dept. of Civil Engineering, Faculty of Science and Technology, Saga University, Japan for over twenty six years. His main research fields of interest are concrete materials and concrete structures. High performance concrete with fly ash and punching shear capacity of prestressed concrete slab are specific areas of research involvement.

INTRODUCTION

Compressive strength and density are the two of the most important parameters of structural lightweight concrete. With suitable lightweight aggregates, structural lightweight concretes can be made with densities which are 25-40 percent lower but with strengths equal to the maximum normally achieved by normal density concretes. While the strength of the coarse aggregate is at par or higher than the range of structural concretes contemplated conventional methods can be employed. In all the cases the strength of the aggregates were not considered since the strength of the concretes to be obtained was lesser than the strength of the coarse aggregate. If the characteristic strength of the coarse aggregate is lesser than the compressive strength of the concrete desired then the conventional methods to proportion the mix are not applicable. The need for considering the synergy between the coarse aggregate and cement mortar matrix in the analysis of strength of concrete has been elucidated by Zahida Banu and Nagaraj [1]. It is necessary to understand the possible modes of failure of concrete in compression to evolve appropriate method.

FRACTURE AND FAILURE OF MORTARS AND CONCRETE

Concrete is a heterogeneous material whose properties depend on the properties of the individual components e.g. coarse aggregates and mortar and their compatibility. In order to process high strength light weight structural concrete with required properties of coarse aggregates must also be known. Galico et. al.,[2] have reported that all the aggregate characteristic effects on concrete intensify only for aggregates sizes above 5 mm i.e., at coarse aggregate level. Concrete can attain the strength of the mortar matrix approximately, only if the strength of the aggregate is higher than that of the matrix, and hence the crack propagation will be across the periphery of the aggregate as shown in figure 1(a). This condition is generally satisfied in the case of normal density concrete. If the strength of the cement mortar matrix is higher than the aggregate crushing strength, due to low modulus of deformation of the aggregate than that of the matrix, the aggregates can no longer transmit the applied stress at the same deformation as that of the mortar. Since the stress in the aggregate is greater than its strength, the failure is through the aggregate as in shown in figure 1(b). A weak aggregate requires higher mortar strength, for identical strength when compared with concrete experiencing bond failure.

CONCRETE AS COMPOSITE MATERIAL

In examining the concrete as composite material, it can be characterized as a two phase material. Mortar is the matrix and coarse aggregate is the distributed phase. According to the rules of the composite material an examination can be made regarding the behaviour of concrete in terms of the properties of the individual phases and their proportions. Once a unit cell model, composed of mortar and coarse aggregate is assumed then effort can be directed to find mathematical relations such that tailor made situation is realized. For the limiting case of no bond, the assumption of identical stresses in the matrix and particles become reasonable, if the particles are more rigid than the matrix. If the particles are less rigid than the matrix, as in the case of lightweight and high strength concretes, the bond is of little significance to composite behaviour. The particles will follow the deformation of the matrix up to its limit compressive strength

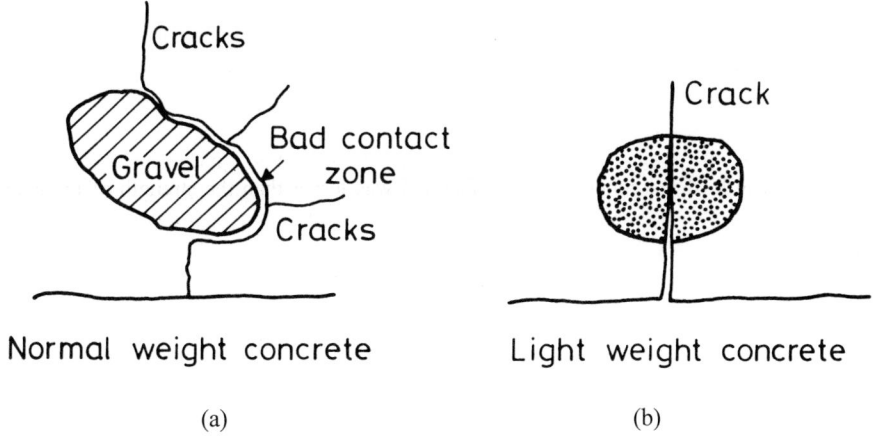

Figure 1 Different modes of failure in concrete under compressive loading

For the assumption of a perfect bonding between the coarse aggregate and matrix without any slippage at the interface, the strains experienced between the coarse aggregate, matrix, and concrete are same. For a unit cell model the relation involving the stress acting on each of the two phases (matrix, σ_m and coarse aggregate, σ_a and their volume fractions, matrix, v_m and that of coarse aggregate, v_a) would be:

$$\sigma_c = \sigma_m v_m + \sigma_a v_a \quad for \quad \varepsilon_c = \varepsilon_a = \varepsilon_m \quad and \quad v_m + v_a = 1 \tag{1}$$

To advance a generalized approach to proportion the concrete mixes taking into account the characteristic strength of coarse aggregate, the possibility of using the above relation for the following assessments merits examination,

(i) From the strength data of concrete where aggregate crushing has been observed along with the compressive strength of constituent mortar matrix of that concrete calculation of the characteristic strength of coarse aggregate.
(ii) Using the same law of mixtures with the characteristic strength of the coarse aggregate known calculation of the required compressive strength of the mortar matrix for the specific compressive strength of the concrete.

For examination of the above possibility the experimental data forms the basis in this investigation.

MATERIALS AND METHODS

Materials

Ordinary Portland cement (OPC) and crushed sand as fine aggregate (specific gravity of 2.56 and fineness modulus of 2.62) has been used. FA-Light is pelletized lightweight coarse aggregate (12.5mm and down size, specific gravity in saturated surface dry condition 1.68), made from fly ash of thermal power station. This aggregate is made of soft pellet using fly ash along with bentonite and powdered coal mixed with sprayed water and burnt in a furnace.

Mix Proportion

Based on the ACI method with 927 kg /m³ the experimentally determined bulk density of the aggregate the corresponding mix proportions are arrived for a water content of 215 lit /m³ with water -cement ratios being in the range of 0.3 to 0.7. The constituent cement mortar matrix mixes were also arrived at for the same water cement ratios. In all the cases the volume fraction of the coarse aggregate, v_a, is 0.314 and that of mortar matrix, v_m, is 0.686.

Experimental Findings and Analysis of data

For w/c ratio of 0.4 the 7day compressive strength of concrete is 38 Mpa. The 7day strength of the constituent mortar at the same water /cement ratio is 45 Mpa. The failure of concrete was observed to be by crushing of the aggregate. The characteristic strength of coarse aggregate accordingly as per law of mixtures is:

$$38 = \sigma_a 0.314 + 45 \times 0.686 \qquad \sigma_a = 22.3 \, MPa$$

The matrix strength developed at an age of 28 days was 61 MPa. Correspondingly the strength of concrete as per law of mixture would be:

$$\sigma_c = 22.3 \times 0.314 + 61 \times 0.686 = 48.9 \text{ as against 49 experimentally observed.}$$

Similarly for water/cement ratio of 0.5 the 7day observed concrete strength was 30 MPa. For the same age the mortar strength developed was 33.5 MPa. This results in a characteristic strength of the aggregate of the same value of 22.3 MPa. At 28day aging the mortar strength increased to 46 MPa. from 33.5 MPa. The predicted strength of concrete

$$\sigma_c = 22.3 \times 0.314 + 46 \times 0.686 = 38.6 \, MPa \qquad \text{as against } 39 \text{ MPa which was}$$
experimentally observed value with failure by aggregate crushing.

With the characteristic value of 22.3 in the case of concrete with a water /cement ratio of 0.6 yielding the strength of constituent mortar of 37.4 MPa. the predicted strength would be 32.7MPa. The expeimental strength value was 32.4 MPa. with aggregate crushing being observed.

It is interesting to examine the two cases of strength of concrete one approaching characteristic strength of the aggregate and the other approaching the limit compressive strength as the strength of mortar being very high due to low water/cement ratios. For a water/cement ratio of 0.7 the 28day compressive strength of constituent mortar was only 30 MPa. With the characteristic strength of coarse aggregate being 22.3 MPa. according to law of mixture the strength of concrete ought to have been 28MPa. The observed value was 24 MPa. with failure partly composite by crushing of the aggregate and partly by interface separation.

On the other hand as the water/cement ratio was reduced to as low as 0.3 obviously the strength developed by mortar rose up to 88 MPa. By law of mixtures the concrete ought to develop a strength of 67 MPa. Actually the developed strength was only 61 MPa. By increasing the water /cement ratio to 0.35 the mortar strength got reduced to 74 MPa. By law of mixtures the strength of the concrete ought to be 58 MPa. The experimental value of concrete strength at this water / cement ratio was 61 MPa. The failure was by crushing of the aggregate. It indicates that the limiting value of concrete strength for the pelletized fly ash aggregates used in this investigation with a characteristic strength of 22.3 Mpa. would be in the range of 61 MPa. for optimum synergy. Still it would not be advocated to obtain this value for practical use since the cement content at such low water cement ratio of 0.3 or 0.35 , apart from cost, would be beyond permissible limits from the durability considerations arising due to very high cement content. Despite this fact the methodology of arriving mix proportions for different strengths in the structural strength range establishes the technical feasibility.

DISCUSSION

Determination of the characteristic strength of coarse aggregate and taking into account of the same by use of law of mixtures to calculate the mortar matrix strength required to obtain the compressive strength of concrete greater than that contributed by coarse aggregate as demonstrated above has to be consistent with;

(i)   the general understanding of the interface characteristics between the coarse aggregate and the mortar matrix and

(ii) the validity of Abrams' law in the analysis of strength mobilization of strength of mortar matrix.

Between the cement mortar matrix and the inclusions there is an intermediary region called the interface. The microstructure of these interfacial zones consists of both larger pore diameters and higher porosity than the microstructure of the bulk mortar [3]. It is well understood that the transfer of forces between two adjacent phases is by far most important factor for strength development of concrete. In the present context since the failure of the concrete is by crushing of the aggregate contrary to bond separation as is observed in normal density concretes. Since the strains in the matrix and that in the aggregate are of the same order, no bond separation takes place at the time of failure. It is interesting to note that with the compressive strength of the mortar matrix the characteristic strength of the aggregate is determined and with this value the strength of the mortar matrix required for mobilization of required concrete strength is assessed. It is likely that the matrix interface strength is likely to be in some proportion to that of the bulk mortar strength for various combinations. It is not possible to assess and take into account the interface mortar matrix strength at the interface. Hence the bulk mortar strength at various conditions itself serve the purpose in the considerations of synergy. This aspect is still to be adequately substantiated possibly by a different approach.

In an independent study by the normalization of strength data of concrete with different cements by the strength data of the same at water cement ratio 0.5 the generalized Abrams' law has been proposed by Nagaraj and Zahida Banu [4]. Subsequently it has been found by Nagaraj [5] that the strength development in the cement mortar matrix with different water cement ratio also follows the same law. By considering the, compressive strength at water - cement ratio, 0.5, $(\sigma_m)_{0.5}$ as the reference mark to reflect the synergetic effects between constituents of concrete, for practical application the appropriate relationships developed are:

$$\left\{\frac{\sigma_m}{(\sigma_m)_{0.5}}\right\} = -0.2 + 0.6\left\{\frac{c}{w}\right\} \quad \text{For } (\sigma_m)_{0.5} \geq 30\, MPa$$

$$\left\{\frac{\sigma_m}{(\sigma_m)_{0.5}}\right\} = -0.73 + 0.865\left\{\frac{c}{w}\right\} \quad \text{For } (\sigma_m)_{0.5} \leq 30\, MPa \tag{2}$$

Using the appropriate relations for compressive strength data at free water -cement ratio, 0.5, i.e., $(\sigma_m)_{0.5}$ obtained from the trial mix, it is possible to compute free water - cement ratios for the entire range of strengths normal concrete mixes within the usual limits of 0.3 to 0.7 and beyond, if necessary. This enables to calculate the water -cement ratio of cement mortar matrix required for synergy such that concrete strength as contemplated is realized. Hence in adopting the method discussed in this paper, the water cement ratio for the cement mortar matrix with strength development required can be arrived by generalized Abrams' law.

CONCLUSIONS

Although this investigation is limited, there are clear indications that the role of fly ash pelletized coarse aggregate in concrete strength development can be accounted for. In addition to the generalized Abrams' law, the law of mixtures can be used to determine both the characteristic strength of pelletized fly ash and as well as the cement mortar strength required to obtain the concrete strength higher than the strength of the coarse aggregate up to its limiting strength.

REFERENCES

1. ZAHIDA BANU AND NAGARAJ, T S. Synergetic roles concrete ingredients in strength development. Construction and Building Materials, Vol.10, No.4, 1996, pp 251-253.

2. GLACCIO, G, ROCCO, C, VIOLINI, D, ZAPPITELLI, J. AND ZERBINO, R. High strength concretes incorporating different coarse aggregates. ACI Mater. J. 1992, Vol. 89. pp 242 - 246.

3. SYNDER, K A, WINSLOW, D N, BENTZ, D P AND GARBOCZI, E J. Interfacial zone percolation in cement - aggregate composites. Interfaces in Cementitious Composites, Proc. RILEM Conf. Toulouse, Oct.1992, E&FN Spon. pp 259 -268.

4. NAGARAJ, T S AND ZAHIDA BANU. Generalization of Abrams' law. Cement and Concrete Research, Vol. 26, No.6, 1996. pp 933 - 942.

5. NAGARAJ, T S. Superfluid microconcretes - possible rational approach for mix proportioning mixes. National Seminar on Advances in Special Concretes, 1998 Bangalore. pp 38- 52.

FINE GRAINED CEMENTLESS CONCRETE CONTAINING SLAG FROM FOUNDRY

S I Pavlenko

Siberian State University of Industry

V I Malyshkin

Khakasenegro Co Ltd

Russia

ABSTRACT. The Department of Civil Engineering of the SSUI under a contract to the Pavlodarsky Tractor Plant (PTP) has developed the composition and technology of fine-grained concrete with ground basic slag as a binding material and acidic slag sand with a particle size of 0 to 5 mm as an aggregate. Both basic and acidic slag are waste products of three foundry departments of the PTP with produces 30 000 tons acidic slag annually. These slags are dumped/polluting the environment and occupying land. Concrete developed included 400 to 700 kg/m^3 (depending on the compressive strength) basic slag, 1000 to 1200 acidic slag sand, 230 to 280kg/m^3 by water and 0,3% by weight of binding materials of plasticizing admixture (technical grade lignosulfonate). The slump of a mixture was 4 to 6 cm and the average density was 1800 to 1900 kg/m^3. The technology of processing acidic slag into sand with a particle size of 0 to 5mm and basic slag into powder has also been developed. Schematic diagrams of this technological process are presented in the paper. The studies showed the possibility and reliability of producing concrete from byproducts and their use in the construction of single, two-story houses as well as in the production of small blocks and unburnt brick.

Keywords: Basic and acidic slags from foundry, Fine-grained cementless slag concrete, Technology of processing slags.

Professor S I Pavlenko is head of the Civil Engineering Department. Siberian State University of Industry, Novokuznetsk, Russia.

V I Malyshkin is Director General of Khakasenergo, Co. Ltd, Abakan, Russia.

INTRODUCTION

The Pavlodarsky Tractor Plant In Kasakhstan has faced the following problems: the problem of utilizing waste products from steelmaking and iron foundry processes and shortage of housing construction for workers due to lack of cement and aggregates for lightweight and heavy concretes.

The objective of the present study was to solve the above problems by investigating slags in order to develop on their basis concrete for load-bearing and non-load-bearing structures for use in the construction of cast in-situ houses. Composition of fine-grained slag concrete and technology of processing slags into sand with a particle size of 0 to 5mm and into powder with a fineness of 4000 to 4500 cm2/g have been developed by the Siberian State Academy of Mining & Metallurgy. The concrete does not contain any natural or artificial porous aggregates.

MATERIALS

The materials used in the investigation were acid slags from the iron foundry department and the steel-making department No I (SD-I), basic slag from the steel-making department No 2 (*SD-2). Physical properties and chemical analysis of the materials are given in Tables 1 and 2, respectively.

Table 1 Physical properties of slags from PTP

CHARACTERISTICS OF SLAG	SLAG FROM SD-1	SLAG FROM SD-2	CUPOLA SLAG
Bulk Density, kg/m^3	1240	1250	1250
Absolute Density, kg/m^3	2450	2750	2800
Crushabilily, %	30	19	13
Cylinder Crushing Strength, MPa	2.44	2.50	2.78
Specific Efficiency, pc/g	0.4	3.1	7.2
Colour	green	yellow-grey	grey

Slag from the Steelmaking Department No 1

Slag from SD-I is an acid slag and from its granulametric composition it is referred to as coarse sand. It is of a green colour, consisting of particles of various shapes including thin-walled spirals. It has an unstable structure and after being boiled thrice, it loses up to 50% weight. It contains up to 20% iron oxides and metallic inclusions.

Therefore, prior to utilization of slag in concrete it was ground to sand, with a particle size distribution of 0 to 5 mm, the metallic inclusions and iron oxides being removed by magnetic separation.

Slag from the Steelmaking Department No 2

Slag from SD-2 is basic. It is yellow-grey In colour and consists of particles 70 mm and even larger in size. It loses 12% of its weight after being boiled three times. The process of slag structure failure is attributable to the polyamorphic transformation of dicalcium silicate from (β to γ-modification. Free lime (up to 6%) which is present in slag in various degrees of overburning is hydrated and the resulting increase in the volume also assists the failure of slag. A decision was made to grind this slag into a powder having a specific surface of 4000 to 4500 cm2, which helped to eliminate the negative properties of slag and utilization of the positive ones (i.e. release a free lime to react with silica and water).

Table 2 Chemical composition of slags

OXIDES, %	SLAG FROM SD-1	SLAG FROM SD-2	CUPOLA SLAG
SiO_2	30.95	30.06	57.76
Al_2O_3	6.88	2.65	4.40
Fe_2O_3	11.97	1.99	13.77
FeO	8.82	0.71	1.93
CaO (total)	8.95	20.00	10.09
CaO (free)	1.76	4.58	2.12
MgO	11.79	22.34	1.62
Na_2O	1.32	2.34	0.95
K_2O	0.10	0.24	0.58
TiO_2	1.01	0.50	0.85
SO_3	0.43	0.76	0.37
MnO	17.67	18.24	9.59
P_2O_5	0.07	0.06	0.10
Loss on ignition	-	0.13	1.94

Cupola Slag

Cupola slag is acid and by its fineness modulus (3,71) is referred to as coarse sand.

Technical grade lignosulfonate (TGL)

TGL is a waste product of pulp and paper industry. It was used as a plasticizer for concrete mixture in order to reduce the amount of mixing water, while the detergent called "Progress", a secondary sodium alkyl sulphate, was used in concrete as an air-entraining admixture.

OBJECTIVE OF STUDY

The objectives of this work were to:

1) Create the optimum composition of cementless finegrained concrete based on a slag waste from a tractor plant which does not contain any natural or artificial porous aggregates;

2) Create a technology of processing of a basic slag into a powder to be used as a cementitious material and of acidic slag into a sand to be used as an aggregate.

EXPERIMENTAL

It is well known [1, 2, 3] that high-calcium ashes and slag have an increased content of free calcium oxide (over 5%) which may result in an irregular volume change and destruction of concrete. To eliminate the irregularity of the volume change (hydration of free CaO after concrete hardening), the following methods were used: slag grinding (breakage of melted strutures and covers of CaO), previous hydration of the material, autoclave treatment, chemical binding.

Therefore, slag from SD-2 was ground in a laboratory ball mill into a powder with a specific surface of 4000 to 4500 cm^2/g which was used as a cementitious material and a fine-dispersed aggregate for the fine-grained slag concrete. Slag sand from SD-I and iron foundry department with a particle size distribution of O to 5mm produced by grinding in a laboratory roller crusher was used as fine aggregate. In selecting optimum compositions of concrete mixtures, the method of rational planning of experiment [4? 5] was used.

The values of primary factors (contents of slag sand, water, plasticizing and air-entraining admixtures) were changed in a given range and their effect on secondary factors (average density, compressive strength of fine-grained slag concrete) was studied.

Concrete was mixed in a laboratory fixed-drum concrete mixer. Test cubes, 100x100x100mm in size, were cured in a laboratory steam-curing chamber at 900°C using a 3+10+3 hour cycle. By changing the values of the primary factors in accordance with Table 3 and by calculating the average resulting values the optimum compositions of finegrained slag concrete mixtures for various strength grades were obtained. The proportioning of the concrete mixtures is summarized in Tables 3,4.

It can be seen that the fine-grained slag concrete has the compressive strengths of 5 to 8 MPa and 10 to 25 MPa for non-load-bearing and load-bearing structures, respectively. The best results were obtained by utilizing as an aggregate acid slag from a steelmaking department.

The data on physic-mechanical, deformation properties, frost resistance and heat conductivity of concrete are given in Table 5, and durability studies of this concrete are given in our work [6]. The results obtained meet the requirements of the Building Code [7] for classes 50 to 200 fine-grained concretes used for the construction of low-rise buildings.

Table 3 Optimum compositions of fine-grained concrete on the basis of ground slag and slag sand for non-load-bearing structures

CHARACTERISTICS	CONCRETE BRAND			
	50	50	75	75
Acid Slag Sand SD-1, kg/m^3	690	-	765	-
Acid Slag Sand from Foundry Department, kg/m^3	-	680	-	750
Basic Fine-Ground Slag SD-2, kg/m^3	500	520	500	515
Water, kg/m^3	290	297	295	298
TGL, %	0,3	0,3	0,3	0,3
Air-Entraining Admixture, %	2	2	1	1
Mixture Slump, cm	4-6	4-6	4-6	4-6
Average Density of Mixture, kg/m^3	1480	1497	1560	1563
Compressive Strength at 28 days, MPa	5,6	5,3	8,1	7,4
Compressive Strength at 180 days, MPa	6,0	5,6	8,6	7,8

Table 4 Optimum compositions of fine-grained concrete on the basis of ground slag and slag sand for load-bearing structures

CHARACTERISTICS	CONCRETE BRAND					
	100	100	150	150	200	200
Acid Slag Sand SD-1, kg/m^3	1077	-	1110	-	1142	-
Acid Slag Sand from Foundry Department, kg/m^3	-	1067	-	1103	-	1140
Basic Fine-Ground Slag SD-2, kg/m^3	500	510	500	510	500	508
Water, kg/m^3	278	283	280	285	268	273
TGL, %	0,3	0,3	0,3	0,3	0,3	0,3
Air-Entraining Admixture, %	-	-	-	-	-	-
Mixture Slump, cm	4-6	4-6	4-6	4-6	4-6	4-6
Average Density of Mixture, kg/m^3	1855	1860	1890	1898	1910	1921
Compressive Strength at 28 days, MPa	10,8	10,1	16,2	15,3	21,0	19,7
Compressive Strength at 180 days, MPa	12,0	11,2	20,2	17,4	25,8	22,3

PROCESSING OF ACID AND BASIC SLAGS INTO SAND AND POWDER

The practical results of this work were the development of technological schemes and regulations [8]. At present, the project for processing acid and basic slags into sand with a particle size of 0 to 5 mm and basic slag sand into powder with a specific surface of 4000 to 4500 crn2/g are being developed (Figures 1,2). The annual output of slags from three departments of the Paviodarsky tractor plant is 80000 tons including 30000 tons of basic slag.

Table 5 Physic-mechanical, deformation and special properties of fine-grained slag concrete

Properties	Concrete Strength Group				
	For Non-Load-Bearing Structures		For Load-Bearing Structures		
	50	75	100	150	200
Axial Tensile Strength, MPa	0.67-0.79	0.88-1.12	0.98-1.23	1.54-1.58	1.94-2.13
Flexural Strength, MPa	0.94-1.11	1.06-1.20	1.39-1.73	2.17-2.33	2.84-3.12
Prism Strength, MPa	3.69-3.83	6.15-6.30	7.07-8.10	11.26-21.72	15.30-15.80
Shrinkage, mm/m	0.74-0.77	0.68-0.72	0.29-0.32	0.31-0.38	0.32-0.37
Initial Modulus of Elasticity, MPa	6.7-7.3	7.9-8.4	18.3–19.6	19.1–20.6	21.8-23.4
Creep minus Shrinkage, mm/m	0.36-0.38	0.31-0.34	0.20-0.22	0.22-0.24	0.24-0.26
Compressibility, mm/m	1.24-1.39	1.02-1.26	0.82-0.91	0.85-0.87	0.97-1.12
Tensibility, mm/m	0.24-0.27	0.18-0.22	0.09-0.11	0.08-0.09	0.11-0.12
Frost Resistance, cycle	55-75	80-100	—	—	—
Heat Conductivity (in a dry state), Kcal/m°C	0.33-0.35	0.36-0.39	—	—	—

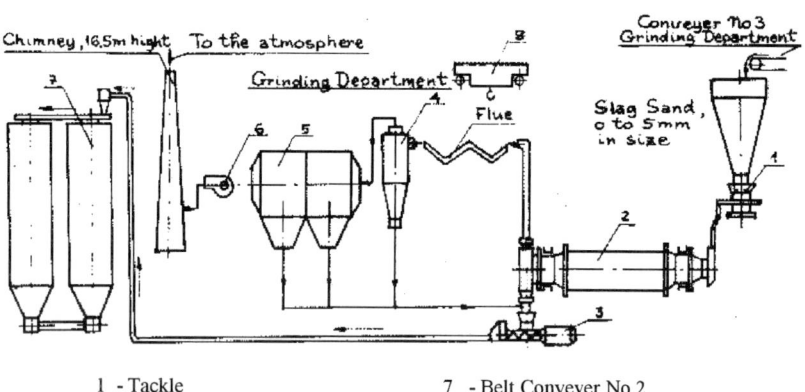

1 - Tackle
2,3 - Alligators
4 - Belt Conveyer
5 - Iron Separator
6 - Roller Crusher
7 - Belt Conveyer No 2
8 - Belt Conveyer No 3
9 - Sector Gate
10 - Electric Dumper
11,12 - Tackles

Figure 1 Schematic diagram of processing slag's into sand

Fine-Grained Cementless Concrete 107

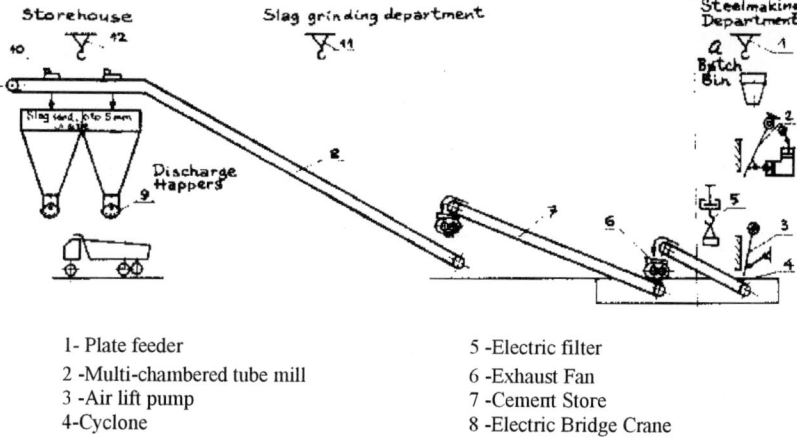

1- Plate feeder
2 -Multi-chambered tube mill
3 -Air lift pump
4 -Cyclone

5 -Electric filter
6 -Exhaust Fan
7 -Cement Store
8 -Electric Bridge Crane

Figure 2 Schematic diagram of processing of basic slag sand into powder

CONCLUSIONS

1. Using the waste products from the foundry department of the PTP (basic and acid slags it is possible to organize the production of fine-grained cementless concrete of 5 to 20 MPa strength grades. Fine-ground basic slag may be used as a cementitious material and acidic slag processed into sand as an aggregate.

2. Five strength grades (50, 75, 100 150 and 200) of the concrete developed for non-load-'bearing and load-bearing structures are in accordance with the requirements of Building Code and State Standard.

3. Technological schemes of processing acid slag into sand with a particle size distribution of 0 to 5 mm and basic slag into a cementitious material with a specific surface of about 4500 cm^2/g have been suggested.

REFERENCES

1. PAVLENKO, S I. Lightweight cementless concrete on the basis of high-calcium fly ash and slag sand from TPP. Blended Cements in Construction. Edited by R.N. Swamy. Elsevier Applied Science London and New York. 1991` pp 95-106.

2. SAVINKINA, M A AND LOGVINENKO, A T. Ashes Produced from the Kansko-Achinsky Brown Coal, Nauka, Novosibirsk, 1979; p. 46.

3. PAPAYIANNI, J. Concrete with high-calcium fly ash. CANMET/ACI International Conference on Advances in Concrete Technology, Athens, Greece 1992) pp. 261-284.

4. PROTODYAKONOV, M M AND TEDDER, R I. Method of rational experiment planning, Nauka Pablishing House, Moscow, 1970. p 70.

5. VOZNESENSKY, V A. Statistical methods of experiments planning in technico-economical investigations. Edition Statistiks, 1974. p 192.

6. PAVLENKO,,S I AND BOGUSEVICH V S. Durability Studies of Concrete Containing Low Cement Content and Fine Particles of Foundry Slag. ACI SP 145-59, Editor by V.M. Malhotra, Detroit, MI, 1994. pp. 1069- 1086.

7. BUILDING CODE,2.03.01-84. Concrete and reinforced concrete structures, Gosstroy USSR, Moscow, 1985.

8. PAVLENKO, S I AND BOGUSEVICH, V S. Technology of fine-grained concrete on the basis of slag sand. CBU/CANMET International Symposium, University of Wisconsin, Milwaukee, USA, 1992~ pp. 1 - 12.

PERFORMANCE OF STEEL SLAG AGGREGATE CONCRETES

M Maslehuddin
M Shameem
M Ibrahim
King Fahd University
N U Khan
Heckett Multiserv
Saudi Arabia

ABSTRACT. This research study was conducted to evaluate the feasibility of using steel slag aggregates in concrete. The strength and durability characteristics of steel slag aggregate concrete were compared with those of crushed limestone aggregate concrete. The performance of steel slag aggregate concretes under thermal variations was evaluated by measuring the reduction in compressive strength and pulse velocity and increase in water absorption. The strength characteristics of steel slag aggregate concrete were marginally better than those of crushed limestone aggregate concrete. The durability characteristics of steel slag cement concretes were, however, better than that of crushed limestone aggregate concrete. The beneficial effect of steel slag aggregate on the durability characteristics of hardened concrete were particularly noted in concrete mixtures with a coarse to total aggregate ratio of 0.55 and above.

Keywords: Aggregates, Concrete, Durability, Mechanical properties, Steel slag, Waste material.

Dr Mohammed Maslehuddin is a Research Engineer in the Center for Engineering Research, Research Institute, King Fahd University of Petroleum and Minerals. He conducts research to improve concrete durability for the hot and arid regions of the world.

Mohammed Shameem is a Civil Engineer in the Center for Engineering Research, Research Institute, King Fahd University of Petroleum and Minerals. His area of specialization is concrete durability and soil mechanics.

Mohammed Ibrahim is a Civil Engineer in the Center for Engineering Research, Research Institute, King Fahd University of Petroleum and Minerals. He conducts research on materials of construction, specially concrete.

Nafees Ullah Khan is a Technical Manager, Heckett Multiserv, Al-Jubail, Saudi Arabia. His research interests are in the use of steel slag aggregate in asphalt and cement concretes.

INTRODUCTION

The steel slag produced during the direct reduction of iron in an electric arc furnace is used in asphalt concrete and for construction of subgrades in pavements. These aggregates which are produced by the disintegration of steel slag, due to rapid cooling, need minimal crushing to produce aggregates of the desired size. Their useage in concrete will be beneficial, particularly in areas where good quality aggregate is not easily available or it has to be hauled from far off distances.

Encouraged by the limited data developed by the authors [1], this study was conducted, firstly, to evaluate the properties of steel slag aggregates and, secondly, to ascertain the strength and durability characteristics of Portland cement concrete made using these aggregates. The effect of aggregate proportions, i.e., the proportion of coarse aggregate in the concrete, on the strength and durability characteristics of steel slag aggregate concrete was also evaluated.

EXPERIMENTAL PROGRAM

Tests on Steel Slag Aggregates

The steel slag aggregates were tested for the following:

a)   Abrasion resistance, according to ASTM C 131 [2],
b)   Specific gravity and water absorption, according to ASTM C 127 [3],
c)   Sulfate soundness, according to ASTM C 88 [4], and
d)   Clay lumps and friable particles, according to ASTM C 142 [5].

Tests on Steel Slag and Crushed Limestone Aggregate Cement Concretes

Concrete specimens were made using steel slag and crushed limestone aggregates. The concrete mixtures were made with a cement content of 400 kg/m^3 and water to cement ratio of 0.40.

Steel slag aggregate concrete mixtures were made with steel slag aggregate to total aggregate ratios of 0.45, 0.50, 0.55, 0.60, and 0.65. These aggregate proportions were utilized to evaluate the effect of coarse aggregate proportions on the properties of hardened steel slag aggregate cement concrete. In the crushed limestone aggregate concrete, the coarse aggregate was 60% of the total aggregate. The weights of the mix constituents are shown in Table 1. Dune sand with a specific gravity of 2.43 and water absorption of 0.57% was used as fine aggregate in both the limestone and steeel slag aggregate concrete mixtures.

Concrete specimens, 75 mm diameter and 150 mm high, were cast using steel slag and crushed limestone aggregate to determine unit weight, compressive strength and water absorption. Beam specimens measuring 100 x 100 x 500 mm were cast to measure the flexural strength. The compressive strength was determined after 7, 14, 28, and 90 days of curing.

Table 1 Mix design parameters used in the steel slag aggregate concretes

PARAMETER	VALUE
Cement content, kg/m^3	400
Water-cement ratio	0.40
Coarse to total aggregate ratio	0.45, 0.50, 0.55, 0.60, and 0.65
Max. aggregate size, mm	12.5

After 28 days of curing both the steel slag and crushed limestone aggregate concrete specimens were exposed to thermal variations. For this purpose, they were exposed to a temperature of 70 °C for eight hours and to 25 °C for 16 hours to complete one thermal cycle. After subjecting the concrete specimens to thermal variations, water absorption, compressive strength and pulse velocity were determined according to ASTM C 642 [6], ASTM C 39 [7] and ASTM C 597 [8], respectively.

Reinforced concrete specimens, 100 x 65 x 300 mm, were cast using steel slag and crushed limestone aggregates. They were reinforced with a centrally placed 12 mm ϕ steel bar. A cover of 25 mm was provided at the bottom of the specimens. After 14 days of curing in water, the concrete specimens were partially immersed in 5% NaCl and reinforcement corrosion was accelerated by impressing an anodic potential of 4 Volts. For this purpose, the steel bars in the concrete specimens were connected in series to the positive terminal of a DC power source. Stainless steel plates were used as counter electrodes. The current supplied to each of the specimens, due to the application of a fixed potential of +4 V, was monitored at 30 minutes interval by measuring the potential drop over a resistor of known resistance. The time-current curves were utilized to evaluate the time to initiation of reinforcement corrosion, which was taken as the point at which a significant increase in current or a change in the slope of the time-current curve occurred. The appearance of cracks on the concrete specimens was also monitored visually.

Shrinkage and expansion characteristics of steel slag cement mortar specimens were evaluated on mortar bars measuring 25 x 25 x 285 mm. These specimens were exposed to dry and moist conditions for four months to simulate extreme environmental conditions and their length change was measured at periodic intervals. The change in length of the mortar specimens was measured using the instrumentation similar to that specified in ASTM C 497 [9].

RESULTS

Properties of Steel Slag Aggregates

Table 2 summarizes the properties of steel slag aggregates. The apparent specific gravity of steel slag aggregate was in the range of 3.44 to 3.48, while the water absorption was in the range of 0.9 to 1.39%. The clay lumps and friable particles were in the range of 0.07 to 0.31%.

Table 2 Results of tests on steel slag aggregate

PARAMETER	VALUE
Apparent specific gravity	3.44 to 3.48
Water absorption, %	0.90 to 1.39
Clay lumps and friable particles, %	0.067 to 0.31
Magnesium soundness loss, %	0.83
Abrasion loss after 100 revolutions, %	4.47
Abrasion loss after 500 revolutions, %	18.46

The abrasion resistance of steel slag aggregate was determined using Grading B of ASTM C 131 [2], after 100 and 500 revolutions. The loss in weight of the steel slag aggregate after 100 and 500 cycles was 4.47 and 18.46%, respectively.

Properties of Steel Slag and Crushed Limestone Aggregate Concretes

The unit weight of steel slag and crushed limestone aggregate cement concretes are shown in Figure 1. These values varied from 2436 to 2769 kg/m^3 for steel slag aggregate concrete, increasing with the proportion of the coarse aggregate. The unit weight of crushed limestone aggregate concrete, with a coarse to total aggregate ratio of 0.6, was 2330 kg/m^3.

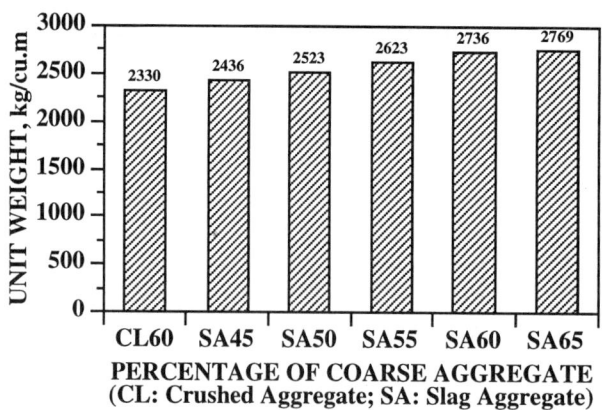

Figure 1 Unit weight of steel slag and crushed limestone aggregate concretes

The compressive strength development in steel slag and crushed limestone aggregate concretes is plotted in Figure 2. As expected, the compressive strength of both these concretes increased with the period of curing. Further, the compressive strength of steel slag aggregate concrete (coarse aggregate content varying from 50 to 65%) was more or less similar to that of crushed limestone aggregate concrete up to about 50 days of curing. At later ages, of more than 70 days, the compressive strength of concrete made with 65% steel slag aggregate was more than that of crushed limestone aggregate concrete. The compressive strength of 45% steel slag aggregate concrete was less than that of crushed limestone aggregate concrete and other steel slag aggregate concretes. This may be attributed to the excess of sand used in these concrete specimens. Further, the compressive strength of concrete with 50% steel slag aggregate and 50% sand was similar to that of concrete with 60% crushed limestone aggregate and 40% sand. Therefore, this proportion of coarse and fine aggregate may be adopted to minimize the weight effect of heavy steel slag aggregate.

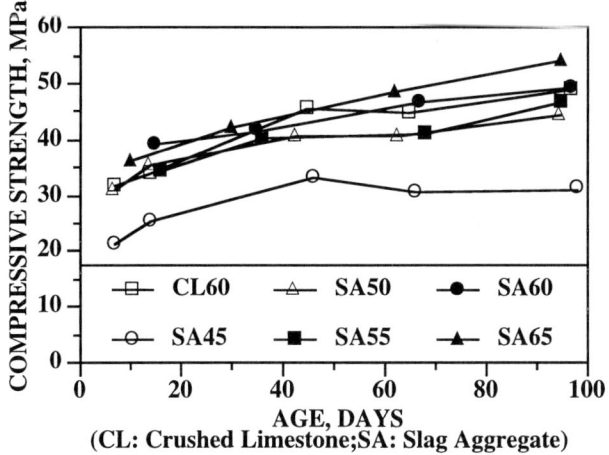

Figure 2 Compressive strength development in steel slag and crushed limestone aggregate concretes

The flexural strength of steel slag and crushed limestone aggregate concretes cured for 90 days is plotted in Figure 3. The flexural strength of steel slag aggregate concrete increased with the proportion of coarse aggregate, varying from 3.47 MPa in 45% steel slag aggregate concrete to 4.21 MPa in 65% steel slag aggregate cement concrete. The flexural strength of crushed limestone aggregate (60% coarse aggregate) concrete was 3.96 MPa. These results indicate that the improvement in flexural strength, due to the incorporation of steel slag aggregate, particularly when the proportion of coarse aggregate was less than 60%, was not very significant.

The 48-hour water absorption in the steel slag and crushed limestone aggregate concretes, tested according to ASTM C 642 [6], is depicted in Figure 4. The water absorbed by the steel slag aggregate concretes was less than that absorbed by the crushed limestone aggregate concrete. The water absorbed by the steel slag aggregate concretes varied from 3.64 to 4.51%, while it was 5.53% in the crushed limestone aggregate concrete. The reduction in the absorption characteristics of steel slag aggregate concrete may be attributed to the impervious nature of the steel slag aggregate compared to the crushed limestone aggregate.

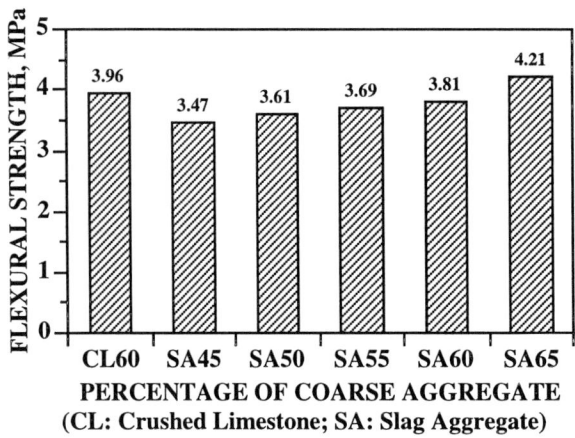

Figure 3 Flexural strength of steel slag and crushed limestone aggregate concretes

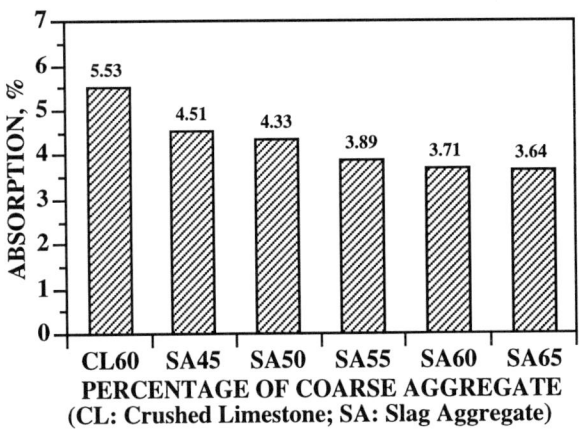

Figure 4 48-hour water absorption in steel slag and crushed limestone aggregate concretes

The effect of thermal variations on pulse velocity in the steel slag and crushed limestone aggregate concretes is shown in Figure 5. The pulse velocity in all the concrete specimens decreased due to exposure to thermal variations. This reduction in the pulse velocity was in the range of 9.7 to 16.7% in steel slag aggregate cement concretes, while it was 17% in crushed limestone aggregate concrete.

Thus, the influence of thermal variations on the pulse velocity is more or less similar in both steel slag and crushed limestone aggregate concretes.

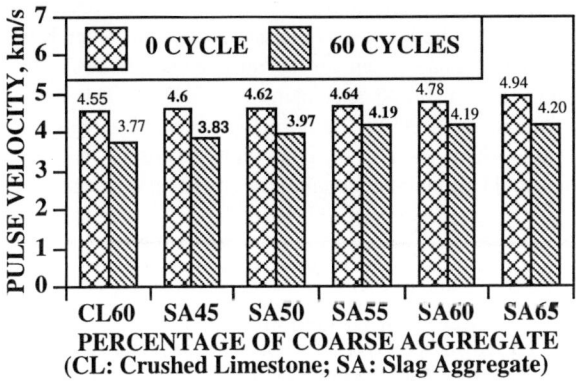

Figure 5 Effect of thermal variations on the pulse velocity in steel slag and crushed limestone aggregate concretes

Shrinkage of steel slag-cement and sand-cement mortar specimens is shown in Figure 6. These measurements were taken after seven days of moist curing and then air drying in the laboratory environment of 25 ± 2 °C and 50 ± 5% RH. As is apparent from these data, shrinkage of steel slag-cement mortar specimens was similar to that of the sand-cement mortar specimens. After 120 days, shrinkage in the former was 0.08%, while it was 0.10% in the latter. These data suggest that shrinkage should not be of concern in the steel slag aggregate concrete.

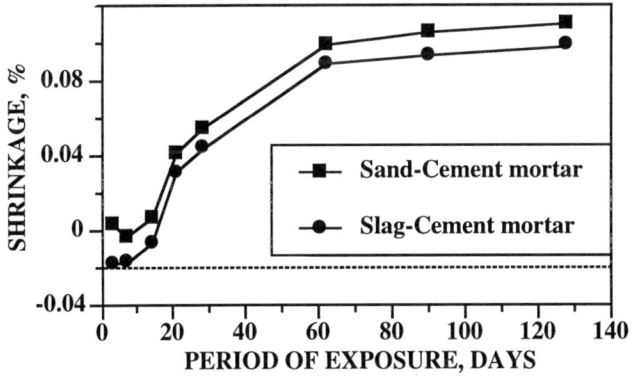

Figure 6 Shrinkage of steel slag-cement and sand-cement mortar specimens

The change in length of the steel slag-cement and sand-cement mortar specimens exposed to a moist environment for about four months is plotted in Figure 7. Initially, shrinkage was recorded in both the steel slag-cement and sand-cement mortar specimens. Thereafter, an increase in length was noted in both types of mortar specimens. After 120 days of exposure, the change in length of the sand-cement mortar specimens was 0%, while the increase in length of the steel slag-cement mortar specimens was 0.04%. This expansion in the steel slag-cement mortar specimens is below the normally accepted value of 0.05% specified in ASTM C 33 [10], after three months of continuous exposure to a moist environment. Therefore, expansion should not be of concern in the steel slag aggregate cement concrete.

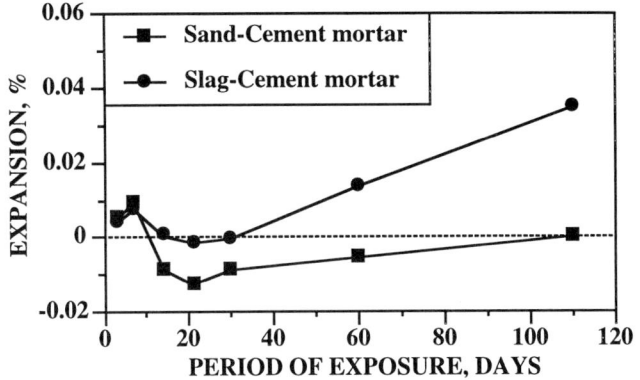

Figure 7 Change in length of steel slag-cement and sand-cement mortar specimens exposed to a moist environment

As stated earlier, corrosion of reinforcing steel bars embedded in steel slag and crushed limestone aggregate cement concretes was accelerated by impressing an anodic potential of 4V. The resulting current was plotted against time.

These figures were utilized to assess the time to initiation of reinforcement corrosion, under accelerated environment. The time to initiation of reinforcement corrosion was taken as a point where an abrupt change in the current requirement was noted. These values are summarized in Table 3.

The time to initiation of reinforcement corrosion in the crushed limestone aggregate cement concrete specimens was 190 hours, while it was in the range of 198 to 367 hours in steel slag aggregate cement concrete. Further, the time to initiation of reinforcement corrosion increased with the quantity of steel slag aggregate.

Table 3 also shows the time when cracks were visually noted on the concrete specimens. Surface cracks were observed on the steel slag aggregate concrete specimens after 509 to 774 hours, while in the crushed limestone aggregate concrete specimens they were noted after 515 hours.

Table 3 Time to initiation of reinforcement corrosion and cracking in steel slag and crushed limestone aggregate cement concretes

AGGREGATE TYPE	COARSE / TOTAL AGGREGATE RATIO	AVERAGE TIME TO INITIATION OF CORROSION, HOURS	AVERAGE TIME TO CRACKING, HOURS
Crushed limestone	0.60	190	515
	0.45	198	509
Steel slag	0.50	280	586
	0.55	325	653
	0.60	329	704
	0.65	367	774

The data in Table 3 indicate that steel slag aggregate concretes perform better than crushed limestone aggregate concrete in resisting reinforcement corrosion, both in terms of longer time to initiation of reinforcement corrosion and cracking of concrete.

CONCLUSIONS

1. The water absorption, abrasion-resistance and magnesium soundness of steel slag aggregate were within the allowable values.

2. The unit weight of steel slag aggregate cement concrete was more than that of crushed limestone aggregate concrete. The increase in unit weight of concrete, due to the incorporation of steel slag aggregate, in lieu of crushed limestone aggregate, was approximately 17%.

3. Though the compressive strength of steel slag aggregate concrete was marginally better than concrete with equivalent proportion of crushed limestone aggregate, no significant improvement in the flexural strength was noted in the steel slag aggregate concrete compared to crushed limestone aggregate concrete.

4. The water absorbed by steel slag aggregate concretes was less than that absorbed by the crushed limestone aggregate concrete. On an average, the water absorbed by steel slag aggregate concretes was 26% less than that absorbed by the crushed limestone aggregate concrete.

5. Shrinkage of steel slag-cement and sand-cement mortar specimens, exposed to a dry environment, was similar. While no expansion was noted in the sand-cement mortar specimens, exposed to moist environment for four months, an expansion of 0.04% was measured in the steel slag-cement mortar specimens exposed to similar conditions. This expansion is, however, less than 0.05% specified by ASTM C 33 [10].

6. The time to initiation of reinforcement corrosion and time to cracking was more in steel slag aggregate concrete specimens compared to crushed limestone aggregate concrete specimens. The superior performance of steel slag aggregate concrete was noted in the concrete specimens made with a coarse to total aggregate ratio of 0.55 and above.

7. The improved properties of concrete made with steel slag aggregate indicate that this material can be beneficially utilized in Portland cement concrete.

ACKNOWELDGEMENTS

The support provided by the Research Institue at King Fahd University of Petroleum and Minerals, Dhahran Saudi Arabia, and Heckett Multiserve, Al-Jubail Saudi Arabia, is appreciated.

REFERENCES

1. TESTS ON SLAG AGGREGATE, A report submitted to Heckett Multiserv, Saudi Arabia, Ltd., by Research Institute, King Fahd University of Petroleum and Minerals, 1996.

2. ASTM C 131, Test Method for Resistance to Degradation of Small-Size Coarse Aggregate by Abrasion and Impact in the Los Angeles Machine, *Annual Book of ASTM Standards*, Vol. 4.02, American Society of Testing and Materials, Philadelphia, 1998.

3. ASTM C 127, Test Method for Specific Gravity and Absorption of Coarse Aggregate, *Annual Book of ASTM Standards*, Vol. 4.02, American Society of Testing and Materials, Philadelphia, 1998.

4. ASTM C 88, Standard Test Method for Soundness of Aggregates by Use of Sodium Sulfate or Magnesium Sulfate, *Annual Book of ASTM Standards*, Vol. 4.02, American Society of Testing and Materials, Philadelphia, 1998.

5. ASTM C 142, Standard Test Method for Clay Lumps and Friable Particles in Aggregates, *Annual Book of ASTM Standards*, Vol. 4.02, American Society of Testing and Materials, Philadelphia, 1998.

6. ASTM C 642, Standard Test Method for Specific Gravity, Absorption, and Voids in Hardened Concrete, *Annual Book of ASTM Standards*, Vol. 4.02, American Society of Testing and Materials, Philadelphia, 1998.

7. ASTM C 39, Standard Test Method for Compressive Strength of cylindrical Concrete Specimens, *Annual Book of ASTM Standards*, Vol. 4.02, American Society of Testing and Materials, Philadelphia, 1998.

8. ASTM C 597, Test Method for Pulse Velocity through Concrete, *Annual Book of ASTM Standards*, Vol. 4.02, American Society of Testing and Materials, Philadelphia, 1998.

9. ASTM C 490, Standard Practice for Use of Apparatus for the Determination of Length Change of Hardened Cement Paste, Mortar and Concrete, *Annual Book of ASTM Standards*, Vol. 4.02, American Society of Testing and Materials, Philadelphia, 1998.

10. ASTM C 33, Standard Specification for Concrete Aggregates, *Annual Book of ASTM Standards*, Vol. 4.02, American Society of Testing and Materials, Philadelphia, 1998.

PROPERTIES OF SELF-COMPACTING CONCRETE WITH SLAG FINE AGGREGATES

M Shoya   S Sugita
Y Tsukinaga   M Aba
Hachinohe Institute of Technology
K Tokuhasi
Chichibu Onoda Corporation
Japan

ABSTRACT. In this study, the ferronickel slag fine aggregates and copper slag fine aggregates, among non-ferrous metal slag fine aggregates standardized in JIS A 5011, were investigated from the point of view the applicability for production of self compacting concrete. In this paper, the self compactability of fresh concretes with slag fine aggregates using a limestone powder were investigated from the slump-flow test, the V-type funnel test and filling vessel test. Properties of hardened self compacting concrete with slag fine aggregates were also investigated by mechanical properties test and durability test. It was judged that both types of non-ferrous metal slag fine aggregate examined in the study were promising as aggregates for production of high performance concrete with self compactability and high durability.

Keywords: Self compacting concrete, Slag fine aggregate, Limestone powder, Durability.

Dr Masami Shoya is a professor in the Department of Civil Engineering at Hachinohe Institute of Technology, Japan. He has written many papers on concrete durability and shrinkage.

Dr Shuichi Sugita is a professor in the Department of Civil Engineering at Hachinohe Institute of Technology, Japan. He has been engaged in research work on cementitious materials for concrete.

Dr Yoichi Tsukinaga is an a professor in the Department of Architectural Engineering at Hachinohe Institute of Technology, Japan. He has been engaged in research work on the durability of concrete structures and non-destructive testing.

Dr Minoru Aba is an assistant professor in the Department of Civil Engineering at Hachinohe Institute of Technology, Japan. He has been engaged in research work on microcracks formed on concrete.

Kazuki Tokuhasi is a technical staff in Chichibu Onoda Corporation, Tokyo, Japan. He has been working on the technical problems of cement and concrete.

INTRODUCTION

Self compacting concrete has been studied and successfully developed in Japan [1]. Self compactability of this type of concrete can be defined as a set of properties including not only high fluidity such as self compaction but also good resistance to segregation. The subcommittee of concrete committee in JSCE published guideline for making self compacting concrete, and in this guideline, it was regulated that the self compactability of fresh concrete could be divided into three ranks assigned according to the combination of slump-flow test, V-type funnel test and filling vessel test [2].

On the other side, recently in Japan, the utilization of industrial by-products and recycled aggregates has been promoted and has proceeded since the enactment of "Recycle Law" in 1992. Among the industrial wastes applicable to aggregates for concrete, ferronickel slag (FNS) fine aggregates and copper slag (CUS) fine aggregates, which were standardized in JIS A 5011 (Japanese Industrial Standard) [3] as slag aggregate for concrete, are examined in this study. These slags are suitable for concrete aggregates because they are almost inert both chemically and physically in concrete, and concretes with them generally show the same strength level and durability as normal concretes with natural fine aggregates. However, These slag fine aggregates have not yet been applied to self compacting concrete.

The work reported in the paper was planned to investigate possibility to make self compacting concretes with FNS fine aggregate and CUS fine aggregate using a limestone powder. Concrete containing FNS fine aggregate and CUS fine aggregate will be denoted conveniently as FNS concrete and CUS concrete. Besides, self compactability and fresh properties, properties of hardened concretes such as, mechanical properties, freezing and thawing resistance, carbonation, permeability and drying shrinkage were tested.

CONCRETE MATERIALS

Materials Used

An ordinary portland cement was used, and its specific gravity, specific surface area by Blaine, 28 day compressive strength and Na_2O equivalent were 3.16, 335 m^2/kg, 42.1 MPa and 0.60 percent, respectively.

The physical properties of fine aggregates used in test are shown in the following. The natural sand for control was a pit sand having a specific gravity of 2.65, water absorption of 1.16 and fineness modulus of 2.57. FNS fine aggregates adopted showed that specific gravity, water absorption and fineness modulus were 2.97, 1.2 % and 2.48, respectively, and was wind granulated and crushed aggregate whose particle shape before crushing was almost sphere with brown color. CUS fine aggregates adopted was a water granulated one with dark color whose specific gravity, water absorption and fineness modulus were 3.63, 0.31 % and 2.20, respectively. The main chemical composition of FNS and CUS were MgO and FeO, and the contents were 34.2 and 36.1 %, respectively. The grading curves of the fine aggregate used shown in Figure 1. These slag aggregates fulfilled the requirements of FNS 2.5 in JIS A 5011-2 (FNS fine aggregate) and CUS 2.5 in JIS A 5011-3 (CUS fine aggregate)[3].

The coarse aggregate used in the test was a crushed limestone with a maximum size of 20-mm, a minimum size of 5-mm, a specific gravity of 2.70 and a fineness modulus of 6.60.

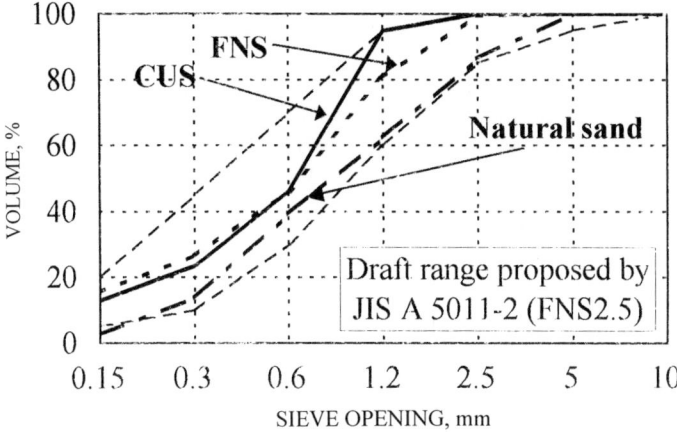

Figure 1 Grading curves of the fine aggregate used

A polycarbonic acid ether based high range air-entraining water reducer as well as two types of air-entraining admixture (TA and YA) essentially made from natural resin, were used to secure the fluidity and the air entrainment. A limestone powder having a specific gravity and specific surface area by Blaine are 2.72 and 570 m^2/kg, respectively, was also used to increase the viscosity of the paste.

TESTING METHOD

Table 1 shows the test plan used when making self compacting concretes with slag fine aggregate. The self compactability of concrete was investigated from the slump-flow test, the V-type funnel test and filling vessel test. It can be seen in this table that target of slump value and V funnel flow time were selected 700-mm and 10 sec. Then, properties of the hardened concretes such as, mechanical properties, freezing and thawing resistance, permeability, carbonation and drying shrinkage were also investigated.

Table 1 Tests made on concrete with slag fine aggregate

VOLUME FRACTION OF SLAG FINE AGGREGATE %	TARGET VALUE			TEST ITEMS
	Slump-flow mm	V funnel flow time sec	Air content %	

0, 50, 100	700 ± 50	10	3	• Evaluation of self compatability
				  Slump-flow test
				  V-type funnel test
				  Filling vessel test
				  Change of properties of fresh concrete
				• Time of setting

			4	• Bleeding • Mechanical properties test • Freezing and thawing test
			5	• Permeability test • Carbonation test • Drying shrinkage test

Evaluation for Self Compactability of Fresh Concrete

Slump-flow test

The slump cone used (upper diameter of 100-mm, lower diameter of 200-mm and height of 300-mm) was regulated in JIS A 1101, and test conducted in accordance with the guideline of JSCE [2]. The slump cone is slowly lifted up vertically, and the diameter, which represents the maximum spread of the concrete is measured as well as a diameter perpendicular to the first one are measured. The average of these diameters was calculated to determine the slump-flow value (1).

$$Fc = \text{slump-flow (mm)} = (d1+d2)/2 \tag{1}$$

Simultaneously with the slump-flow test, measurement of time taken by slump-flow value to reach 500-mm and naked eye observation on segregation was also performed.

V-type funnel test

V-type funnel test was conducted using a funnel with the shape and dimensions shown in Figure 2. The fresh concrete was placed in the funnel, and the bottom of the funnel opened allowing the concrete to flow down. The time taken by the fresh concrete to flow out of the funnel was measured.

Filling vessel test

Filling vessel test was conducted using a box regulated in the guideline of JSCE [2]. Figure 3 shows the testing method of filling vessel test. The fresh concrete was casted in the left box, and the partition gate opened. The filling height in the right box was measured. The obstacle used in this test was R1 that was most severe condition.

Test for Time of Setting and Bleeding Test

Test for the time of setting was conducted in accordance with ASTM C 403 using Proctor penetration resistance needles. The initial time of setting and the final time of setting were determined when the Proctor penetration resistance reached 3.5 MPa and 28.4 MPa, respectively. The samples used for test was the wet-screened mortar using a 5-mm sieve.

Bleeding tests were conducted in accordance with JIS A 1123 using 250-mm diameter by 285-mm high cylindrical container.

Mechanical Properties Test

Tests for compressive strength and splitting tensile strength were done in accordance with JIS A 1108 and JIS A 1113, respectively. Compressive strength and tensile strength were tested using 100 by 200-mm cylindrical specimens at the ages of 7, 28 and 91 day. Static modulus of elasticity was measured using 100 by 200-mm cylindrical specimens with compressometer. Static modulus of elasticity was also measured.

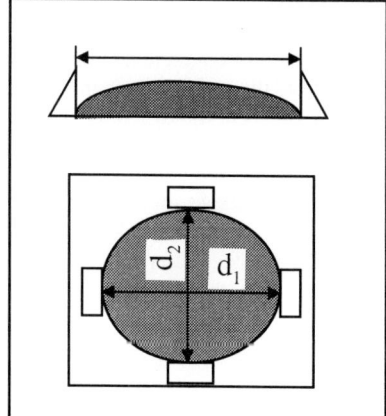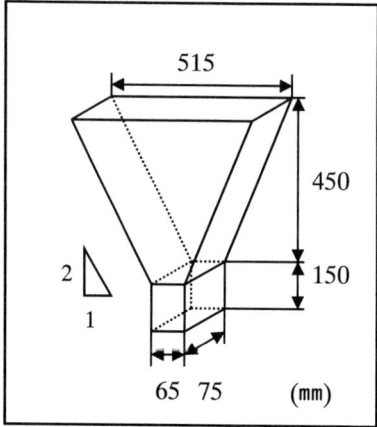

Figure 2 Method of slump-flow teswt and V-type funnel test

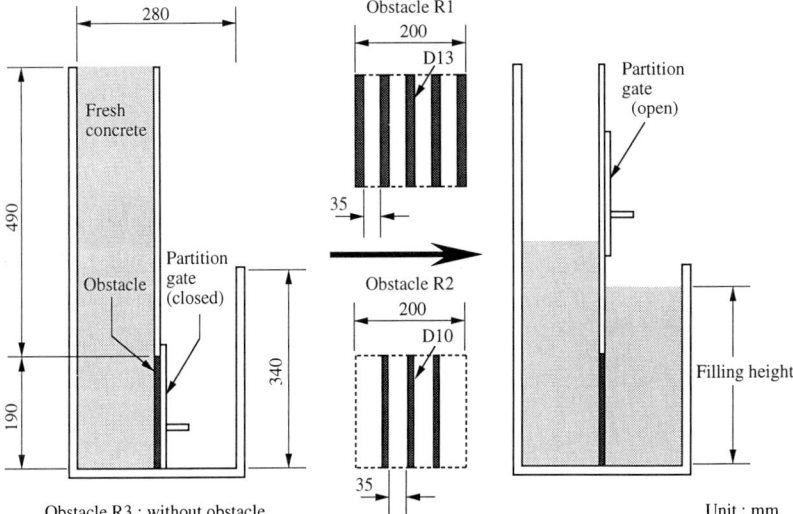

Figure 3 Method of filling vessel test

Freezing and Thawing Test

Freezing and thawing test was conducted in accordance with ASTM C 666-92 procedure A (freezing and thawing in water). The test started at an age of 28 days using 100 by 100 by 400-mm prisms. Three specimens were tested simultaneously representing each mixture and testing conditions. The change in the relative dynamic modulus of elasticity and mass, were measured every 30 cycles up to the 300th cycle. Then, durability factor DF was calculated.

Permeability Test

Water permeability was measured by the in-put method proposed by Murata. The experiment was made applying a water pressure of 980 kPa to an area of concrete having a diameter of 150-mm for 48 hours. After the completion of the test, the specimens were split, and the average depth of the penetration of water was measured. Then, the diffusion coefficient was calculated by the theoretical equation proposed by Murata.

Air permeability was measured by the constant pressure method proposed by Nagataki and Ujike. The experiment was made holding a load bearing pressure of 490 kPa. The amount of air penetrating the concrete specimens was captured using the water replacement method. The coefficient of air permeability was calculated by the theoretical equation applying Darcy's Law.

Carbonation Test and Drying Shrinkage Test

The carbonation test was made using 100-mm diameter by 50-mm high circular specimens cut from 100-mm diameter by 200-mm high cylindrical specimens. The test was made in the condition of 5 % CO_2 concentration, 30°C and 60 % R.H.. The neutralized depth was measured by spraying 1 % phenolphthalein ethanol solution to the split specimen surfaces.

Drying shrinkage tests were conducted according to JIS A 1129. The test was started at an age of 28 days using 100 by 100 by 400-mm prisms in a controlled room with 60 % in relative humidity and 20°C in temperature. The loss of moisture was also measured during drying shrinkage test.

EXPERIMENTAL RESULTS AND CONSIDERATION

Mixture Proportioning

Table 2 shows the mixture proportions of concrete while the target slump-flow value and V funnel flow time being kept at 700-mm and 10 sec, respectively, at a fixed water cement ratio of 0.55. It can be seen that the water demand of FNS concrete decreased with the increase of the volume fraction of FNS fine aggregates. It was decreased by 10 kg/m3 when the volume fraction of FNS fine aggregates was 100 % as compared to that of control concrete. The reason was guessed to be due to the influence of round shaped FNS particle.

On the other side, the water demand of CUS concrete a bit increased with the increase of the volume fraction of CUS fine aggregates. The reason why the water demand of CUS concrete increased, may be attributed that the small particles in the CUS fine aggregates used in this examination behaved as same a fine powder, probably due to containing a lot of small particles under 0.06-mm.

The optimum sand percentage of concretes were almost constant regardless of volume fraction of slag fine aggregate in this testing conditions.

Self Compactability of Fresh Concrete with Slag Fine Aggregates

Table 2 Mixture proportions of concrete

SYMBOL	GMAX mm	W/C*	W/P	S/A %	C/L**	VSS*** %	UNIT CONTENT, kg/m³						SP*2, c x %	AE, c x %
							W	C	L	S (NS)	S (SS*1)	G		
Control						0	165	300	260	880	-	740		0.4*3
FNS-50						50	160	291	252	448	502	746		2.2*3
FNS-100	20	0.55	0.86	55	1.15	100	155	282	245	-	1017	756	1.15	2.5*4
CUS-50						50	167	304	243	442	606	737		0.2*4
CUS-100						100	170	309	247	-	1199	730		0.45*4

* Volumetric water-powder ratio,   ** Volumetric cement-lime stone powder ratio,   *** Volume fraction of slag fine aggregate,
*1 Slag fine aggregate, *2 High range air-entraining water reducer,   *3 Air-entraining admixture TA,   *4 Air-entraining admixture YA

Table 3 showed the results of the test for self compactability of fresh concrete recommended by JSCE. Table 4 shows the rank of self compactability and corresponding target values [3]. From the table 3, it can be seen that the V funnel flow time became somewhat slower when the volume fraction of FNS fine aggregates increased. However, these results of test for self compactability almost were fallen in the range of rank 1 from Table 4. And, that in CUS fine aggregate concrete is also assigned to the rank 1. From these results, FNS and CUS fine aggregate examined in the study were judged as a promising aggregates for producing high performance concrete with self compactability.

Time of setting and bleeding property

The initial time of setting and the final time of setting of FNS concretes was almost at the same level as that of control concrete. However, those values of CUS concretes became a bit slower with the increase of the volume fraction of CUS fine aggregates. The amount of bleeding of both slag concretes was almost nothing as not more than 0.003 ml/cm2.

Mechanical Properties

As shown in Table 5, the compressive strength of slag concretes tends to be a bit grater than that of the control concrete, at a fixed air content of 5 %. The tensile strength of slag concretes were almost similar to those of the control concrete. The relationship between static modulus of elasticity and volume fraction of slag fine aggregates was a bit larger to that of compressive strength. The reason can be due to the higher hardness of slag fine particle. From these results, the mechanical properties of concretes with slag fine aggregates were found to be similar to those of the control concrete made with natural sand.

Freezing and Thawing Resistance

Table 5 shows the test results for freezing and thawing resistance of concretes with slag fine aggregates. The durability factor DF showed the high value when the volume fraction of slag fine aggregates increased, at a fixed air content while spacing factor was likely to increase. Especially, the durability factor of CUS concrete became relatively higher comparing to that of FNS concretes. This will be perhaps due to the powder effect of small particles in the CUS fine aggregates.

Table 3 Test results for self compactability of fresh concrete

TYPE OF CONCRETE	CONTROL CONCRETE	FNS CONCRETE		CUS CONCRETE	
		FNS-50	FNS-100	CUS-50	CUS-100
Slump-flow value, mm	700	695	690	700	705
V funnel flow time, sec	14	17	21	11	9
Time taken by slump-flow value to reach 500-mm. sec	6	8	18	5	5
Filling height, mm	300	300	340	340	340
Unit coarse aggregate content by absolute volume, m^3/m^3	0.274	0.276	0.280	0.273	0.270

Table 4 Rank of self compactability and corresponding target values

RANK OF SELF COMPACTABILITY		1	2	3
Structual Conditions	Minimum clearance of bar, mm	35 ~ 60	60 ~ 200	Not less than 200
	Mass of reinforcement, kg/m^3	Not less than 350	100 ~ 350	Not more than 100
Filling height by filling vessel test[*1], mm		Not less than 300 (Obstacle R1)	Not less than 300 (Obstacle R2)	Not less than 300 (Obstacle R3)
Unit coarse aggregate content by absolute volume, m^3/m^3		0.28 ~ 0.30	0.30 ~ 0.33	0.32 ~ 0.35
Flowability	Slump-flow value, mm	600 ~ 700	600 ~ 700	500 ~ 650
Segregation resistance	Funnel flow time, sec	9 ~ 20	7 ~ 13	4 ~ 11
	Time taken by slump-flow value to reach 500 mm, sec	5 ~ 20	3 ~ 15	3 ~ 15

Table 5 Results of strength test and freezing and thawing resistance test

SYMBOL	VOLUME FRACTION OF SLAG FINE AGGREGATES, %	28 DAYS STRENGTH, N/mm^2	AIR*, %	AIR**, %	SPACING FACTOR $\bar{L}$, mm	DF	MASS AFTER (300 CYCLE), %
Control-5	0	41.2	4.5	5.4	0.260	71.6	97.7
FNS-50-5	50	45.7	5.2	4.3	0.350	84.7	98.3
FNS-100-5	100	41.7	5.0	5.0	0.270	86.4	97.4
CUS-50-5	50	42.1	4.9	2.9	0.269	92.8	99.1
CUS100-5	100	43.8	5.1	4.0	0.326	81.4	96.8

Based on the test results, it is seen that the slag fine aggregate is a good fine aggregate for making durable concrete with self compactability when a suitable air content is included.

Permeability

Table 6 shows the permeability results for the concretes with slag fine aggregates. As shown in this figure, both the coefficient of water diffusion and the coefficient of air permeability of slag concretes showed little change with an increase of volume fraction of FNS and CUS fine aggregates.

Table 6 Results of permeability test, drying shrinkage test and carbonation test

SYMBOL	DIFFUSION COEFFICIENT, (x 10^{-13}), m^2/sec	COEFFICIENT OF AIR PERMEABILITY (x 10^{-13}), m/sec	DRYING SHRINKAGE STRAIN AFTER 100 DAYS, x 10^{-5}	NEUTRALIZED DEPTH AFTER 91 DAYS, mm
Control	2.43	2.51	51	6.97
FNS-50	3.38	2.45	45	5.72
FNS-100	2.70	2.44	42	6.49
CUS-50	1.98	4.17	53	6.83
CUS-100	1.89	4.07	49	6.74

From these results, it was found that the permeability of FNS fine aggregate concrete and CUS fine aggregate were similar to those of the control concrete made with natural sand.

Drying Shrinkage and Carbonation

Table 6 shows the test results of drying shrinkage and carbonation of concretes. As shown in this figure, the drying shrinkage strain of the FNS concretes showed a somewhat smaller value due to the reduction of unit water content. That of CUS concretes showed the almost at the same level as that of the control concrete. And, the neutralized depth of slag concretes was almost similar or a bit lower than that of the control concrete.

CONCLUSIONS

Experimental examination of self compacting properties, mechanical properties and durability of concrete with slag fine aggregates were made in detail. The results are summarized as follows:

1. The water demand of self compacting concrete made with FNS fine aggregates decreases when the volume fraction of FNS fine aggregates increases. Also, the water demand of CUS concrete increased with the increase of the volume fraction of CUS fine aggregates. The optimum sand percentage of slag fine aggregate concretes was found to be the same as that of control concrete made with natural sand.

2. The fresh properties and properties of hardened concrete made with the FNS fine aggregate and CUS fine aggregate were found to be almost similar or a bit higher than those of the control concrete made with the natural sand.

3. FNS fine aggregates and CUS fine aggregates examined in the study was found to be promising as an aggregate for the production of high performance concrete with self compactability.

REFERENCES

1. NAGATOMO, N and OZAWA, K. Mixture properties of self compacting high performance concrete. Proceedings of ACI international conference on high performance concrete, SP 172-33, 1997, pp 623-636.

2. JSCE. Guideline on Making Self Compacting Concrete. Concrete Library 93, 1998 (In Japanese).

3. JAPANESE INDUSTRIAL STANDARD. Slag Aggregate for Concrete : Ferronickel slag aggregate (JIS A 5011-2), Copper slag fine aggregate (JIS A 5011-3), 1997 (In Japanese).

ELECTRO-SURFACE PROPERTIES OF AGGREGATES FROM WASTE PRODUCTS AND THEIR INFLUENCE ON THE QUALITY OF FINE GRAINED CONCRETE

V A Matviyenko
N M Zaichenko
Donbass State Academy
S M Tolchin
Concrete and Reinforced Concrete Works, Makeyevka
Ukraine

ABSTRACT. To obtain the qualitative products and structures on the base of fine-grained concrete it is necessary to carry out such circumstances: 1- optimum granularity composition of aggregate; 2 - combination of aggregates with different integral charges of a surface. On an example of fine-grained concrete with aggregates from waste products is shown, that the carrying-out the second circumstance will make it possible to speed up the structure formation, to increase concrete strength, frost resistance, to reduce the size of deformations at drying shrinkage. The possibility of using of aggregate from crushed burnt coal pit rock in a mixture with Martin slag (in the ratio 2:1 on weight) for production of concrete B25 is shown. As a result saving of the natural aggregates and cement for concrete as well as a certain solution an ecological problem take place.

Keywords: Fine-grained concrete, Aggregate, Waste products, Electro-surface properties, Structure formation, Strength.

Professor Vasiliy A Matviyenko is a Lecture in Donbass State Academy of Civil Engineering and Architecture, a Head of the Department of Academy of Higher School, Ukraine. He develops the scientific school of electric activation forces in concrete technology. Professor Matviyenko has published widely and serves on many Technical Committees.

Dr Stanislav M Tolchin is Director of the Concrete and Reinforced Concrete Works, Makeyevka, Ukraine. His main research interests are connected with the problem of maximum using of waste products in the concrete structures industry.

Dr Nickolai M Zaichenko is a Lecture in Donbass State Academy of Civil Engineering and Architecture. He specialises on the problems of early concrete strength and rheology of concrete mixes.

INTRODUCTION

The production of fine-grained concrete with aggregates from waste products is an actual problem. A singularity of such concrete is the high values of interface of phases on the boundary a matrix - filler. The properties of fine-grained concrete depend on granulometric and mineral composition of aggregates, properties and consumption of cement, on technology factors. The use as aggregate of various waste products is connected with additional costs on their preliminary processing, which is necessary for increasing their quality. The investigations of various waste products (slags of power stations, blast and Martin slags, burnt coal pit rock) have shown that each from them at identical granularity and structure of concrete mix gives fine-grained concrete with sharply distinguished properties. It is possible to explain by influence of electrosurface properties of filler on the rheological properties of concrete mixes and on the process of concrete structure formation. In development of this position a hypothesis about a possibility of improving of physical-mechanical properties of fine-grained concrete due to the combination of fillers with various acid-basic properties has been brought up [1]. That kind of fillers combination makes it possible not only to optimise their granularity composition, but also to polarise a cement-water phase between opposite charged particles.

Polarisation of double electrical layer (DEL) of cement grains in space between such fillers will change conditions of their hydration in favour of heavily interaction with water. Structuring of water, modifying of adhesive contacts on the boundary cement paste - filler will promote the formation of more homogeneous and dense microstructure of fine-grained concrete. Concrete will have the higher physical-mechanical properties as a result [2].

EXPERIMENTAL DETAILS

Materials

Ordinary Portland cement (OPC-500) was used. The fine aggregates (fillers) consisted of dumping Martin slag (MS), blast granulated slag (BGS), crushed burnt coal pit rock (BCPR), ash-slag mix of thermal power stations (ASM), in 10 mm and 0,14 mm single size. The chemical composition of aggregates used is summarised in Table 1.

Curing Environment

The following curing conditions were used: air curing of concrete samples in air at 20-22°C/90-96%RH 28 days.

Methods and Equipment

The electrosurface properties of fillers have been determined with the help of various methods: indicator [3], sedimentary, pH-method [4]. Structure formation of concrete mixes has been studied under the data of plastometry. The main physical-mechanical properties of fine-grained concrete were established on standard techniques. A degree of probability of experimental outcomes is 95%.

Table 1 The chemical composition of aggregates from waste products

MATERIAL	CONTENT OF OXIDES, %							
	SiO_2	Al_2O_3	$FeO+Fe_2O_3$	CaO	MgO	MnO	SO_3	R_2O
Argillite burnt coal pit rock	62,8	25,1	3,6	4,0	1,6	-	0,1	2,8
Quartz burnt coal pit rock	78,4	11,1	1,8	4,2	1,1	-	0,3	3,1
Ash and slag mix	50,5	23,2	13,5	3,2	2,0	-	3,2	4,4
Martin slag	20,6	4,9	18,6	39,4	9,1	7,4	-	-
Blast granulated slag	37,6	6,8	1,1	47,0	2,1	3,3	2,1	-

RESULTS AND DISCUSSION

Determination of Electrosurface Properties of Aggregates (fillers)

The electrical relief of a solid body surface has a mosaic structure, which is characterised by presence of negatively and positively charged local sites. The sum of surface charges determines integral polarity of a particle surface. The differentiated analysis of concentration of active centres on a surface of various mineral waste products carried out under the data of an adsorption of Hammet's indicators, has shown the following. The lowest concentration of acid centres (pKa=-4,4) has Martin slag, and the highest - quartz ingredient of burnt coal pit rock. In comparison with other materials blast and Martin slags have an increased concentration of moderately sour Brensted's centres (pKa=2,1). Burnt coal pit rock and ash and slag mix have on a surface the higher concentration of alkaline (pKa=12,8) active centres. The interpretation of active centres of Brensted's type as centres of an adsorption (physical and chemical) of hydroxyl groups allows to divide researched waste products into two groups:

1. With explicitly expressed acid properties (burnt coal pit rock and ash-slag mix);
2. With a dominance of moderately and poorly sour active centres (blast and Martin slags).

This division will be agreed the data of suspension effect (Table 2). The display of suspension effect in 0,1-n solution of potassium chloride allows also to differentiate researched materials on sour (burnt coal pit rock and ash-slag mix) and basic (blast and Martin slags). However, in a solution of calcium hydroxide this effect is exhibited in a smaller degree, that is possible connected with high concentration and high alkalinity of solution. The influence of dispersing environment on a long-range action of active centres was investigated on sedimentary deposit density of dispersions of various materials. Its higher values in a saturated solution of calcium hydroxide (with the exception of dispersion of argillite ingredient of burnt coal pit rock) are stipulated by compression of a double electrical layer and decreasing of a long-range action of repulsive forces between the same charged particles. If the particles have opposite charged surfaces, the electrostatic forces of attraction between them are exhibited and the more dense deposit of dispersions will be formed. It is confirmed on an example of dispersion of materials consisting of burnt coal pit rock and Martin slag. Maximum density of a deposit in a solution of calcium hydroxide is achieved at the contents in an initial mix of dispersion of burnt coal pit rock 65-90%, and in 0,1-n solution KCl 50-70%. These data confirm a hypothesis about expediency of a combination of fillers with various electrosurface properties.

Table 2 Suspension effect ($\Delta\varepsilon H$) and density (ρ) of a sedimentary deposit of dispersions of fillers (fraction less than 0,14 mm) from waste products

MATERIAL	IN 0,1-n SOLUTION KCl			IN SATURATED SOLUTION Ca(OH)$_2$		
	$\Delta\varepsilon H$, mV, in			$\Delta\varepsilon H$, mV, in		
	10 s	3 min	ρ, kg/m^3	10 s	3 min	ρ, kg/m^3
Argillite burnt coal pit rock	-14	-18	800	+3	+2	680
Quartz burnt coal pit rock	-27	-39	950	0	-1	990
Ash and slag mix	-26	-19	1075	0	-3	1110
Martin slag	+17	+112	1100	+6	-1	1110
Blast granulated slag	+14	+85	1040	+5	-5	1100

Structure Formation of Concrete Mixes

It is important in the technology of concrete products formation the fact that the structuring of concrete mix on combined aggregates begins to be exhibited only under condition of sufficient approach of particles. The researches of concrete mixes workability on separate fillers and their mixes have shown, that their water consumption submits "to a rule of mix" and is not increased with transition to the combined composition of fillers at C/W=(1,3-2). The structuring begins to have an effect in 2,5-3 hours of concrete mix hardening for the stage of condensed-crystallised structuring (Figure 1). By that time as a result of hardening responses water is linked in hydrates and is structured by their developed surface. It affects in decreasing forces of wedge pressure between particles.

The Strength of Concrete

The strength of concrete has explicitly expressed area of maximum strength at the composition of filler mix: burnt coal pit rock 67% and Martin slag 33%. Obtained on the criterion of strength, an optimum ratio between these kinds of waste products coincides the data of the sedimentary method. Taking into account that fact, that granularity structure of the selected tests of rock and Martin slag does not differ considerably (modulus of coarseness M_c=4,1 and M_c=4,3 accordingly), the presence of an optimum ratio is explained by the structure formation role of mix fillers with various acid-basic properties. For a mix of fillers of ash and slag mix and Martin slag the optimum on strength criterion is the ratio 2:1-1:1.

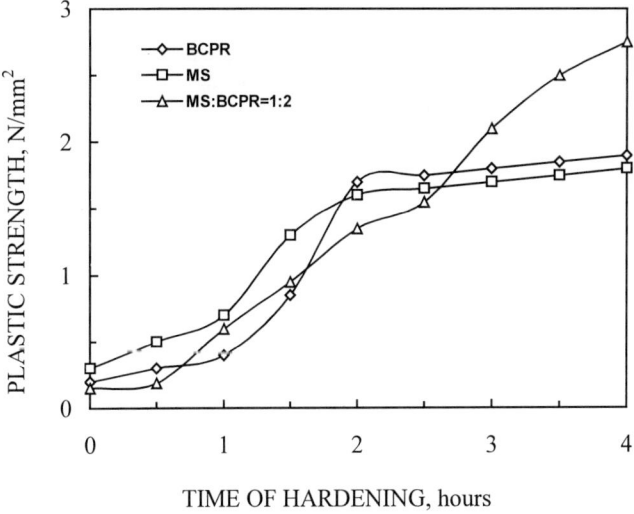

Figure 1 Dependence of plastic strength of concrete mix on hardening time

On the base of combined fillers of optimum composition the main physical-mechanical properties of fine-grained concrete have been investigated. By a variation of the consumption of cement within the limits of 250-500 kg/m^3 fine-grained concrete of classes B7,5-B25 were obtained. The properties of concrete B15 are adduced in a Table 3.

Table 3 Mix proportions and properties of concrete B15 on the base of aggregates from waste products

AGGREGATE CONTENT IN CONCRETE MIX, %				W/C	CEMENT CONTENT, kg/m^3	STRENGTH, N/mm^2	ELASTIC MODULUS, N/mm^2
Burnt coal pit rock	Martin slag	Ash and slag mix	Quartz sand				
100	-	-	-	0,60	470	19,8	1,84·10^4
67	33	-	-	0,54	406	20,2	2,26·10^4
-	50	50	-	0,53	400	21,5	2,40·10^4
-	75	-	25	0,52	270	20,4	2,44·10^4

From these data is obvious, that at the expense of a combination of fillers with opposite integral charges of a surface the required strength of concrete is reached at the smaller consumption of cement.

PRACTICAL APPLICATION OF RESULTS

In Donbass region the sizeable quantity enterprises of coal-extractive, metallurgical, chemical industry is concentrated, that creates high-power load on an environment. In this connection special urgency is gained with problems on salvaging of various waste products. At Makeyevsky plant of prefab reinforced concrete during the time of production of concrete structures the various slags of an iron and steel industry, ashes of thermal power stations, burnt coal pit rock are used as fillers of concrete. In addition, the determination of an optimum composition of fillers is made under the following scheme [6].

The evaluation of a degree of conformity of granularity each from initial fillers for optimum is previously made. If there are large divergences, the enrichment of filler is possible. Especially it is necessary at the high contents in it of dusty fractions, sharply raising the water consumption of concrete mix. A following stage of decision making is the determination of electrosurface properties of fillers according to the data of suspension effect ($\Delta\varepsilon H$) or by the other ways (electrokinetic potential etc.). By this parameter the fillers with opposite charges of a surface are selected. The determination of an optimum composition of filler mix is made in two stages. At the first stage the ratio between fillers according to the data of sedimentation method (on the size of maximum deposit of dispersions of fillers mix) is determined. In the future obtained ratio is updated on by the criterion of the strength of concrete samples.

According to the adduced scheme at the plant the optimum composition of combined filler consisting of burnt coal pit rock and elimination of splitting of Martin slag in the ratio 2:1 was established. At the consumption of cement 410 kg/m^3 the fine-grained concrete (in the age of 28 day) is characterised by the following parameters: compressive strength 31,5 N/mm^2, module of elasticity 2,26×10^4 N/mm^2, drying shrinkage 7 mm/m.

CONCLUSIONS

1. The hypothesis about a capability of regulation of structure formation process and improving of physical-mechanical properties of fine-grained concrete is confirmed by the combination of a composition of aggregates (fillers) with various electrosurface characteristics.

2. The positive influence of mix of fillers with opposite charges of a surface on the structure formation and concrete strength was established.

REFERENCES

1. MATVIYENKO, V A, TOLCHIN, S M, VYSOTSKY, Y B, AND MALININA, Z Z. Electrosurface properties of aggregates from waste products [in Russian]: the Bulletin DGASA, Vol.96-3 (4), 1996, pp 117-121.

2. TOLCHIN, S M, ZAICHENKO, N M, AND GUBAR, V N. Concrete on the base of aggregates from waste products [in Russian]: the Bulletin DGASA, Vol.96-3 (4), 1996, pp 125-128.

3. NECHIPORENKO, A P, AND SHEVCHENKO, G K. Research of influence of heat treatment and dispersibility of a sample on acid-basic properties of a surface of silica [in Russian]: the Log-book of common chemistry, 1985- Vol.55 (2), pp 244-253.

4. NECHIPORENKO, A P, AND KUDRYASHOVA, A I. Researches of an acidity of rigid surfaces by the pH-method. Journal of Applied chemistry, No.9, 1987, pp 1957-1961.

5. MATVIYENKO, V A, ZAICHENKO, N M. Hardening of Cements Polarized in Electrical Field: 10-th International Congress on the Chemistry of Cement, Sweden, 1997, pp 785-789.

6. TOLCHIN, S M, MATVIYENKO, V A, ZAICHENKO, N M AND GUBAR, V N. The principles of combination of aggregates from waste products for fine-grained concrete [in Russian]: Proceedings of the Conference "Modern problems of building", Donetzk, 1997, pp 103-105.

THE USE OF RECYCLED CONCRETE AND MASONRY AGGREGATES IN CONCRETE: IMPROVING THE QUALITY AND PURITY OF THE AGGREGATES

J Desmyter

J Van Dessel

S Blockmans

Belgian Building Research Institute

Belgium

ABSTRACT. The BBRI is since the seventies involved in research on recycling in the construction industry. On the one hand, the optimisation of separation-purification process for the recycling of C&D waste is investigated. Technologies commonly used in the metallurgy industry are evaluated and optimised in order to be used for C&D waste. Representative samples coming from several crushing plants were characterised in order to realise a statistical overview of the current quality of the recycled aggregates. On the other hand, efforts were directed towards the production of ready mixed concrete and concrete products with recycled aggregates. Concrete specimens were fabricated with several kinds of C&D aggregates and tested for them workability, strength, durability, creep and shrinkage. A summary of the most important and available results of the current research programs of the BBRI is given.

Keywords : C&D waste aggregates, Separation-purification technologies, Recycled concrete

J Desmyter is head of the laboratory Structures at BBRI (Belgian Building Research Institute). He is in charge of several research projects regarding the use of recycled construction and demolition waste as aggregates in concrete. Ir. J. Desmyter is chairman of a technical committee of CRIC-Certification, which is elaborating a national standardisation document for recycled aggregates (PTV).

J Van Dessel is a project co-ordinator of the division Structures at BBRI (Belgian Building Research Institute). He is a member of several working groups on waste handling, recycling and the environment. J. Van Dessel is responsible for the elaboration of policy studies for the regional authorities in this field and for the diffusion of knowledge with regard to relevant recycling and sorting technology towards the industry.

S Blockmans is a geologist researcher at BBRI (Belgian Building Research Institute). She was in charge of a research project regarding the optimisation of the purification and separation process for the recycling of C&D waste. At the moment she is mainly working on the technological aspects of the use of recycled construction and demolition waste as aggregates in concrete. She is a member of the above-mentioned technical committee of CRIC-Certification.

INTRODUCTION : HISTORIC OVERVIEW OF RECYCLING IN BELGIUM

General

The first construction and demolition (C&D) waste recycling plant in Belgium started to operate already back in the fifties. However, the recycling industry was only developed on a broader scale from the seventies on, amongst others thanks to scientific and technical research (as by example [1]). This resulted in the eighties to important research and pilot projects ([2] [3]). Since then, the interest for recycling has increased continuously.

Nowadays about 90 recycling plants are operating all over the country. About 75 % of the installations are fixed or mobile with a fixed location, the remaining 25 % are mobile plants. The plants are generally equipped with a weighing bridge, equipments for pre-processing (bull, crane, ...), a preliminary sieve to eliminate the finest materials, a primary crusher, electrical magnet systems and a sieve installation to separate the materials in accordance with the specified aggregate sizes. The most advanced plants can also be equipped with an air sieve or a washing installation and a secondary crusher and sieving installation.

An important part of the recycled aggregates is used in road construction. This sector takes the crushed concrete and mixed aggregates for use as unbound base-course and sub-base material. Moreover, crusher sand, concrete and mixed aggregates are already for some years used as aggregates for treated or stabilised sand and lean concrete. The masonry aggregates and to a lesser degree the sieve sands, are used in earthworks and raising, i.e. for low-grade applications. All of these uses are recognised and allowed by the authorities through their technical specifications for public works. Since such documents have a relevant exemplary function, recycled products are also frequently used in private construction works.

Waste Policy in Belgium

Belgium is a federal state with three regions, each responsible, with few exceptions, for their own environmental and waste policy and legislation. Table 1 gives a short overview of the elaborated waste management plans and the target-levels specified for the recycling of C&D waste. The different technical specifications in force for each region are also specified in Table 1.

An important element in the acceptation process of recycled aggregates is the voluntary certification scheme for their use in unbound applications, cement treated sand and gravel and lean concrete, which was developed on the initiative of an association of recycling plants. COPRO is the responsible party for the certification of these aggregates. A recent development, which should lead to still a wider acceptance of recycled products, is the elaboration by a CRIC (Certification Organism for cement, aggregates, concrete, fly ash and additives) working-group of a technical specification for recycled concrete, masonry and mixed aggregates (PTVxxx [12]). This document would enable to grant a BENOR quality label to these aggregates.

To protect the environment against possible hazards caused by the use of recycled products, most of the authorities are preparing or have prepared an environmental hygiene legislation to prevent such accidents.

Table 1 Overview of the waste management plans and of the different technical specifications regulating the use of recycled aggregates in public works

	PLANS AND TECHNICAL DOCUMENTS	TARGETED RECYCLING LEVELS
WALLONIA	Walloon Waste Management Plan "*Horizon 2000*" [4]	* 60 % recycling for 2000 * 65 % recycling for 2005 * 75 % recycling for 2010 * Disposal of C&D waste in C.E.T.'s [1] should decrease to a level of 10 % for 2010
	* *Technical specifications*	* Circular AWA/178-95/150 [5] * CCT 300 [6] * CCT W10 [7] *These documents will be merged in the near future.*
FLANDERS	Implementation plan for construction and demolition waste [8]	* 75 % recycling for 2000 * Landfilling within environmental hygienic conditions of the remaining 25 %. * Gradual reduction of the quantity of waste.
	Technical specifications	Standard Specifications 250 [9]
BRUSSELS	Waste management Plan 1991-1996 Waste management Plan 1998-2002 [10] *Technical specifications*	70 % recycling for 1996 95 % recycling or reusing for 2002 Circular of 9th of May 1995 (annex to General Technical Specification Document 150) [11].

[1] Centres d'Enfouissement Techniques

The VLAREA legislation in Flanders [13] establishes clear rules, mainly based on chemical composition and lixiviation characteristics, for the recognition of waste materials as secondary products. According to this legislation most of the construction and demolition rubble can be processed into secondary products.

Amount and Types of Waste Materials Produced in Belgium

The year production of construction and demolition waste in Belgium is estimated at 8 million tons (i.e. 25 % of the total production of waste) of which about 3,6 million tons are processed (i.e. 45 % recovery of C&D waste). This represents approximately 6 % of the primary aggregate consumption [14]. The production and recycling levels are, however, slightly different from Region to Region (Table 2)

OPTIMIZATION OF THE QUALITY OF THE RECYCLED AGGREGATES

In order to realise the target recycling levels, the market for the recycled aggregates has to be widened. This requires an improvement of the quality of the recycled aggregates. In this context, the BBRI in collaboration with the ULg (MTM) Université de Liège (service de métallurgie et traitement des minerais) has initiated a research project [15] regarding the optimisation of the purification and separation process for the recycling of C&D waste.

Table 2 Differences in the recycling rate and facilities between the three regions

	C&D WASTE PRODUCTION	RECYCLING RATE	RECYCLING - CRUSHING PLANTS / DUMPSITES	COMMENTS
WALLONIA	2,600,000. T/Year	37 % (1997) = 960.000 T	- 30 C.E.T. - 10 recycling plants	- Quarries: 75 % of the natural aggregates production. - Lower population density: allow C.E.T.-sites.
FLANDERS	4,600,000. T/Year	65 % (1998) = 3,000,000 T	80 crushing plants (total capacity: 5 MT / year)	- Lack of natural resources. - High population density: limited capacity for dumpsites.
BRUSSELS	850,000. T/Year	75 % (1995) = 640.000 T	No recycling plants nor dumpsites; C&D waste processed by Flemish recycling plants or landfilled in Wallonia.	- Priority to selective demolition. - Very high population density: limited capacity for dumpsites and recycling plants.

The aim of this research project was to study the possibilities offered by ore treatment and mining technologies in the processing of C&D waste aggregates. The first phase of the study consisted in a detailed characterisation of different types of recycled aggregates, which were carefully selected based upon the diversity of impurities contained. In a second phase different techniques were applied to purify the recycled products, i.e. to remove impurities such as wood, plastic, paper, glass, plaster and other weak materials.

Characterisation of Recycled Aggregates Before Purification

The aim of this part of the study was to identify the critical aspects related to the purity of C&D waste aggregates. 19 samples of C&D aggregates from different nature (sieve and crusher sand, crushed masonry, mixed and concrete aggregates) were selected and analysed for the characteristics shown in Table 3.

Aggregate grading

Most of the analysed samples did not fit into the usual (or future) technical specifications for aggregate grading. This can be partly explained by the fact that most of the samples chosen were not the higher quality products of the recycling plants. As far as aggregate grading is concerned, recycling plants should, improve their production process and quality control.

Visual analysis of the recycled aggregates

A visual analysis with naked eye and/or (stereo)microscope was realised in order to determine in detail the composition of the sampled recycled aggregates. This analysis was executed on several granulometric fractions of each sample (i.e. 0/0.125 mm, 0.125/0.0250 mm,..., 31.5/∞). The results were combined and cumulated to have an overall overview of the composition, purity and quality of the different samples. The materials found in the samples were classified in 7 categories defined in the Table 4. Results were expressed in percent by mass and in percent by number of element present in the defined categories.

Table 3 Recycled aggregate characteristics measured before purification

CHARACTERISTIC	MEASURED IN ACCORDANCE WITH
Grading	NBN B 11-001 - 1978 / NBN B 11-002 - 1979 / NBN B 11-012 - 1984
Detailed visual analysis, based upon petrography, stereoscopy and microscopy	Method developed at BBRI
Apparent specific gravity by heavy liquids separation (for the crushed masonry and mixed aggregates)	Method developed at Ulg-MTM
Amount of fine particles < 80 µm	NBN B11-209 - 1991
Sand equivalent value	NBN 589-208 - 1969
Sand equivalent value at 10 % fines	NF P18-597 - 1990
Methylene blue test	NBN B11-210 - 1989
Ignition loss	NBN B11-253 - 1975
Sulphate content	NBN B11-254 - 1975
Chloride content	NBN B11-202 - 1973
Organic material content ($K_2Cr_2O_7$-method)	NBN B11-207 -1969
Lixiviation test 3 x 24 h	NEN 2489 - 1976

Table 4 Description of the categories of material that can be encounter in recycled rubbles aggregates (appendix 3 of the PTV XXX, method developed by the Ministère de l'Equipement et des Transports (M.E.T.)

CATEGORY N°	LIST OF MATERIALS ENCOUNTER PER CATEGORY
1	*Crushed concrete and natural stone material:* Concrete rubble, aggregates with mortar gangue, natural stones, gravel,...
2	*Masonry type material:* Bricks, mortar, roofing tiles,...
3	*Other artificial stony material:* Tiles, slates, slag, cellular concrete, expanded clay, asbestos cement,...
4	*Hydrocarbonated material:* Hydrocarbonated coatings, bitumen, tar, roofing,...
5	*Non stony material:* Gypsum, rubber, plastic, insulation material, glass, metals, lime, plaster,...
6	*Organic maters:* Wood, rests of plants, paper, cellulose cement,...
7	*Special materials.*

The analysed concrete and mixed aggregates respect in general the requirements of the technical prescriptions (SB 250, CCT 300 and W10) regarding the composition in stony elements (categories 1,2,3 and 4). The presence of non-desired elements from the categories 5 and 6 is limited to an acceptable level for this kind of aggregates.

Aggregates currently sold by some recycling plants as masonry aggregates, have a composition rather similar to the definition of mixed aggregates. Their amount of impurities from the categories 5 and 6 is generally too high. Masonry aggregates, such as defined in the technical specifications, are in fact not yet available on the market.

The same conclusion can be made when considering the amount of impurities in the fraction > 4 mm of the recycling sands.

Chemical parameters and properties of the fines (< 80 μm)

Most of the recycled sands and all-in aggregates studied have a fine particles content higher than 8 %, the upper limit prescribed by the PTVxxx for recycled aggregates. However, the upper limit of 30 % imposed by the PTV 401 for traditional construction sand is clearly respected by these recycled materials.

The sand equivalent 10 %- and methylene blue- values are variable for aggregates of the same nature, but fall generally within the categories specified in the technical specifications. The parameters measured by these methods on recycled aggregates have to be studied more deeply in order to be able to explain the observed variations and to know if low quality values for SE and MB are automatically associated with low performance of the aggregates.

The obtained values for the chemical parameters (i.e. chlorides, sulphur, sulphates content, and ignition loss) are excellent. The only problem identified concerns the organic material content, which is sometimes higher than the limit of 0.5 % admitted for natural aggregates. This is due to the fact that the used method is based upon experience with natural aggregates, and is not adapted to the chemical composition of recycled aggregates. Therefore, in future this method will no longer be retained in the PTVxxx, and will be replaced by a simple visual analysis based upon the principles described above.

Apparent density characteristics of recycled aggregates

The apparent density characteristics of eleven samples, separated in three granulometric fractions (7/10, 4/7 and 0.300/4), were analysed. By immersing successively the plunging part of the samples into solutions of increasing density, their densimetric distribution curve was determined. The composition of the floating part of each solution was visually analysed, according to the procedures described above. A typical repartition of the principal categories of recycled materials in function of the apparent density classes is shown in figure 1, which shows that it is possible to group the C&D materials in three density classes (δ):

1. $\delta < 1.6$: the main materials are non-desirable elements from the categories 5 and 6 in addition to highly porous elements and/or elements with a lower mechanical resistance from the categories 2 and 3 (cellular concrete, highly porous bricks, slag, flat materials like bitumen and slates).

2. $1.6 < \delta < 2.1$: practically no non-desired elements are present. Most of the materials belong to the category 2 (masonry aggregates), few elements belong to the category 1 (concrete rubble). These products have a pure character and an intermediate mechanical resistance.

3. $\delta > 2.1$: aggregates with a high mechanical resistance and materials coated with hydrocarbons are present. These materials belong mainly to the category 1. Concrete rubbles are found at densities from 2.1 to 2.46. Above 2.46, natural stone elements are found.

These results show that choosing techniques based upon density to purify the C&D rubble would be very reasonable: if one is able to eliminate the density fraction < 1.6, aggregates poor in non-desired elements can be produced.

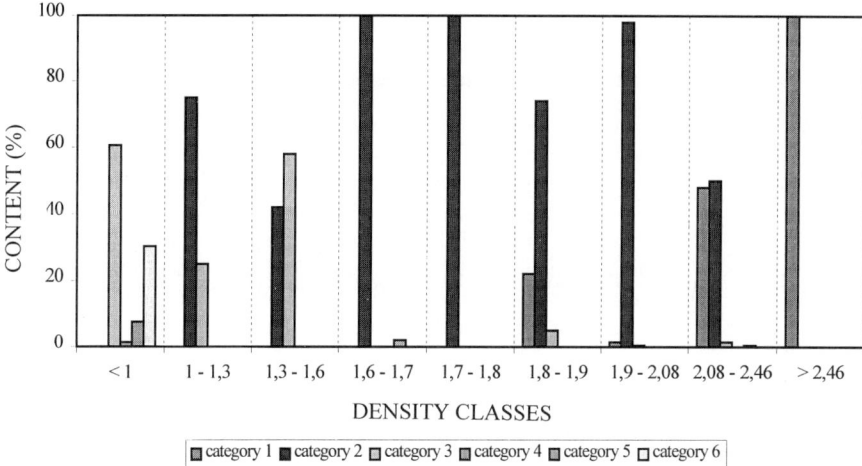

Figure 1 Repartition of materials as a function of apparent density

Optimisation in Terms of Purity of the Aggregates

Different gravimetric methods frequently used in the ore recovery and mining industry were tested at the laboratory and pilot scale. The selected techniques were optimised in view of the processing of C&D waste. The studied techniques, of which the performances depend on the granulometric fractions, are shown in Table 5. Taking into account economical and technical constraints to which recycling plants are submitted, the most interesting techniques to purify the aggregates appear to be the pneumatic table and the jig.

Table 5 Selected techniques and task to realise on each granulometric fraction of the recycled aggregates

GRANULOMETRIC FRACTION	TESTED TECHNIQUES	TASK
< 0,075	nil	nil
0,075/0,3	attrition	elimination of the cement covering the grains of quartz
0,3/4	shaking table hydrocyclone	gravimetric separations at different values
4/7	Jig, pneumatic table heavy media	gravimetric separations at different values
7/10	Jig, pneumatic table heavy media	gravimetric separations at different values

For the explanation of the results it is important to note that :

- the separation tests were realised on a standard samples having a really bad quality
- the presented results are obtained after one separation operation.

Therefore, it is clear that these techniques will yield far better results when treating materials with a common and better quality.

Results of the separation test with the pneumatic table and the jig

Several tests were realised on lab- and industrial pneumatic tables and on lab- and pilot jigs. Depending on the sieve fraction analysed (4/7 or 7/10), the separation tests provided several products, which are more or less heavy. The heaviest products are of excellent quality but only account for some % of the not-yet-processed product when considering the pneumatic table (Table 6). A visual analysis of the different products separated with the industrial pneumatic table was made according to the previous described method for both sieve fractions 4/7 and 7/10. The results for the heavy and light products are given in Table 6. Similar conclusions can be made for the products separated by jig.

Table 6 Recuperation rates of the different products after separation test

		% OF THE ORIGINAL MATERIAL		
		Heavy	Mixed	Light
Industrial Pneumatic Table	4/7	18	58	24
	7/10	26	60	14
Pilot Jig	4/7	42	40	18
	7/10	40	50	10

Table 7 Visual analysis of the heaviest and lightest products separated by pneumatic table

CATEGORIES	SIEVING FRACTION	% WEIGHT		
		Original Sample	Heavy	Light
1	4/7	48	80	29
	7/10	54	75	37
2	4/7	43	13	63
	7/10	36	16	56
3	4/7	2	1	3
	7/10	3	2	2
4	4/7	2	6	1
	7/10	2	4	2
5	4/7	3	0.5	3
	7/10	3	2	2
6	4/7	0.5	0	0
	7/10	0.5	0	0
7	4/7	1	0	0
	7/10	1	0	0

From the heavy to the light products we notice:

- A diminution in stony elements from the category 1 parallel to an enrichment in elements of the category 2.
- An enrichment in porous elements from the category 3.
- A diminution of the amount asphalt coated materials.
- An enrichment in impurities of the category 5 for the fraction 4/7. The elements of the category 5 present in the heavy product consist essentially of glass (which is not harmful) and iron (which could be easily removable by the optimisation of the magnet systems already used in most of the recycling plants); in the light product however we find harmful elements like plastic, paper and plaster.
- No elements of the categories 6 and 7 were found in both of the separation products. Therefore, no conclusions about the behaviour of these elements can be made.

RECYCLED AGGREGATE IN CONCRETE AND CONCRETE ELEMENTS

A research project regarding the use as aggregate in concrete and concrete products is currently running at the BBRI, this in collaboration with the Magnel Laboratory of the University of Ghent [16]. Besides the concrete technology aspect, this research focuses mainly on the study of the durability and of the creep and shrinkage behaviour of recycled concrete. Results concerning the creep and shrinkage behaviour were presented at the International Symposium "Sustainable Construction: Use of Recycled Concrete Aggregate" organised by the University of Dundee in London in November 1998.

With regard to the concrete technology of recycled concrete the research allowed to demonstrate that producing recycled concrete with crushed concrete and masonry aggregates with an acceptable compressive strength (i.e. C20/25) is technically feasible. However, one has to take into account the rather high porosity of these aggregates, which make the production of workable concrete somewhat more complicated than for the traditional natural aggregates. In most of the cases higher water content is necessary for the production of workable recycled concrete. Since a part of the water is, however, absorbed by the porous aggregates, the effective water content of the cement paste is lower than the total water content of the concrete. Therefore, one has to make a clear distinction between the total and effective water-cement-ratio of recycled concrete. This is clearly illustrated in Figure 2: for a same total water-cement-ratio concrete types fabricated with more porous aggregates result in higher compressive strength values. Based upon these findings, the research is currently working on tools that would enable concrete producers to produce a stable concrete quality with recycled aggregates with variable porous properties.

As far as concerns the durability of recycled concrete, the research demonstrated that preventive measures regarding the alkali-silica-reaction are useful. Expansion tests executed following the procedures of the French Standard NF P18-587, demonstrated that recycled aggregates could possess a residual reactivity. Since in most cases the source of the recycled aggregates is diverse and per definition unknown, preventive measures such as the use of Low Alkali cements, are recommended. The research is currently concentrating on the study of the effectiveness of different preventive techniques (i.e. use of low W/C-ratios, of Low Alkali and/or blast furnace slag cements and of admixtures such as air entrainers). Different concrete prism expansion tests such as a modified NF P18-587 test and the ATILH-LCPC Annex G procedure are used.

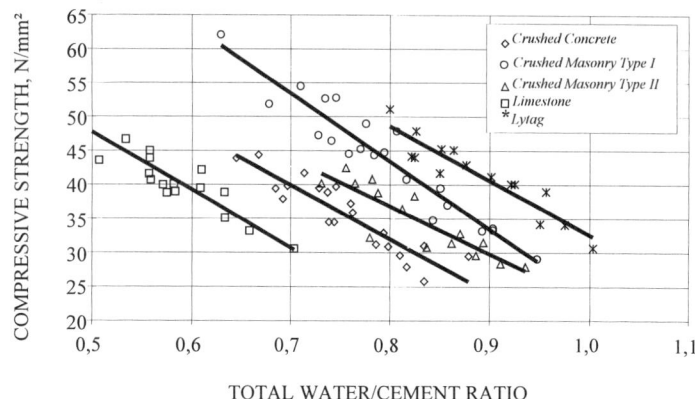

Figure 2 Compressive strength vs total water/cement ratio of concrete produced with recycled concrete aggregate, with two types of recycled masonry aggregate, limestone aggregate and lytag

CONCLUSIONS

The research projects of the BBRI with regard to the purification and separation process of C&D waste and with regard to the production of recycled concrete show clearly that the concrete market offers a lot of opportunities for the recycling of C&D waste. However, although the quality of the recycled products is already acceptable for the nowadays authorised applications, recycling plants have to invest in new separation and purification techniques, which enable the production of higher quality recycled aggregates suitable for the production of concrete. These investments, which would also improve indirectly the market acceptance and the quality image of these products, are probably only economically viable, if the use of recycled aggregates in such high quality applications is further stimulated by the authorities.

REFERENCES

1. DE PAUW C., Béton recyclé, CSTC - Revue, 1980, n° 2, June, pp 2 - 15.

2. DE PAUW C., Recyclage des décombres d'une ville sinistrée, CSTC - Revue, 1982, n° 4, December, pp 12 - 28.

3. SEMINAR KVIV-TI-BBRI, Hoe 80.000 m³ gewapend beton veilig laten springen en recycleren, Antwerp (Belgium), 1987.

4. GOUVERNEMENT WALLON, Horizon 2010 Plan wallon des déchets, 1998, 15th January.

5. MINISTÈRE WALLON DE L'EQUIPEMENT ET DES TRANSPORTS, Circulaire AWA/178-95/150: Utilisation des matériaux de réemploi dans les travaux routiers, Nivelles, 1995, July.

6. MINISTÈRE DE LA RÉGION WALLONNE, Cahier des charges type 300, 1994.

7. MINISTÈRE WALLON DE L'EQUIPEMENT ET DES TRANSPORTS, DIRECTION GÉNÉRALE DES AUTOROUTES ET DES ROUTES, Cahier des charges type W.10, 1991.

8. OVAM, Uitvoeringsplan Bouw-en sloopafval, 1995.

9. MINISTERIE VAN DE VLAAMSE GEMEENSCHAP, DEPARTEMENT ALGEMENE ZAKEN & FINANCIËN, Standaardbestek 250 voor de wegenbouw, 1996, 19th of December.

10. BIM, Plan betreffende de preventie en het beheer van afvalstoffen 1998-2002, 1998.

11. MINISTÈRE DE LA RÉGION BRUXELLES-CAPITALE, Circulaire relative à la réutilisation de débris dans les travaux routiers et d'infrastructure, 1995, 9th May (Moniteur Belge: 1995, 22nd September).

12. CRIC, PTVxxx: Granulats de débris de démolition et de construction recyclés - granulats de débris de béton, de débris mixtes et de débris de maçonnerie - Projet de version 10.0 -, 1998, 1st July.

13. OVAM, Vlaamse Reglement inzake Afvalvoorkoming en -beheer (VLAREA), D/1998/5024/2, 1998, January.

14. DESMYTER J., LAETHEM B., SIMONS B., VAN DESSEL J., VYNCKE J., Towards sustainability with construction and demolition waste in Belgium?, Environmental Aspects of Construction with Waste Materials, J.J.J.M. Goumans *et al.*, Elsevier Sciences, 1994, pp 759-773.

15. Optimalisation du processus de purification-séparation pour le recyclage des déchets de construction et de démolition, Final report, biennial 1996-1998.

16. Béton recyclé: technologie du matériau, aspects de durabilité et cditères de conception, Final report and Activity reports, biennials 1994-1996 and 1997-1999.

RECYCLED AGGREGATES FROM OLD CONCRETE HIGHWAY PAVEMENTS

W Fleischer
M Ruby
Heilit & Woerner BAU-AG
Germany

ABSTRACT. Due to an economic crushing and screening technology old concrete pavements can be repeatedly recycled to the same high quality concrete pavement. Coarse recycled aggregates from an old concrete pavement which does not show any damage due to the attack of frost and de-icing salt or due to alkali-aggregate reactions can be used in the lower course concrete of new pavement, but also in the upper-course concrete and in single course concrete pavement. Recycled aggregates which are not used in new pavement concrete can be used in cement stabilized bases or in unbound bases. Then unbound bases are covered directly with a thicker concrete pavement. Mixing and placing of concrete with recycled aggregates can be done with existing modern batching-plants and slipform pavers. It is not possible to transfer the demands for unused aggregates directly to recycled aggregates. In each single case it has to be decided which testing methods and demands are adequate for recycled aggregates and for concrete with recycled aggregates. The performance concept is more suitable for concrete with recycled aggregates than the prescription concept.

Keywords: Concrete pavements, Stabilized bases, Unbound roadbases, Recycled aggregates, Frost resistance, Alkali-aggregate reaction, Slipform paver, Crushing, Screening

Dr-Ing Walter Fleischer is Department Head of Construction Technology, Research and Development of Transportation Infrastructure in the Head Office of HEILIT+WOERNER BAU-AG in Munich, Germany. He earned his doctorate at the Institute of Building Materials of the Technical University in Munich for research in the influence of cement on shrinkage and swelling of concrete. His main interest includes concrete technology, concrete pavements, cement stabilized and unbound bases, equipment for construction of roads, aircraft movement areas and rigid tracks, design of concrete pavement.

Dipl-Ing Michael Ruby studied civil engineering at the Technical University of Munich. He is working as a project leader in the Department of Construction Technology, Research and Development of Transportation Infrastructure of HEILIT+WOERNER BAU-AG, Head Office in Munich, Germany. His main areas of responsibility are design of concrete pavement for highways and aircraft movement areas, disposal of special machines for construction of roads, aircraft movement areas and rigid tracks.

INTRODUCTION

The idea of environmental protection is - and, it must be said, very often justifiably - gaining increasing importance on an almost international scale in the construction industry also. There are indisputable economic reasons for this, apart from the idea of environmental protection in itself. For example, natural raw materials are becoming increasingly scarce, it is getting more and more difficult to obtain permission to extract them, and the recultivation measures required by law are becoming more and more extensive. Also for a considerable time now there have been signs of restrictions in landfill capacity, especially in conurbation areas and the landfill charges are increasing continually. In building construction and civil engineering there is at the moment, especially in countries which still have large quantities of natural supplies of aggregate, a relative reluctance to make use of recycled aggregates. On the other hand, the situation is different with regard to road-construction concrete. Several years of experience in the use of recycled aggregates in road-construction concrete have already been obtained, especially in the USA, Austria, Switzerland and Germany [1-6]. Of course there is a reason for the interest shown by the construction industry and road-construction authorities in a reuse of old road-construction concrete at as high a level as possible. Both are, first of all, interested in the large amounts of old road-construction concrete - and thus of recycled aggregates of a uniform quality - which are available at the job-site itself. If, for example, 10 km of carriageway on an old concrete motorway are taken up and suitably processed, more than 30,000 tonnes of high-quality recycled aggregates can be obtained, depending on the width and thickness of the old concrete pavement.

FIELDS OF APPLICATION

The essential question to be clarified by a state authority when inviting tenders for a reconstruction or by a company in preparing an alternative proposal is where the old road construction concrete can be reused. This will depend above all on the state of the old concrete pavement and on the properties of the old concrete. What basic possibilities are open?

- If the old concrete pavement has become defective simply for design reasons (e.g. is it underdimensioned for current traffic levels, are there fracture cracks and no structure damage, is the roadbase destroyed?), then the recycled aggregates can be reused in new road construction concrete. This is regulated in a current instruction sheet of the German Road and Transportation Research Association (Forschungsgesellschaft für Straßen- und Verkehrswesen) [7]. In Germany today recycled aggregates of this kind e.g. in the grain sizes 2-8, 8-16 and 16-32 mm are employed in the new construction of two-course concrete roads in the lower-course concrete [1,6,8]. The recycled sand is used in the raodbase. Tests are currently being made to determine to what extent recycled sand can also be used in the lower-course concrete. In Germany its use in lower-course concrete therefore is still limited to test areas. In the USA, for example, good results in this field have already been obtained [3]. The employment of recycled aggregates in single-course construction or in upper-course concrete is also not very widespread in Germany, in particular because two-course or two-layer construction is preferred for reasons of quality and cost effectiveness. Up to now in Germany has been one single-course construction lot in which the grain size of 8-16 mm consisted of recycled aggregate [6], but there is basically no reason why further grain sizes should not be used [7,9].

- If the old concrete pavement has been damaged by the effects of frost and de-icing salt, a reuse in new road construction concrete is usually not possible. At the moment extensive preliminary tests would be necessary to allow parts of the lower-course concrete to be replaced by this kind of recycled aggregate. An employment of recycled aggregate in hydraulically bound or in unbound roadbases (Figure 1) is, however, possible [10].

- If the old concrete contains structural damage caused by alkali-aggregate reactions or similar expansive reactions, recycled aggregates should not, according to current thinking, be reused in cement-bound courses. Old concretes of this kind are used in unbound road bases [7,10].

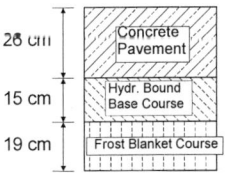
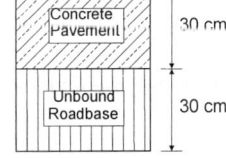

"Standard Superstructure"   Alternative: "Thick Concrete Pavement" on an Unbound Roadbase of Crushed Aggregates

Figure 1 German concrete pavement construction methods for the heaviest traffic loadings and a frost-proof superstructure of 60 cm, for hydraulically bound base courses and unbound roadbases recycled aggregate can be used

In the USA good experience has been obtained with recycled aggregates from old traffic movement area concrete, which is characterized by damage caused by aggregates with inadequate frost resistance (durability cracks) or as a result of alkali-aggregate reactions [3]. For this purpose, however, careful preliminary testing and specific technical properties of the concrete are necessary (e.g. maximum recycled aggregate size must be 19 mm, use of flyash and Portland cements with a total alkali content below 0.6 %). There are even favourable results in the USA concerning the reuse of road-construction concrete already produced using recycled aggregates. These show that road-construction concretes using "re-recycled" aggregates also have the necessary functional properties and durability [3].

TESTS, PROPERTIES AND REQUIREMENTS

Basic Principles

Some cautious colleagues would prefer to wait before using recycled aggregates until everything has been laid down permanently in technical rules and regulations. To them it may be said that to act in this way would be to show little confidence in our engineering competence. Of course there are risks inherent in every new development that have to be reduced to a minimum by thorough research and testing. In the case of new developments, such as the use of recycled aggregates, it may be asked to what extent the relevant regulations - in this case those relating to aggregates - can be taken as any basis at all. Each limiting value should be examined to establish whether it is at all the right one for recycled aggregate or whether other criteria are needed. The decisive factor for new construction components is to achieve a good standard in use over as many decades as possible.

For this purpose engineering thinking and experience are necessary. The basis for this is formed by research results and material testing. Technical regulations are important aids for the experienced engineer to achieve maximum performance in practical use, but it is important for them to be continually revised and brought up to date.

Most testing of building materials is still carried out today according to the "prescription concept". That is to say, base materials have to meet certain requirements and limiting values determined for the composition of the concrete (e.g. aggregate properties, water-cement-ratio or air entrainment values). Long years of experience suggest that, if these limiting values are met, the desired concrete properties will be achieved. When new base materials are used, however, it is becoming more and more important to be able to assess the properties of the concrete in service on the basis of direct tests - of course in good time before it is put to practical use. The term used in this case is "performance concept".

Old Concrete Pavements and Old Concrete

A first assessment of the condition of the old concrete pavement is made after a visual inspection. Fracture cracks together with pumping pavement slabs or slabs lying over voids are familiar to every expert. Damage due to the attack of frost and de-icing salt is made visible either through surface weathering or by structural damage going deeper into the concrete. If damage caused by expansive reactions is suspected, chemical mineralogical tests are recommended to identify the causes. Tests on this subject have been made, especially in the Scandinavian countries and the USA.

If the answer to the question of whether the old concrete has survived the attack of frost and de-icing salt during its service life without damage is not sufficient to assess the resistance to frost and de-icing salt, laboratory tests carried out on drilling cores may be of assistance. If the recycled aggregate is to be used in upper-course concrete or in single-course construction, the resistance to frost and de-icing salt in the old two-course concrete pavement of the lower-course concrete should be subjected to separate tests, since in comparison to the concrete surface, it was subject to only a low level of attack of frost and de-icing salt and therefore there may have been hardly any damage during its service life even if resistance to frost and de-icing salt was inadequate.

Although the strength of the old concrete is only of minor importance in determining the strength of the new concrete [3,11], the compressive strength of the old concrete is commonly taken into account to assess the state of the old concrete. It can be easily examined, e.g. in drilling cores, and within certain limits permits conclusions to be made regarding the other strength characteristics possessed by the old concrete. For example, the compressive strength to be found in old hard shoulders is of interest, since they were previously made of concrete with a cement content of only 325 kg/m^3 and average compressive strength of 35 N/mm^2, in contrast to driving lane concretes with a cement content of at least 340 kg/m^3 and average compressive strengths of 40 N/mm^2 [9]. The compressive strengths determined today for driving-lane concretes of this kind are, because of concrete re-hardening, between 80 and over 100 N/mm^2, and those of hard-shoulder concretes between 40 and 50 N/mm^2 [6]. According to the new instruction sheet on this subject [7], recycled aggregates from old road-construction concrete are allowed to be reused for new road construction concrete if the compressive strength of the old concrete amounts to at least 45 N/mm^2 and to at least 50 N/mm^2 on average.

Recycled Aggregates

Since our testing requirements to be met by aggregates - as is the case for most construction materials - are predominantly based on empirical values obtained only with these construction materials, conventional aggregate tests in the case of recycled aggregates can in any case supply valuable information. Under no circumstances can it be assumed that, through an uncritical transferring to recycled aggregates of the requirements to be met by conventional aggregates, the same performance in service can in principle be safely achieved. Taking the matter so simply may result in damage or in the need for additional measures which are not necessary from a technical point of view [9]. Which properties of conventional aggregates can also be of importance for recycled aggregates?

The same requirements as in conventional aggregates must be met by recycled aggregates with regard to grain shape, the grain-size grading curve, volume stability, and the ultra-fine material, since in this case the same concrete technology conditions apply. These properties are substantially influenced by the strength of the old concrete and by the processing method.

Especially in the case of recycled aggregates that are to be used for upper-course concrete or in single-course construction the same requirements are to be met with regard to expansive solids as in the case of conventional aggregates, since expansive solids can lead to surface damage (pop-outs). Such substances are, for example, wood from old expansion joints or old joint seals.

Old bituminous concrete content from repair areas, etc. must be considered separately. Tests have shown that, the lower the compressive strength of concrete with recycled aggregate, the higher the share of old bituminous concrete. Compressive strength with 4 % old bituminous concrete can, however, still be higher than that of concrete only with gravel aggregate [5]. In the case of recycled aggregates for lower-course concrete an extraneous matter content of up to 20% is not a problem, according to these tests. This represents a limiting value that should not, however, be made full use of.

The strength of recycled aggregate depends on the strength of the old concrete and on its processing. If the processing is correct and the strength of the old concrete is above that of the above-mentioned limits, then there is no need to test it again in the recycled aggregate. The impact crushing value, which is now and again used to assess strength, has no relation to real practice in the case of road-construction concrete, in contrast to bituminous concrete surfaces compacted with rollers, since during concrete compaction in which internal vibrators are used no impact loads occur.

Resistance to frost or de-icing salt requirements to be met by recycled aggregates are being discussed at the moment by experts. Views are to the effect that either recycled aggregates have to meet the same requirements as conventional aggregates, or it is assumed that recycled aggregate has passed the test if it has survived for decades the hardest frost and de-icing salt test without damage, i.e. the loads experienced in practice. The last view makes more sense. In most cases - especially if the old concrete has an adequate air void system - recycled aggregates meet the requirements with regard to aggregates for lower-course concrete, and sometimes even those of upper-course concrete aggregate. If doubts exist as to whether a concrete with adequate frost-road salt resistance can be produced using recycled aggregates, then the prescription concept should be abandoned and a direct measurement of the resistance to frost and de-icing salt of the concrete made.

This may be the case if, e.g. recycled aggregates are to be used in upper-course concrete.

A further characteristic of the quality of recycled aggregate is its old mortar content, since the processing of coarse aggregates and mortar cannot be completely separated from each other. Coarser recycled aggregates may also consist exclusively of mortar. It may be taken as a rule of thumb [12] that the larger the size of the recycled aggregates, the smaller the share of mortar (Figure 2). What is also of importance for the technology of a concrete which makes use of recycled aggregate is that in particular the very fine constituent parts of the recycled aggregate have a high content of old hardened cement paste [2,12], while the coarse grain groups often occur as almost pure aggregate [6]. The old mortar content in recycled aggregate can have a considerable influence on the fresh and hardened concrete properties of the new concrete. A high content, for example, leads to high water requirements for the new fresh concrete and to poor workability. The water absorption of the recycled aggregate can therefore be of importance for the mixing formula for a concrete using recycled aggregate. In the case of coarse recycled aggregates from old road construction concrete with water absorption values between 2 and 4 % water absorption is still relatively small, since these grain groups occur predominantly as pure aggregate and the old road-construction concrete had a low water-cement value, that is to say, the capillary void content of the old hardened cement paste is low. In the case of recycled sand, water absorption is of course higher [2,3,5,6,8,12,13]. During concrete production any sharp fluctuation in water absorption or in the natural moisture of the aggregates is inconvenient. It is therefore advisable to determine the water absorption of the recycled aggregates not only for preliminary testing but possibly also during construction work, or as an alternative to its natural moisture content. If fluctuations occur, the concrete mixing personnel can react accordingly, as in the case of sand moisture. As a water absorption testing procedure the usual method in the case of lightweight aggregate of 30-minute or even 24 hour water immersion tests for aggregates dried at 105°C is used. Water absorption capacity can be taken into account in the same way as with lightweight concrete through the effective water-cement value (i.e. the only decisive factor is the water that is effectively available for the cement paste).

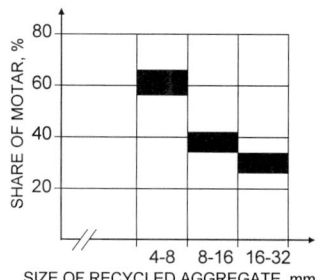

Figure 2 The larger the size of the recycled aggregate, the smaller the share of mortar [12]

Road-Construction Concrete With Recycled Aggregates

When recycled aggregates are used, one thing must be carefully remembered: concrete using recycled aggregates must not be awarded a quality bonus! That is to say, the same requirements with regard to performance capability and durability must be met as for concrete with conventional aggregates.

The tests in current use before and during building construction to ascertain directly the properties of road-construction concrete are also to be used without restriction in the case of concrete with recycled aggregate (e.g. compressive and bending tensile strength, air void content, pavement thickness). In cases of doubt it may be advisable to make use of the "performance concept" e.g. to make additional direct laboratory tests of the resistance to frost and de-icing salt of the concrete.

The mixing formula for concrete with recycled aggregate does not differ in principle from that used for concretes with unused aggregates. If only coarse recycled aggregates from old concrete pavements are used which have no structural damage, the requirements for road-construction concrete with a cement content (normally a CEM I 32,5 R Portland cement) of 340 kg/m³ and a water-cement-ratio of not more than 0,45 can be met. Only in the use of recycled sand is the attitude in Germany somewhat more cautious. Basic tests are being carried out at the moment. Of interest in this context are tests carried out in the USA [3], according to which the maximum degree of strength in new concrete was achieved when the sand consisted of 20 to 30 % of recycled sand (Figure 3).

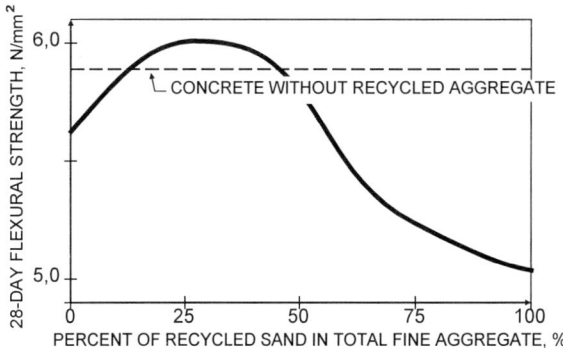

Figure 3 The maximum degree of flexural strength in new concrete is achieved when the sand consists of 20 to 30 % recycled sand [3]

PROCESSING OF RECYCLED AGGREGATES

Old concrete pavements are broken up into lumps, loaded onto trucks and transported to the processing site. There the lumps are crushed into aggregates and separated into grain sizes by screening plant. Depending on the number and type of crushers and screens in use, up to 250 t/h of old concrete can be crushed and more than 200 t/h of high-grade recycled aggregate obtained. The reinforcement contained in old road-construction concrete is broken away from the concrete during the crushing process and removed by magnetic separators from the crushed material. Other extraneous matter, such as wood from expansion joints or joint seals should be as far as possible separated before breaking the old concrete pavement or when the concrete lumps are loaded. If the permissible amounts of extraneous matter, especially wood, are strictly limited when recycled aggregates are used in upper-course concrete or in single-course construction, it has been found practicable to separate them from the individual fractions by using wind sifting [6].

When recycled aggregates are used in load-bearing courses, it is not necessary to separate into grain sizes after crushing, if after crushing the crushed recycled aggregates are already within the desired grading envelope for unbound and bound roadbases [10].

At the moment various crushing and processing concepts are being employed. Today after prescreening the crushing of the more or less large old concrete lumps is carried out mainly using impact breakers, jaw crushers or impellers. We know today that the quality of recycled aggregates - as indeed that of any natural rock or gravel chippings - is largely determined by the crushing process. When impact breakers are used, a reduction in size takes place without forces being exerted on grain shape, in contrast to jaw crushers. If jaw crushers are used, the feed material is squeezed, which means that incipient cracks in the chippings may remain, with the result that strength and, above all, frost resistance are reduced. On the other hand, when the size of the grain is reduced by impact, it breaks at its weakest points, such as incipient cracks or intercalations, so that the result is a grain with better properties. Reduction by impact also causes the less solid mortar to be knocked off to a greater extent, resulting in a cubic crushed grain, which consists predominantly of pure rock. The old concrete is also subject to a selective process, with the less solid areas being more reduced in size and the brittle constituents accumulating in the sand.

Wet processing of recycled aggregates is for technological reasons not necessary with a modern treatment plant [7]. In our experience a prespraying of recycled aggregates is not necessary, as long as the natural moisture of the aggregates is properly taken into account during mixing. Also, given the great amount of crushing work needed, it is hardly possible to prespray the recycled aggregates uniformly at a level of work that can still be justified. An artificial spraying of aggregate piles is in no sense sufficient, since this can keep damp only the aggregate at the top of the piles and can have no effect on the dampness of the core of the pile.

CONSTRUCTION OF CONCRETE HIGHWAY PAVEMENTS FROM RECYCLED AGGREGATES

Mixing The Concrete

Road construction concrete with recycled aggregates can be mixed in the same mixing plants as conventional road-construction concrete. For modern concrete pavement construction, very mobile high-capacity mixing plants, which can be resited within a few days, are needed. In addition to the normal batching plants, continuously working mixing plants have proved their worth. Modern mixing plants can produce up to 300 m^3/h of concrete of one kind. The two kinds of concrete needed for two-course concrete installation can also be produced thanks to a quick change in the mixing formula in one of these mixing plants [1].

Concrete Placement

The placement of concrete with recycled aggregates takes place in exactly the same way as when unused aggregates are used. Normally slipform pavers are employed. With the use of modern slipform pavers, as for example designed and also produced by HEILIT+WOERNER itself, it is by no means rare to place 800 m and more per day [1].

Placement widths of more than 15 m are possible. Because of their modular design, these slipform pavers can be easily adapted to the desired placement width. Thanks to their short length no problem is involved in transferring them on a low-loader vehicle in normal road traffic without the need for special authorisation.

In the HEILIT+WOERNER slipform system [1], two separate slipform pavers are used for two-course placement (Figure 4). The lower-course concrete is placed to the desired height by the first slipform paver and compacted with interior vibrators and the pressure plate. Then dowels and tiebars are vibrated into the lower-course concrete. The upper-course concrete is distributed over the compacted lower-course concrete. The second slipform paver places the upper-course concrete to the desired thickness and height, compacting again being carried out by means of vibrating cylinders and pressure plate. In this case it is important to place upper and lower-course concrete "fresh on fresh" to obtain a permanent bond between both courses.

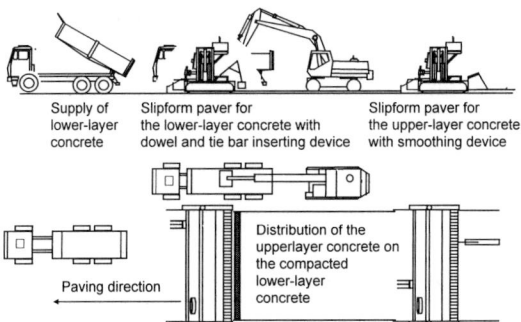

Figure 4 Slipform paving train according to HEILIT+WOERNER for construction of concrete pavement in two courses or layers (schematic)

After the concrete has been compacted, the concrete surface is smoothed transversely by means of a smoothing screed and then longitudinally using a "smoother". To make the concrete pavement permanently skid-resistant and also with low-noise levels right from the start, the fresh concrete surface is given its own texture e.g. by dragging a jute cloth over it longitudinally. Immediately afterwards it is sprayed with a liquid curing agent. If necessary additionally the concrete pavement is cured with water.

CONCLUDING REMARKS

It can be taken as state-of-the-art practice today that, for two-course concrete road-traffic pavement construction, suitable recycled aggregates (with the exception of sand components) from old road-construction concrete are being used to produce high quality concrete for the lower course of a concrete highway pavement. In Germany and the neighbouring countries about 1.5 million m² of two-course concrete pavements for heavily trafficked motorways and aircraft movement areas have now been built by HEILIT+WOERNER alone using recycled aggregates in lower-course concrete. Recycled aggregates which are not used in new road construction concrete can be used in new hydraulically bound base courses or in unbound road bases.

The latter are then covered with a thicker concrete pavement than is the case in bound roadbases (Figure 1). This is a construction method that, because of its advantages, is already being regarded by experts as the system of the future [10]. HEILIT+WOERNER has used this system to build about 2 million m² of heavily trafficked motorway.

Today old road traffic pavements made of concrete can be processed several times to become the same high-grade highway concrete pavement, so that in the future road construction concrete which has already been produced using recycled aggregates can also be recycled again. Close co-operation between the construction industry, research institutions and road construction authorities will make considerable progress possible, especially in the fields of processing and construction material technology.

REFERENCES

1. VON WILCKEN, A. Moderne Verkehrsflächen aus Beton. Beton, 1995, No.8, pp 547-552.

2. HANSEN, T C. Recycled aggregates and recycled aggregates concrete, second state-of-the-art report, developments 1945-1985. Matériaux et Constructions, 1986, Vol.19, No.111, pp 201-246.

3. YRJANSON, W A. Recycling of portland cement concrete pavements. National Cooperative Highway Research Program, Synthesis of Highway Practice 154, Transportation Research Board, National Research Council, Washington, D.C., 1989.

4. WERNER, R, HERMANN, K. Recycling von Bauschutt: neue Normen. Cementbulletin, 1995, Vol.63, No.2, pp 3-7.

5. SOMMER, H. Beton aus Altbeton und lärmarme Betonoberflächen auf Autobahnen in Österreich. Straße + Autobahn, 1992, Vol.43, No.3, pp 160-167.

6. FRANKE, H J. Recycling von Betondecken im Autobahnbau. Straße + Autobahn, 1993, Vol.44, No.10, pp 615-621.

7. FORSCHUNGSGESELLSCHAFT FÜR STRASSEN- UND VERKEHRSWESEN. Merkblatt zur Wiederverwendung von Beton aus Fahrbahndecken. Köln, 1998.

8. FRANKE, H J. Recycling von Betondecken im Autobahnbau. Beton, 1994, No.9, pp 504-509.

9. SPRINGENSCHMID, R. Möglichkeiten und Grenzen der Wiederverwendung von Beton aus Fahrbahndecken. Straße + Autobahn, 1996, Vol.47, No.4, pp 203-208.

10. BLESSMANN, W, FLEISCHER, W, WIPPERMANN, D. Concrete pavement on a crushed aggregate unbound roadbase, a new design for heavy-traffic motorways. Paper to be presented at the 8th International Symposium on Concrete Roads, 13 -16 Sep. 1998, Lisbon, PIARC.

11. WESCHE, K, SCHULZ, R R. Beton aus aufbereitetem Altbeton, Technologie und Eigenschaften. Betontechnische Berichte 1982/83, Ed. G Wischers, Beton-Verlag, Düsseldorf, 1984, pp 17-40.

12. HILSDORF, H K. Recycling von Beton im Straßenbau. Proceedings of Betonstraßentagung 1985, Schriftenreihe der Arbeitsgruppe "Betonstraßen" Heft 17, Kirschbaumverlag, Bonn, 1986, pp 55-58.

13. SPRINGENSCHMID, R, FLEISCHER, W. Zur Technologie der Wiederverwendung von altem Straßenbeton. Straße + Autobahn, 1993, Vol.44, No.10, pp 715-718.

EARLY AGE PROPERTIES OF RECYCLED AGGREGATE CONCRETE

F T Olorunsogo

University of Durban-Westville

South Africa

ABSTRACT. Early age properties of concrete made using recycled concrete aggregate as coarse aggregate were investigated. The properties studied are workability, compressive strength; measured at 3, 7 and 28 days, flexural strength and abrasion resistance which were determined at 28 days. The results showed that, in general, there was improvement in workability of concrete mix (i.e. increased in value of slump) with increases in the amount of recycled aggregate (RA) in the mix. As for compressive strength development, the general tendency was for reductions in compressive strength with increases in the proportion of RA in the mixes. No specific trend could be established in the other properties investigated. Nonetheless, the results indicate that there are potentials for use of RA in manufacture of concrete, especially for use in ordinary concrete applications such as domestic garages, driveways and pathways.

Keywords: Abrasion resistance, Aggregate, Compressive strength, Flexural strength, Recycled aggregate, Workability.

Dr Folarin T Olorunsogo is a Senior Lecturer in Civil Engineering, University of Durban-Westville, Durban, South Africa. His main areas of research include studies of properties, structural characteristics and product development of concrete materials and the prediction of failure modes in reinforced concrete structures. He also carries out research on response of concrete structures to aggressive environments.

INTRODUCTION

With the ever increasing world population there is a growing need for facilities which in turn require the use of finite natural resources. For this reason, many industries, backed by government's support and regulations, are now looking for ways of re-using materials in manufacture of new products. This process has been in operation for a number of years and the construction industry worldwide is no exception. In Europe and other developed countries, recycling of building materials started about the end of the Word War II when bricks and other materials that were recovered from the ruins of the war were utilised for reconstruction of amenities. Although, in those days, the use of recycled materials in this manner may be regarded as a means of solving an economic problem, recycling as a means of sustainable use of materials did not actually start until fairly recently.

In South Africa, very little is known about the use of recycled aggregate in manufacture of concrete. This is probably due to lack of knowledge about the behaviour of the material. Therefore, there is a need to investigate and understand the behaviour of concrete using recycled aggregate (RA) (with or without natural coarse aggregate (NA)) as such material is likely to provide both environmental and economic advantages. In this study, a preliminary investigation has been carried out to quantify properties of concrete made by partially or fully replacing natural coarse aggregate with RA. The properties investigated are workability, compressive strength development, flexural strength and abrasion resistance.

EXPERIMENTAL DETAILS

Materials

The materials used in this study are ordinary Portland cement (PC), natural fine (Umgeni sand) and coarse (crushed granite rock) aggregates with recycled aggregate. All the materials including the recycled aggregate were obtained from local suppliers. The recycled aggregate which was processed to 26.5 mm maximum aggregate size using the conventional method used in preparation of natural coarse aggregate was obtained from concrete and brick recyclers who are situated in Clairwood, Durban. Constituents of the finished recycled aggregate are shown in Table 1. Due to the nature of the original demolished structure it can be seen that the finished product (i.e. recycled aggregate) consists of materials such as dust, mortar, brick and stone/mortar conglomerate. The dust and stone/mortar constituted the lowest (1.9%) and highest (84.6%) proportions respectively. Table 2 shows the physical properties of all the aggregates in terms of relative density, moisture content, fineness modulus, compacted and loose bulk densities.

Mix Design

Five concrete mixes, as shown in Table 3 were prepared. Since, there is no existing standard method of designing concrete mixes incorporating recycled aggregate, the method of mix design proposed by the Cement and Concrete Institute (C&CI) was employed to design a concrete mix containing 100% natural coarse aggregate. The mix was designed to have a 28-day target compressive strength of 30 MPa. The RA mixes were derived simply by partially replacing (by mass) the natural coarse aggregate proportion in the control PC mix with RA at 30, 50, 70 and 100% replacement levels.

Table 1 Constituents of recycled aggregate (% by mass)

DUST	MORTAR	BRICK	STONE/MORTAR
1.9%	7.1%	6.4%	84.6%

Table 2 Physical properties of aggregates

TYPE OF AGGREGATE	RELATIVE DENSITY	MOISTURE CONTENT (%)	FINENESS MODULUS	COMPACTED BULK DENSITY (kg/m^3)	LOOSE BULK DENSITY (kg/m^3)
Natural Fine Aggregate (**FA**)	2.60	4.53	2.9	1441	1200
Natural Coarse Aggregate (**NA**)	2.61	5.13	--	1458	1344
Recycled Coarse Aggregate (**RA**)	2.60	5.32	--	1397	1362

Table 3 Mix proportions

MIX No.	CEMENT (kg/m^3)	% OF RECYCLED AGGREGATE	FINE AGGREGATE (kg/m^3)	COARSE AGGREGATES (kg/m^3)		WATER (litre)
				Recycled	Natural	
1	395	0	563	0	1196	198
2	395	30	563	361	835	198
3	395	50	563	598	598	198
4	395	70	563	835	361	198
5	395	100	563	1196	0	198

TESTING PROCEDURES

Properties of Aggregates

Classification of the constituents of RA was carried out visually after determining the percentage of dust content by sieve analysis. The sieve analysis and determination of fineness modulus of fine aggregates were carried out in accordance with SABS 829 [1]. Bulk density and relative density of the aggregates were determined using the methods suggested by SABS 845 [2] and BS 1377: Part 2 [3] respectively.

Properties of Concrete

All concrete properties investigated were monitored following standard procedures. Workability, compressive and flexural strengths of the mixes were measured in accordance with the procedures prescribed in SABS Methods 862 [4], (861 [5] & 863 [6]) and BS 1881 [7] respectively. Abrasion resistance of the concrete mixes was determined using the method of assessment proposed by C&CI [8]. Reported observations are averages of three measurements for all the properties.

RESULTS AND DISCUSSIONS

Workability

Workability of all mixes (1-5) was studied by carrying out slump tests at 0, 15 and 30 minutes after completion of concrete mixing. Figure 1 shows the results of the slump tests. As expected, it can be seen that, slump values for all mixes decreased with time, with all mixes achieving a nominal slump of 75+25 mm. The effect of increasing RA at certain proportion on the slump of concrete is shown in Figure 2, which indicates that there was a tendency for improved workability (i.e. increase in slump value) with increasing proportions of RA in the mixes. Mix 5 (100% RA) exhibited 45, 63 and 312% higher slump values than mix 1 (0% RA) at 0, 15 and 30 minutes after mixing respectively. Similar slump improvements for mix 3 (50% RA) were 35, 37 and 188% more than mix 1, respectively. The results obtained in this study are contrary to the findings of de Vries [9] and di Niro, $et\ al$ [10]. de Vries reported that because of the more angular shape and higher water absorption capacity of RA, total water demands of concrete using RA were higher than those without. In their investigation, di Niro $et\ al,$ observed that an optimum proportion of RA existed which produced significant improvement in workability.

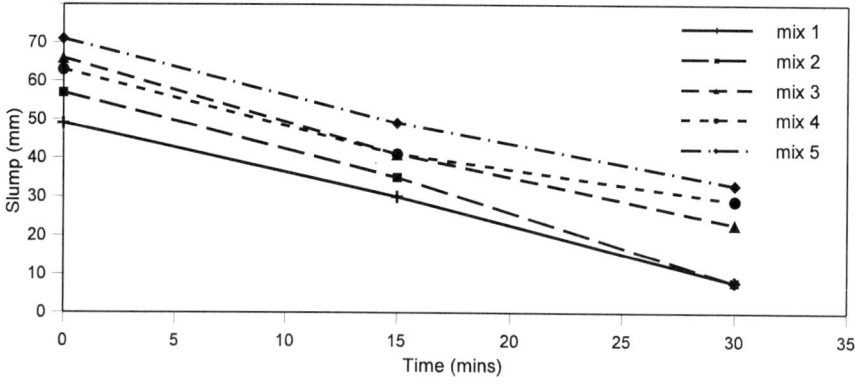

Figure 1 Workability

The mixes investigated in the present study contained RA which were relatively more round in shape than the NA. Although, both aggregates had about the same moisture content (5.13% for NA and 5.32% for RA) the water absorption capacity was not determined. The observed tendency for improved workability with increasing proportion of RA may, therefore be explained partially, with the round shape of the RA compared to the more angular shape of the NA. Another factor which may be responsible for the observation made in this study could be the higher percentage of fines and well graded nature of the RA compared to NA.

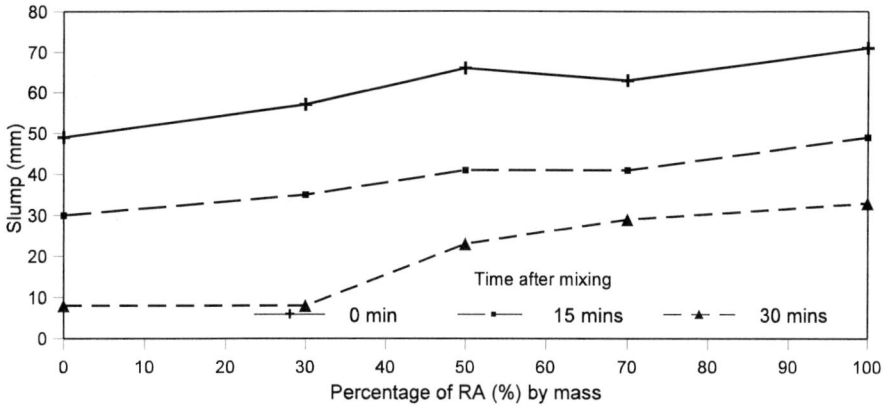

Figure 2 Effect of proportion of RA on workability

Compressive Strength

Figure 3 shows the compressive strength development of mixes 1-5. All mixes attained the 28 day design target strength of 30 MPa with the highest and lowest strengths associated with mixes 1 (36.0 MPa) and 5 (31.1 MPa) respectively.

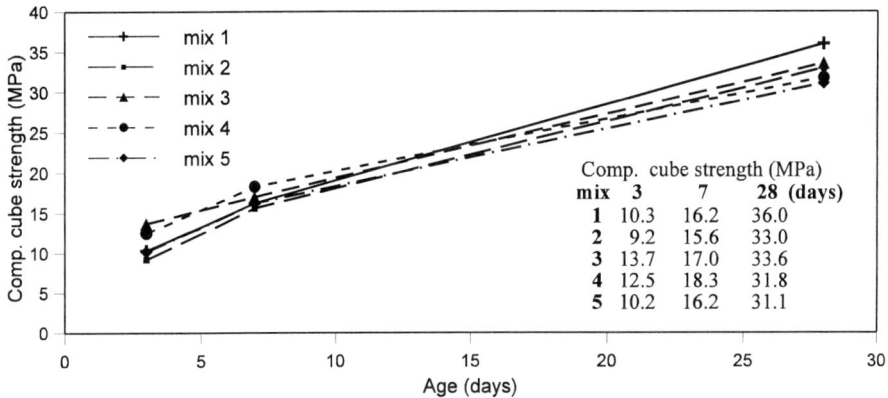

Figure 3 Compressive strength development

Mix 3 which contained 50% RA developed a 28-day compressive strength of 33.6 MPa. Considering the effect of proportions of RA used in each mix, (as illustrated in Figure 4) it is observed that there was a tendency for slight reductions in compressive strength the higher the proportion of RA in the mixes at all the ages of testing. This is expected since the RA used in this study contained certain proportions of materials such as dust, mortar and brick which are weaker in strength than the actual stones in the mix. Another reason for reductions in compressive strength with increase in proportions of RA in the mix is probably the smoother texture and rounder shape of the RA used when compared with the NA.

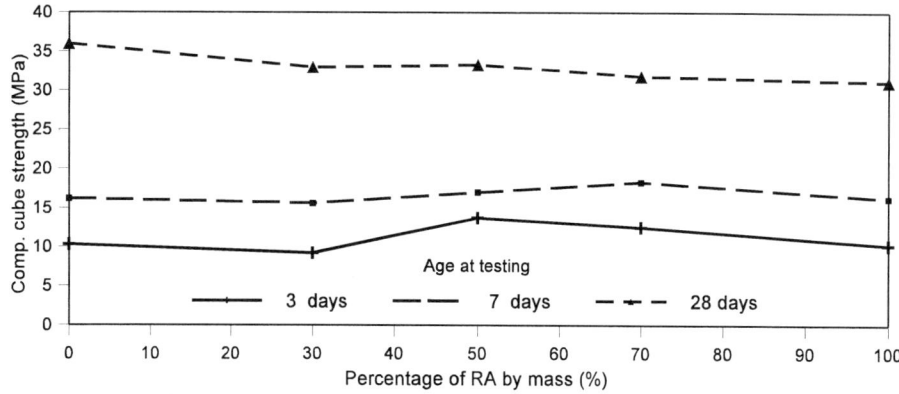

Figure 4 Effect of proportion of RA on compressive strength

The results obtained in this study are similar to the findings of previous investigators such as di Niro et al, [11] and de Vries [9]. di Niro et al, also reported slight decreases in compressive strength with increases in the proportions of RA. They found that of all the replacement levels considered (i.e. 0, 30, 50, 70 and 100% RA) only the 30% RA mix achieved the design target strength of 40 MPa (43 MPa). Comparing the 0% and 100% RA mixes, di Niro et al, [11] and de Vries [9] reported reductions in compressive strengths of up to 20% while in this study the similar value was 14%. With the results of compressive strength reported in this and other studies [9,11,12] it is evident that there is potential for use of RA in production of medium strength concrete. Further study is, however, required to establish the possibility of preparing high performance concrete using RA.

28 Day Flexural Strength

Results of the 28-day flexural strength test are presented in Table 4. All mixes, except mix 4 (70% RA) which exhibited a flexural strength of 6.25 MPa, attained a minimum flexural strength of 7 MPa at 28 days. There is no specific trend in the relationship between the proportion of RA included in concrete mix and flexural strength. However, the results obtained here on the mixes in which RA were included are comparable with flexural strength of the concrete containing 100% NA. This signifies that, on the basis of flexural strength only, concretes in which RA are included will perform satisfactorily (i.e. to similar standard as achieved by those concretes containing only NA as coarse aggregate).

Abrasion Resistance

Abrasion resistance is the ability of a concrete element to resist wear which may arise as a consequence of attrition by sliding, scraping or percussion [13]. Evaluation of concrete resistance to abrasion is somewhat difficult since the destructive action differs depending on the nature of the actual cause(s) of the wear. In this study, the wire brush method of assessment proposed by C&CI [8] was employed to evaluate abrasion resistance of all the mixes investigated at 28 days. The results obtained are shown in Table 4. Mixes 3 and 4, which contained 50 and 70% RA respectively performed the best with penetration depths of

1.60 mm and 1.61 mm respectively, followed by mix 1 (0% RA) which had a penetration depth of 1.85 mm. The poorest performance in terms of ability to resist abrasion was recorded for mixes 2 (2.26 mm) and 5 (2.25 mm) with RA contents of 30% and 100% respectively. As can be seen in Table 4, there is no clear relationship between abrasion resistance and proportion of RA included in the mixes.

Table 4 Abrasion resistance and flexural strength at 28 days

MIX No.	% of RECYCLED AGGREGATE	ABRASION RESISTANCE (penetration depth, mm)	FLEXURAL STRENGTH (MPa)
1	0	1.85	7.8
2	30	2.26	7.8
3	50	1.60	7.3
4	70	1.61	6.3
5	100	2.25	7.8

CONCLUSIONS

Inclusion of recycled aggregate led to improvement in workability and reductions in compressive strength of concrete mixes. The reasons for this are attributed to the relatively round shape of, and higher percentage of fine particles in the recycled aggregate compared to natural coarse aggregate. As for the other properties (i.e. flexural strength and abrasion resistance) investigated, no distinct relationship could be established between any of the properties and proportion of recycled aggregate in concrete mixes. Nevertheless, the results indicated that the performance of mixes incorporating recycled aggregates is comparable to the concrete mix in which 100% natural coarse aggregate was used.

From the results obtained in this study there are potentials for use of recycled aggregate in manufacture of concrete. However, clearly, further research must be carried out in order to fully quantify and understand the behaviour of recycled aggregate concrete.

ACKNOWLEDGEMENTS

The research programme of which findings are reported in this paper is part of a broader research programme titled; *"Use of Recycled Aggregate in Manufacture of Concrete"*. The programme is jointly funded by the University of Durban-Westville and Foundation for Research Development (FRD).

The author acknowledges with gratitude, assistance of Bruce Raath (formerly of C & CI, Durban) of CONTEST Concrete Technology Services, Westmead, by making available for use, certain laboratory facilities at his disposal, for continuing implementation of the research programme.

REFERENCES

1. SOUTH AFRICAN BUREAU OF STANDARDS. The Sieve Analysis of Aggregates, *SABS*, Pretoria. SABS Method 829, 1990.

2. SOUTH AFRICAN BUREAU OF STANDARDS. The Determination of the Fineness Modulus of Fine Aggregates. *SABS*, Pretoria. SABS Method 845, 1990.

3. BRITISH STANDARDS INSTITUTION. Particle Density Tests. *BSI*, London. BS 1377: Part 2, 1990.

4. SOUTH AFRICAN BUREAU OF STANDARDS. The determination of slump of freshly mixed concrete. *SABS*, Pretoria. SABS Method 862, 1990.

5. SOUTH AFRICAN BUREAU OF STANDARDS. The Making and curing of concrete test cubes. *SABS*, Pretoria. SABS Method 861, 1990.

6. SOUTH AFRICAN BUREAU OF STANDARDS. The determination of compressive strength of concrete test cubes. *SABS*, Pretoria. SABS Method 863, 1990.

7. BRITISH STANDARDS INSTITUTION. Method for determination of flexural strength. *BSI*, London. BS 1881: Part 118, 1983.

8. CEMENT AND CONCRETE INSTITUTE. General Assessment Criteria for Concrete Surfaces. *C & CI*, Midrand. PCI TM 7. 11, 1990.

9. DE VRIES, P. Concrete Recycled: Crushed Concrete as Aggregate. *In Proc. of Int. Conf. on Concrete in Service of Mankind,* Dundee, Scotland. 24-28 June, 1996, pp 121-130.

10. DI NIRO, G., DOLARA, E. AND CAIRNS, R. Recycled Aggregate Concrete (RAC): A Mix Design Procedure. *In Proc. of FIP Sym. on Concrete Way to Development*. Johannesburg, South Africa. 9-12 March 1997, pp 675-682.

11. DI NIRO, G., DOLARA, E. AND RIDGWAY, P. Recycled Aggregate Concrete (RAC): Properties of Aggregate and Beams made from RAC. *In Proc. of Int. Conf. on Concrete in Service of Mankind,* Dundee, Scotland. 24-28 June 1996, pp 141-149.

12. COLLINS, R.J. AND SHERWOOD, P. The Use of Waste and Recycled Materials as Aggregates: Standards and Specifications. HMSO, London, 1995.

13. PRIOR, M.E. Abrasion Resistance. ASTM Special Technical Publication, No. 169A, 1966, pp 246-60.

PROCESSED CONCRETE RUBBLE FOR THE REUSE AS AGGREGATES

G Mellmann

U Meinhold

M Maultzsch

BAM (Federal Institute for Materials Research and Testing)

Germany

ABSTRACT. In the context with the participation in a BRITE/EURAM project and further studies, comprehensive investigations were carried out on recycled concrete. The aim was the total substitution of natural aggregates with a particle size of 2 – 32 mm. The investigations focus on industrially processed building rubble with concrete as the main constituent as received from different plants. Tests on the material's properties delivered the basis for the experiments on concrete and the respective evaluation. Besides the composition and grading, the porosity and water absorption proved to be decisive. The properties of concrete were assessed by test results of workability, time-dependent strength and elasticity modulus, creep and shrinkage, freeze-thaw resistance, carbonation and water absorption. Considering the results, the manufacturing of high-grade concrete with reuse of processed building rubble seems to be feasible.

Keywords: Recycling concrete, Processed building rubble, Aggregate, Porosity, Strength, Elasticity, Creep, Shrinkage, Durability.

Dipl-Ing G Mellmann is civil engineer and project leader in the Building Materials division of BAM (Federal Institute for Materials Research and Testing), Berlin, Germany. He is experienced in investigations on concrete properties and the assessment of structures on site, particularly applying NDT methods.

U Meinhold received his practical experience by working in the ready-mixed concrete industry before joining the BAM division Building Materials. He is responsible for the concrete laboratory for a long time and has further interests in concrete repair systems.

Professor Dr-Ing M Maultzsch is head of the BAM division Building Materials. As an engineer in materials science, his particular interests are in the fields of concrete technology and durability aspects such like chloride ingress and carbonation.

INTRODUCTION

Building rubble from demolition work arises more and more in Europe. In Germany the amount is about 70 million tonnes per year with an increasing tendency. A major part of the rubble is concrete that usually is either dumped or taken for sub-grade use. The reuse as aggregate in high-grade concrete is up to now restricted by missing standards and a lack of experience and knowledge. On this background studies were carried out particularly within the frame of a BRITE/EURAM Project [1] with the aim to demonstrate that the recycling of the crushed concrete for manufacturing new high-grade concrete might be possible without an increased risk concerning strength and durability. An important precondition for these studies was the exclusive use of industrially processed rubble with respect to the application of the findings in practice.

EXPERIMENTAL DETAILS

Selected Material

The material designated as concrete rubble was selected from 3 processing plants in the Berlin region. These plants have stock pile capacities of more than 10,000 t and a turnover of about 100,000 t per year. Two of them used an impact crusher, the third one a jaw crusher. Separation before crushing and finally air-sifting was applied. In each case the material was taken in the fractions 2/8, 8/16 and 16/32 mm. The origin of the rubble was unknown, and the fractions were mostly from different batches. The total amount was about 30 t.

Tests on the Material

All tests were carried out on each fraction of the recycled concrete (RC) material. The first test was the visual assessment of the composition by separating the constituents and weighing the portions. The sulphate and chloride content was chemically analysed. Furthermore the thermogravimetric analysis was applied. The physical tests comprehended the grading, particle shape i.e. the flakiness, particle strength, density and water absorption after 10 minutes as well as after 24 or 48 hours.

Concrete Mix Design and Procedure

The aggregate with particle size 2 - 32 mm was totally substituted by the RC material. Only the sand fraction 0/2 mm was natural river sand. The grading was fit to the "B_{32}-line" according to DIN 1045 [2] which represents the border line between the recommended "advantageous" and "less advantageous" area of grading. With regard to the different processing plants, the materials and concrete mixes were marked with the letters "X", "K" or "W".

An ordinary Portland cement CEM I - 32.5 R was used in proportions of 280, 310 and 340 kg per cubic meter concrete.

The water addition was determined in preliminary tests on 280-kg-mixes. As the "worst case" in practice should be covered, the water addition was increased during mixing until any segregation could just avoided. This amount was taken for all subsequent mixes leading to decreasing w/c ratios if the cement content increased. The target line was principally the workability assessed by the initial flow in the range of 50 - 60 cm. A superplasticizer was used for limitation of the water addition. The resulting effective water/cement ratio was predominantly in the range of 0.47 - 0.57 that is appropriate for structures exposed to weathering [2].

The concrete was mixed in forced action pan mixers in batches of 40, 60 or 190 litres. After giving the sand and aggregate into the mixer, 2/3 of the mixing water was added and immediately hereafter the cement. During the 2 minutes mixing period the residual mixing water including the admixture was added. 5 - 6 minutes later the concrete was agitated again for 0.5 minutes before the tests on fresh concrete and the casting of specimens were started.

Control mixes with only natural aggregates, i.e. dense siliceous gravel, without and with superplasticizer addition were made in the same way.

Tests on Concrete

Fresh concrete was tested on workability using the flow table and on density and air content.

Comprehensive tests were performed on the hardened concrete. The strength development was determined using 15 cm cubes. They were usually stored under water until the age of 7 days and then in air at 20 °C and 65 % r.h. The development of the dynamic E- and G-Modulus was determined on prisms 10 x 10 x 30 cm^3, cured under water as well as in the mentioned climate. The static E-Modulus was measured as well on the prisms as on cylinders 15 cm in diameter and 30 cm high. For the tests on creep, shrinkage and carbonation the prisms were applied. In the creep test the compressive load was set to a third of the failure value, starting at the age of 28 days. The creep and shrinkage values were continuously recorded.

With regard to durability aspects, the capillary absorption and the freeze/thaw resistance with salt impact was determined on 15 cm cubes. For the latter property a temperature shock method was applied [3]. The carbonation in the ambient atmosphere of laboratory rooms is assessed on prisms.

TEST RESULTS AND DISCUSSION

Properties of Aggregates

In all RC material fractions, concrete was the main constituent. Material K contained considerable amounts of natural stone in the coarse fractions which can be accounted to the concrete proportion. This ranged from 87.1 % to 94.1 % by weight. Mortar was found in proportions of 0.3 - 8.3 %. Masonry and asphalt amounted to 0.9 - 4.4 % and 2.1 - 5.6 %, respectively. Other impurities such like wood and plastics were found in proportions of 0 - 1.4 %. Glass and gypsum plaster did not play any role.

The latter was confirmed by the analysis of sulphate content that ranged from 0.2 to 0.4 %, in only one case up to 0.55 %. So the absence of gypsum renderings and mortars at least in the fractions with more than 2 mm particle size must be assumed.
The chloride content was determined to 0.012 - 0.026 %. There was no evidence of enrichment of sulphate or chloride in any fraction. This points to an uniform statistical distribution of cement stone and original aggregate in the crushed concrete particles, which could be qualitatively confirmed by visual inspection of polished cross sections of concrete specimens made with the RC aggregates.

The results of grading tests were more or less in accordance with existing standards. Nevertheless the achieved scatter of values of undersize proportions seems to be insufficient in case of RC material as the water demand is more affected than in case of dense natural aggregates. The applied screening technique at the processing plants might be improved. The particle shape is obviously affected by the crusher type. While the impact crusher material contains usually about 90 % round-shaped particles, this value decreases to approximately 55 - 75 % if a jaw crusher is used, the more the finer the fraction. Similar results were found by the project partners in other regions.

The particle strength achieved in a standardised compaction test [4] correlated only with the particle size but nothing else. Though this test was supposed to be crucial for quality assessment of recycling material, the results did not deliver any use.

The apparent density and water absorption, however, proved to be decisive for most of the fresh and hardened concrete properties. An excellent correlation was found for the total water absorption and porosity by the evaluation of results of about 100 different samples as reported before [5]. For the results of 17 samples of partners in another region in Germany and in England exactly the same correlation was valid. The certainty index R^2 reached 0.997. The particle size did obviously not affect the water absorption.

The average particle density was 2,180 kg/m^3 for material X and slightly higher for material K with 2,280 kg/m^3. Here the considerable amount of natural stone in the coarse fractions of material K causes this increase. The density only of the concrete proportion would reach a value about 2,200 kg/m^3 too. In case of material W, however, the medium fraction had a density of only 1,950 kg/m^3 and consequently an average value for the whole material of 2,050 kg/m^3. Depending on the origin of the building rubble, the density values might scatter within a more or less wide range. For the material of suppliers X and K, the scattering range was 160 and 200 kg/m^3, respectively, with the main proportion in a range of 100 kg/m^3. In both cases the material proved to be appropriate, while the material W appeared to be less good. Its density values scattered within a range of 400 kg/m^3. So the uniformity of the recycled aggregate can be assessed by this measure.

The initial water absorption affects the fresh concrete properties and the effective water/cement ratio. With this respect the value measured after 10 minutes is usually accepted [6]. In Figure 1 these values measured on the material as received are plotted versus the total water absorption of the oven-dried material. Unfortunately there is only a poor correlation with a certainty index $R^2 = 0.84$ for linear regression. The correlation factor is in this case 0.638, i.e. 64 % of the total water absorption capacity is reached in average. The correlation might be better if the initial suction is also measured on oven-dried aggregates, but this would be apart from practice. Therefore the additional determination of the initial water absorption in the actual state of the aggregate is indispensable for estimating the amount of the additional mixing water and for calculating the effective w/c.

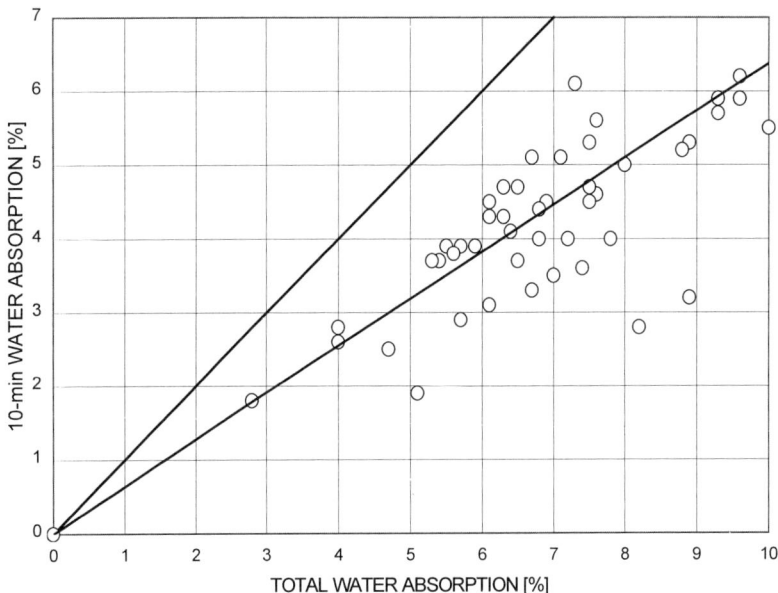

Figure 1 Initial and total water absorption of recycled concrete aggregates

Properties of Fresh and Hardened Concrete

The flow 10 minutes after adding the water was predominantly in the range 50 – 60 cm. Within the subsequent 20 minutes stiffening was observed resulting in a decrease of flow values up to 15 cm. After that the concrete was well workable for at least 30 minutes.

The initial stiffening might be avoided if the superplasticizer would be added later. The air content increases slightly if RC material with low density substitutes the dense aggregate.

The development of strength of RC concrete follows that of conventional concrete, but on a lower level if comparable mixes are considered.

In Figure 2 the compressive strength after 28 days is plotted versus the cement/water ratio. There is an approximately linear correlation for each type of aggregate as was found earlier [7].

The level depends evidently on the density: the values for RC materials X and K with very similar density are about 5 MPa lower than those of the reference concrete, while much lower values are achieved with RC material W.

But it must be emphasised that all RC concrete samples deliver strength values in the range of 35 – 50 MPa though the grading and water addition are not optimised.

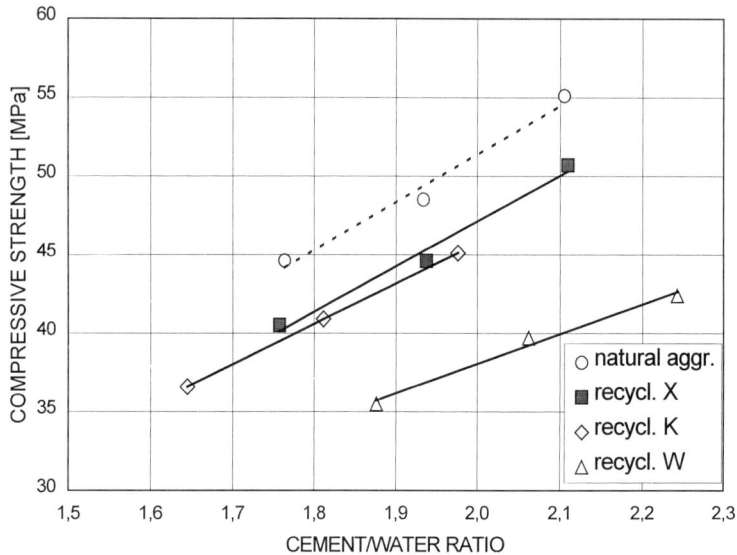

Figure 2 Compressive strength after 28 d depending on the cement/water ratio

The development of E- and G-Modulus in the early age does not principally differ if the aggregate is substituted by RC material. The level is, again depending on the density, lower as that of the reference concrete. Figure 3 shows a correlation of the E-Modulus of all RC concrete specimens with the compressive strength. The values of the reference concrete follow a different correlation on a higher level but with a lower gradient, i.e. the higher the strength, the lower is the difference of the E-Modulus. Same results were received by partners within the project using totally different RC aggregates and mixes [8].

The creep and shrinkage increases if RC aggregate is used, depending on its density too. In case of the apparently appropriate materials X and K the shrinkage after 90 days increased from 0.44 to 0.50 and 0.54 mm/m, respectively, and the creep strain from 0.63 to 0.88 and 0.92 mm/m, respectively. Material W produced considerably higher values.

The capillary water absorption is slightly increased in the magnitude of 0.1 - 0.2 % by weight. This difference complies in first approximation with the increase of the volume of mortar matrix, if RC aggregates are used, and the respective porosity. The freeze/thaw resistance in saturated sodium chloride solution decreased in case of the RC materials X and K, while it remarkably increased with material W.

The higher porosity and the rather low E-Modulus of the W concrete should be responsible for this surprising result. Nevertheless should mentioned that even the X and K concrete shows a resistance comparable with usual concrete of C30 grade. The carbonation will be tested over a long period, but the previous results do not unambiguously point to any effect of RC aggregates.

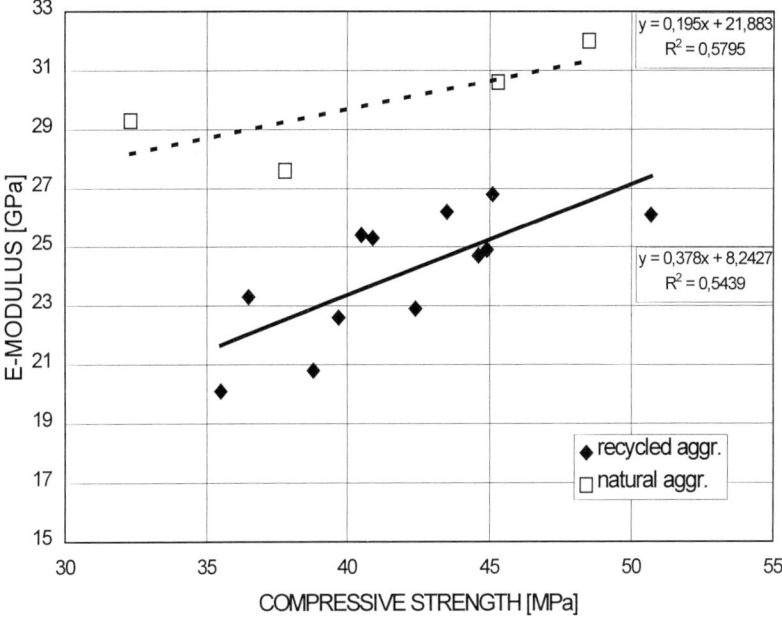

Figure 3 Elasticity modulus depending on the compressive strength

CONCLUSIONS

The investigations on industrially processed crushed concrete from demolition work have proved that the production of high-grade aggregate for the reuse in concrete is possible. According to the pre-separation, processing and screening the content of impairing constituents can be kept small and the quality high. A large stock-pile capacity can ensure sufficient uniformity of the material that can be assessed at first by the apparent density, porosity and total water absorption. The minimum density should be 2,100 kg/m³ with regard to the results achieved from concrete tests. For low-grade concrete, however, material with lower density might be appropriate too.

For the concrete mix design with total substitution of natural aggregates by RC aggregates, the water addition shall take into account the initial absorption. With regard to practice the real short-term absorption of the material in the state just before use should be taken. This value cannot sufficiently be derived from the total water absorption and therefore it must be tested. A superplasticizer should be provided.

The strength depends at first on the cement content and the water/cement ratio. If the density of the RC aggregate is in the usual magnitude, the strength is slightly decreased. The elasticity is affected as well by the particle density as by the strength; the correlation with the strength values differs from that of conventional concrete. The durability is obviously influenced only if the impact of cyclic freezing and thawing in salt solution is considered. Water absorption and carbonation are according to previous results not crucially impaired.

In summary it can be concluded that high-grade concrete can be manufactured with the reuse of processed concrete rubble. Though some properties of fresh and hardened concrete can deviate from those of comparable concrete mixes with natural aggregates, it must not be inevitable that these differences impair the suitability of RC concrete.

ACKNOWLEDGEMENTS

The authors would like to appreciate the financial support of the reported work by the European Commission. The excellent collaboration with the partners of the BRITE/EURAM Project shall be emphasised. The commitment of the laboratory staff in BAM is gratefully acknowledged.

REFERENCES

1. BRITE/EURAM Project BE 2145 "Construction Recycling Technologies for High Quality Cement and Concrete", 1996 – 1999, Co-ordinator LEMONA Ind., Bilbao (Spain).

2. DIN 1045: Reinforced concrete structures; design and construction. July 1988.

3. MAULTZSCH, M; AND GÜNTHER, K. Temperature shock test for the determination of the freeze/thaw resistance of concrete. In: Setzer, M.J.; Auberg, R. (Ed.): Frost Resistance of Concrete. Proceedings of the International RILEM Workshop, Essen 22 - 23 September 1997. London: 1997, pp 314 - 320.

4. DIN 4226-3: Aggregates for concrete; testing of aggregate of compact or porous structure. 1983.

5. MAULTZSCH, M; AND MELLMANN, G: Properties of large-scale processed building rubble with respect to the reuse as aggregate in concrete. In: Dhir, R K; Henderson, N A; Limbachiya, M C (Ed.): Sustainable Construction: Use of Recycled Concrete Aggregate, Proceedings of the International Symposium, London 11 - 12.Nov.1998, pp 99 - 107.

6. GERMAN COMMITTEE FOR REINFORCED CONCRETE (DAfStb). Concrete with recycled aggregate. DAfStb-Guideline, Edition August, 1998.

7. HANSEN, T C: Recycled aggregates and recycled aggregate concrete. Third state-of-the-art report 1945 – 1989, in: Recycling of Demolished Concrete and Masonry, RILEM Report of Technical Committee 37-DRC, ed. by T.C. Hansen. London 1992, pp 1 – 160.

8. KNIGHTS, J: BRITE/EURAM Project BE 2145, 6 Month Progress Report, Taywood Engineering Ltd., Southall, 1998 (not published).

STRENGTH AND ELASTICITY OF BRICK AND ARTIFICIAL AGGREGATE CONCRETE

M Zakaria
Bangladesh University of Engineering and Technology
Bangladesh

ABSTRACT. Crushed brick and artificial aggregates from pfa and clay have a great potential for use in concrete construction. Crushed brick and artificial pfa-clay aggregates were used to produce dense moderately high strength concretes and compared with normal weight concrete using quartzite gravel aggregates. Compressive and tensile strengths and modulii of elasticity (both static and dynamic) were determined for concretes made with experimental aggregates. The strength and elasticity behaviour of these concretes were compared with those of gravel aggregate concrete. Test results reveal that at later curing period and with lower water/cement ratio, there is no significant difference between the strengths of gravel aggregate concrete and brick and artificial aggregate concrete. The compressive strength and modulus of elasticity of brick aggregate concrete at early ages are significantly lower than those of gravel aggregate concrete and the gap decreases with the curing period. Although the compressive strength of manufactured aggregate concrete is lower than gravel aggregate concrete, their tensile strengths are found to be higher. It is found that the modulus of rupture of brick aggregate concrete is on average 10-12 percent higher than that of gravel aggregate concrete. The static modulus of elasticity of brick aggregate concrete is 30 percent lower than its dynamic modulus of elasticity and, on average, 30 percent lower than that of gravel aggregate concrete. On the basis of the results reported in this paper it was concluded that the concretes containing crushed brick or artificial pfa-clay aggregate can be utilised for general structural use and are substitutes for gravel aggregate concrete.

Keywords: Concrete, Natural aggregate, Crushed brick aggregate, Artificial aggregate, Compressive and tensile strength, Elasticity, Modulus of elasticity, Modulus of rupture.

Professor M Zakaria is a Professor in the Department of Civil Engineering at Bangladesh University of Engineering and Technology, Dhaka, Bangladesh. He obtained his PhD at the University of Birmingham, UK in 1986 and was a visiting research fellow at Leeds University in 1994-95. His main research interest include the performance of civil and highway engineering materials especially for developing countries.

INTRODUCTION

Concrete, when prepared following a particular technical 'know-how', should meet two overall criteria by which it can be defined : it has to be satisfactory in its hardened state and also in its fresh state while being transported from the mixer and placed in the form work. In order to solve the problem of either depletion of natural aggregates or reuse of demolition wastes, aggregates from bricks and artificial sources are to be used in concrete construction. Since aggregates occupy roughly three fourth of the volume of concrete, its properties have considerable importance to the quality of concrete. The research reported in this paper was programmed to investigate the performance and durability aspects of brick and artificial fly ash-clay aggregates in structural concrete. The use of these aggregates in concrete constructions was reviewed and a laboratory-testing programme was formulated in order to assess (i) the characteristics of these aggregates in terms of their intrinsic properties. (ii) the strength development behaviour, (iii) the stress-strain behaviour and (iv) the modulii of elasticity.

LITERATURE REVIEW

Schulz and Hendricks [1] reported the properties of aggregates obtained from crushing building bricks from rubbles, their characteristics when mixed with cement for concrete production. According to researches in Germany, as reported by Schulz and Hendricks [1], the increase in compressive strength of recycled brick rubble concrete from 28 days to 90 days was 30 to 40 percent. For recycled concrete produced with crushed hard burnt brick (clinker), the increase in compressive strength was as much as 67 percent. Akhteruzzaman and Hasnat [2] tested four grades of concrete with crushed brick as aggregates and reported that for the same grade of concrete the modulus of elasticity is about 30 percent lower and the tensile strength is about 11 percent higher than normal weight concrete. Khaloo [3] presented the results of three types of crushed clinker brick concrete compared with those of stone aggregate concrete and reported that, the 28 days compressive, tensile and flexural average strength of crushed clinker brick concrete to those of crushed stone concrete are 0.93, 1.02, 1.15 respectively. The use of crushed brick as coarse aggregate has also been studied by researchers of India and Pakistan [4,5].

From laboratory and field studies on non-bloated synthetic aggregates it was concluded that these aggregates have great potential for use in both highway and building construction in many areas of the world where high quality aggregates no longer exist but where the clays required for the production of synthetic aggregates are plentiful [6]. Ledbetter [7] reported the evaluation criteria for synthetic aggregates from clay and shale while Bonifay et al [8] examined rotary kiln-fired synthetic aggregates manufactured from Texas lignite fly ash. It is reported that most of kiln-fired aggregates are suitable for asphaltic concrete, unexposed Portland cement concrete and Portland cement concrete pavements [8].

With the development of reinforced concrete and owing to the rarity of natural porous aggregate deposits and their non-existence in most developed countries, research for the manufacture of artificial aggregate commenced [9]. In the 1950s, the Building Research Establishment, UK, started to produce high quality lightweight aggregate based on palletised pulverised fuel ash (pfa), a ash resulting from burning pulverised bituminous or hard coals [10]. Research on structural concrete has established that appropriate lightweight aggregates do not have the deficiencies in performance compared to concrete made with natural aggregates. Newman [11] discussed the strength, density and modulus of lightweight aggregate concrete. It is reported that the lower modulus of elasticity and higher tensile strength capacity of lightweight aggregate concrete provides better impact resistance than normal weight concrete.

MATERIALS AND LABORATORY TESTS

Type of coarse aggregate was the main variable for the concrete mixes. Naturally occurring quartzite gravel was used to manufacture the control concrete to compare with the concretes made with crushed brick aggregate and artificial aggregates from pulverised fly ash and clay. An uncrushed quartzite gravel aggregate, from commercial deposits within the county of Nottinghamshire, was used in this investigation. Brick aggregates were obtained by crushing BS 3921 -class 7 bricks, sieving to required sizes. Pfa- clay artificial aggregates were manufactured by mixing 50 percent pfa and 50 percent clay and firing to a temperature of 1150°C. The fine aggregate, used in this investigation was natural quartzite sand conforming to the medium grading of BS 882 [12] and the cement was ordinary Portland cement. The physical and mechanical properties of coarse and fine aggregates are shown in Table 1.

The mix design, used in this investigation, followed the method of 'minimum porosity proposed by Cabrera [13].. The mix design yielded a proportion of 1 : 2.33 : 3.5 for cement, fine aggregate and coarse aggregate respectively. Relatively lower slump was observed for crushed brick and pfa-clay aggregates for all the mixes which can be attributed to the angularity of these aggregates. After measuring the workability, cube, prism (beam), slab and cylinder moulds were compacted in two layers on a vibrating table. The vibration of the compaction table was kept constant throughout the casting programme. The specimens were cured in this standard curing environment until their respective testing time.

100 mm test cubes were prepared, cured and tested as per the requirements of BS1881 [12] Compressive strength of the cubes prepared from initial mixes was determined after 7-day moist curing in the 'fog room'. From the 7-day results, 28 day strengths were predicted using an empirical formula and as per Code of Practice CP 114 [12]. The test results and the predicted values are shown in Table 2. The results of compressive strength with curing time for further mixes are reported in Table 3 with the variations due to curing period for different concretes. The relative gain of strength of experimental concretes in terms of 28 days strength is depicted in Figure 1. In the present study, 100 x 100 x 500 mm beams were cast and cured as per the requirements of BS 1881 and tested for different curing periods. The results of the flexure test is shown in Table 3 which also contains the variation of modulus of rupture with curing period. Cylinders were cast, cured and tested for splitting tensile strength as per the requirements of the BS 1881:1983 [12] and the results are shown in Table 3 with other values.

Attempts were made, in this investigation, to determine the static modulii of elasticity in accordance with BS 1881:1983 where the prism specimen is loaded up to 1/3rd of the ultimate stress. The stress-strain data at different loads up to 1/3rd of the ultimate load were recorded and stress strain behaviour of the experimental concretes were investigated.

Figure 2 shows the stress-strain behaviour of concretes using experimental aggregates for 28 days curing period. The modulii results and the variation of static modulus with curing period for different types of concrete are contained in Table 4. The specimens for flexural test were used, in this investigation, for the determination of dynamic modulus. At the end of a particular curing g period the 100 x 100 x 500 mm prism specimens were used for dynamic modulus before the flexure test. Table 4 gives the results of moduli tests while Table 5 shows the relations between different parameters.

Table 1 Properties of experimental aggregates

PROPERTY	TEST DESIGNATION	VALUES		
		GRAVEL	BRICK	PFA-CLAY
Sp. Gravity, O-D basis	BS 812 : 1975	2.61	2.00	2.03
Sp. Gravity, SSD basis	"	2.62	2.08	2.07
Sp. Gravity - Apparent	"	2.65	2.19	2.11
Absorption, percent	"	0.54	4.58	2.06
Sp. Gravity by Helium Pycnometry	Micromeritics	2.67	2.66	2.52
Sp. Gravity by Mercury Intrusion Pycnometry		-	2.61	2.51
Aggregate Crushing Value	BS 812:1975	16	28	19
Aggregate Impact Value	"	15	25	16
Los Angeles Abrasion Value – Grading B	ASTM C131-87	18	30	22
	Sand			
Sp. Gravity – Apparent	BS 812:1975		2.71	
Absorption, percent	"		0.1	
Sp. Gravity by Helium Pycnometry	Micromeritics		2.71	

Table 2 Compressive Strength of Cubes from Initial Mixes

CURING PERIOD	AGGREGATE TYPE	COMPRESSIVE STRENGTH, MPA FOR WATER/CEMENT RATIO OF		
		0.40	0.50	0.70
7 – day experimental value	Gravel	60.95	43.50	28.75
	Brick	61.00	41.10	25.53
	Pfa-clay	61.30	43.20	28.08
28 – day predicted value*	Gravel		65.25	
	Brick		61.65	
	Pfa-clay		64.80	
28 – day observed value**	Gravel		55.90	
	Brick		53.70	
	Pfa-clay		54.90	

* $f_{28} = 1.5\, f_7$ ** from further mixes

Table 3 Strengths in MPa of different concretes for a water/cement ratio of 0.50

CURING PERIOD, DAYS	STRENGTH TYPE, MPA	GRAVEL	BRICK	PFA-CLAY
3	Compressive	40.2	33.0	36.8
	Flexural	4.65	4.95	4.45
	Splitting	3.52	3.61	3.51
7	Compressive	43.4	41.2	43.1
	Flexural	4.85	5.38	5.14
	Splitting	3.60	4.06	4.05
28	Compressive	55.9	53.7	54.9
	Flexural	6.15	6.82	6.20
	Splitting	4.87	5.65	4.90
56	Compressive	60.4	56.7	59.1
	Flexural	6.41	7.05	6.52
	Splitting	5.05	5.62	5.25
90	Compressive	63.5	58.8	61.6
	Flexural	6.62	7.18	6.80
	Splitting	5.37	6.05	5.55

Table 4 Moduli of concretes with a water/cement ratio of 0.50

CURING PERIOD, DAYS	STRENGTH, MPa	CONCRETE WITH AGGREGATE TYPE		
		GRAVEL	BRICK	PFA-CLAY
3	f_c	40.20	33.00	36.80
	E_c	33.37	17.72	25.81
	E_d	39.65	25.98	31.79
7	f_c	43.40	41.20	43.10
	E_c	34.80	22.70	29.09
	E_d	41.39	28.60	34.15
28	f_c	55.90	53.70	54.90
	E_c	37.71	26.82	33.81
	E_d	43.10	31.91	37.05
56	f_c	60.40	56.70	59.10
	E_c	38.85	28.74	35.85
	E_d	43.80	32.85	39.08
90	f_c	63.50	58.80	61.60
	E_c	40.12	30.60	38.10
	E_d	44.40	33.35	40.85

f_c = Compressive strength, MPa
E_c = Modulus of elasticity (static), Gpa
E_d = Modulus of elasticity (dynamic), Gpa

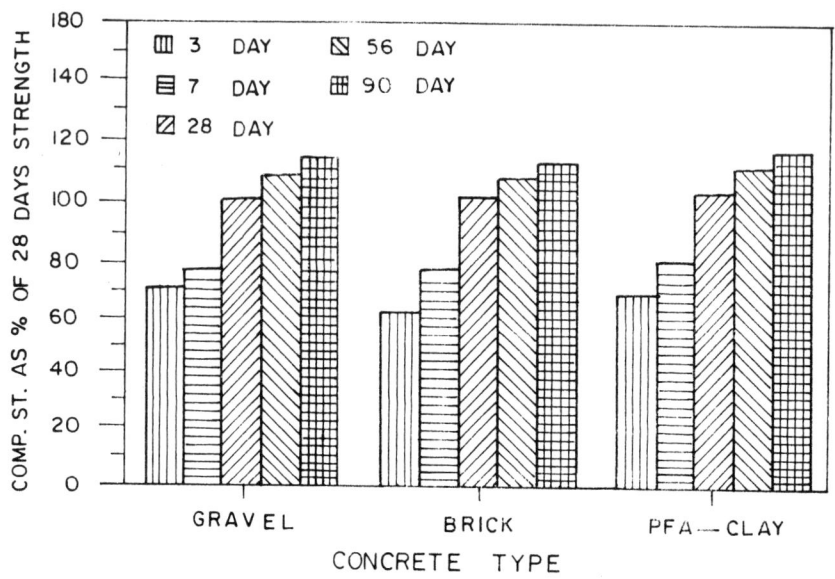

Figure 1 Relative gain in strength in terms of 28 day strength

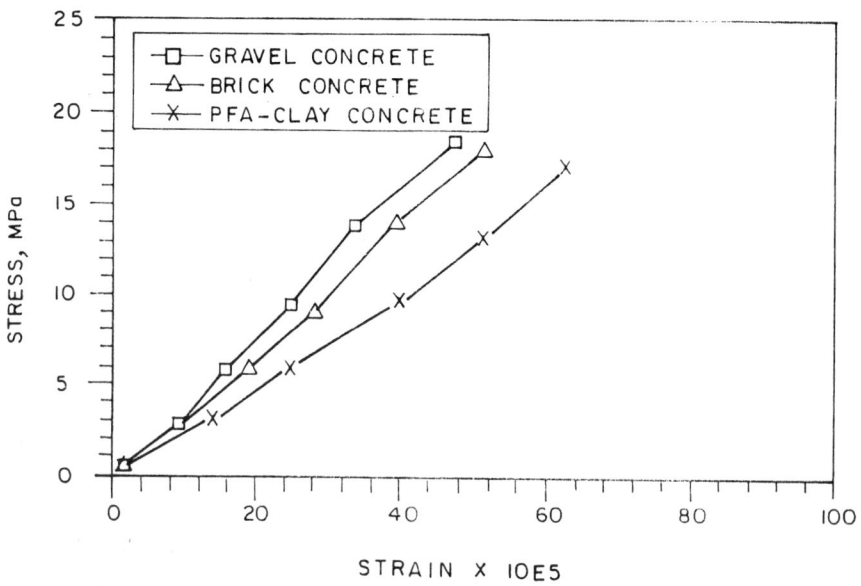

Figure 2 Stress-strain curves of concretes using different types of aggregate

Table 5 Relationship between different parameters with their correlation ratios for different concretes

PARAMETER	CONCRETE TYPE	CORRELATION EQUATION	CORRELATION RATIO (r^2)
MR and CS	Gravel	MR = 1.10 + 0.09 CS	0.99
	Brick	MR = 1.78 + 0.09 CS	0.98
	Pfa-clay	MR = 1.31 + 0.09 CS	0.98
f_t and CS	Gravel	f_t = 0.25 + 0.08 CS	0.98
	Brick	f_t = 0.54 + 0.09 CS	0.97
	Pfa-clay	f_t = 0.60 + 0.08 CS	0.98
E_c and CS	Gravel	E_c = 22.72 + 0.26 CS	0.99
	Brick	E_c = 2.92 + 0.46 CS	0.98
	Pfa-clay	E_c = 8.65 + 0.47 CS	0.99
E_d and CS	Gravel	E_d = 32.84 + 0.18 CS	0.95
	Brick	E_d = 17.84 + 0.28 CS	0.95
	Pfa-clay	E_d = 19.33 + 0.34 CS	0.98
E_d and E_c	Gravel	E_d = 17.32 + 0.68 Ec	0.98
	Brick	E_d = 16.56 + 0.54 Ec	0.92
	Pfa-clay	E_d = 12.96 + 0.73 Ec	0.99

MR = Modulus of rupture, MPa
CS = Compressive strength, MPa
f_t = Splitting tensile strength, MPa
E_c = Static modulus of elasticity, Gpa
E_d = Dynamic modulus of elasticity, Gpa

DISCUSSION

Experimental Aggregates

The physical and mechanical properties of the experimental aggregates contained in Table 1 reveal that both brick and pfa-clay aggregates are semi-lightweight type. The water absorption capacity of crushed brick aggregate is significantly high in comparison with pfa-clay aggregates. With regard to the strength of aggregates, natural gravel and pfa-clay artificial aggregates are comparable while brick aggregates have low values for strength parameters.

Strength of Concretes

The compressive strength of 100 mm cubes after 7 days curing in standard conditions (20°C and 100% RH) for different water/cement ratios are shown in Table 2. In order to compare the relative gain of strength at different ages, the 28-days strength of the mixes having same proportion and curing conditions are also reported in Table 2. It is seen from Table 2 that the 28-day strength for all concretes are less than 1.5 times the 7-day strength for a water/cement ratio of 0.50. It is seen that the strength at lower water cement ratio is dependent largely on the aggregate type.

For brick and artificial aggregate concrete, the strength at lower water/cement ratio (0.40) is relatively higher than for gravel aggregate concrete. Higher strengths at lower water/cement ratios are also reported by Akhteruzzaman and Hasnat [2] and Khaloo [3] for crushed brick aggregate concrete. Higher compressive strengths at lower water/cement ratios were also reported by Schulz and Hendricks [1] from the works of researchers in Germany but they explained the reason to be the higher absorption of brick aggregates.

Table 3 shows variations of the average cube strengths for different concretes at different curing periods. The variation of individual values from the mean was observed to be within 10 percent for all curing periods and smaller at early ages. This is, probably, due to the variation of the degree of hydration in different concretes which is higher at later ages. It is also seen that high absorptive brick aggregate concrete attains relatively lower strengths at early ages. This relative gain of strength is more clear from Figure 1 where the strengths in all curing periods are expressed as the percentage of 28 day's strength. The relatively lower strength at early curing period for brick and artificial aggregate concrete can be attributed to the higher absorption values for these aggregates. Higher strengths of crushed brick aggregate concrete at later ages were also observed by early researchers in Germany [1] who attributed this to the pozzolanic effect of the finely ground portion of the burnt brick.

The tensile strengths of the experimental concretes are reported as modulus of rupture (MR) or flexural strength and splitting tensile strength (f_t) and are shown in Table 3. Both flexural and splitting tensile strengths of brick and artificial aggregate concretes follow a pattern similar to the compressive strength but the strengths are higher than gravel aggregate concrete. Especially the flexural strength of brick aggregate concrete is significantly higher than that of gravel aggregate concrete. Similar findings were reported by Akhteruzzaman and Hasnat [2] and Zakaria [4]. It is seen from the results in Table 3 and Table 5 that the ratio of tensile to compressive strengths largely depends on the aggregate type used and method of testing. For gravel aggregate concrete, the ratio of MR to f_c varies from 11.5 percent to 10.5 percent while the ratio of f_t to f_c varies from 9.0 to 8.5 percent, the higher ratio being for the concretes at early ages. For crushed brick aggregate concrete the same ratios vary from 15 to 14 percent and 11 to 10 percent respectively. The pfa-clay artificial concrete has value in between brick and gravel aggregate concrete.

Elasticity of Concretes

The stress-strain behaviour of different concretes made with experimental aggregates are shown in Figure 2 for a curing period of 28 days. It is seen from the stress-strain curves that they are sufficiently straight to determine the initial tangent modulus. Akhteruzzaman and Hasnat [2] reported stress-strain curves for brick aggregate concrete and found that for higher strengths the curves are more linear. Similar trend is also seen from the results of this investigation. Results in Table 4 show that the patterns for brick and artificial aggregate concrete are similar to that for gravel aggregate concrete but the values are in lower magnitudes. The static modulus of elasticity of brick aggregate concrete, especially at early ages, is approximately half of the same for gravel aggregate concrete. The pfa-clay artificial aggregate concrete is halfway between the gravel and brick aggregate concrete. At later ages, the static modulus of brick and artificial aggregate concrete increase sufficiently but up to two third of those for gravel aggregate concrete. Schulz and Hendricks [1] also reported similar findings for dense crushed brick concrete.

The dynamic modulus of elasticity of brick aggregate concrete, in comparison to the static modulus, is relatively higher in all curing periods ranging from 65 percent at early ages to 75 percent of gravel aggregate concrete at later ages. In this case also the pfa-clay artificial aggregate concrete gives values in between gravel and brick aggregate. The relationships of static and dynamic modulus of elasticity with compressive strengths are shown in Table 5. Good correlations as shown in Table 5 can be attributed to the good quality control in all phases of concrete production and testing [13]. It is seen that the correlation ratios for dynamic modulus for all varieties of concrete are lower than those for static modulus of elasticity. The results in Table 5 also reveal that, in all cases, the dynamic modulii are higher than the static modulii. This is because of the test technique adopted for dynamic testing where negligible stress is applied and the specimens are unaffected by the creep. The relation of static and dynamic modulus of elasticity for gravel aggregate concrete are in good agreement with that recommended by CP 110 [12], although the magnitude varies slightly with the curing periods. For all types of concrete, the relation of E_c with E_d invariably is dependent on the age of curing which confirms the findings of early researchers [14].

CONCLUSIONS

In addition to the conclusions reached in the discussion of the results, the following specific conclusions can be drawn from the research reported in this paper.

1. Pulverised fly ash (pfa) is a suitable material for the production of preshaped and fired artificial aggregates. 50 percent pfa with 50 percent clay as binder material and fired to 1150°C produced suitable aggregates for use in Portland cement concrete.

2. Concrete of moderately high strength can be produced either by using crushed brick or pfa-clay artificial aggregates. At later ages of curing and for lower water/cement ratios there is no significant difference between the compressive strength of artificial aggregate concrete and gravel aggregate concrete.

3. The tensile strength of brick and pfa-clay aggregate concrete is higher than gravel aggregate concrete. The modulus of rupture of brick aggregate concrete is, on the average, 10 percent higher than that for gravel aggregate concrete.

4. The static and dynamic modulus of brick and artificial aggregate concrete is lower than those of gravel aggregate concrete. Good correlations exist between compressive strength and tensile strength and elastic moduli.

5. Brick aggregate and artificial pfa-clay aggregate concrete are recommended to be used for general structural constructions and are deemed to be the best alternatives for areas where natural aggregates are depleted and/or industrial wastes like fly ash are to be utilised.

REFERENCES

1. SCHULZ, R. R. AND HENDRICKS, CH. F. Recycling of masonry rubble. 'Recycling of Demolished Concrete and Masonry', Part 2, RILEM Technical Committee Report No. 6, 1992, E & FN SPON, London.

2. AKHTERUZZAMAN, A. A. AND HASNAT, A. Properties of concrete using crushed brick aggregates. Conc. Int. : Design and Construc., Vol. 5, No. 2, Feb. 1983.

3. KHALOO, A. R. Properties of concrete using crushed clinker brick as coarse aggregate. ACI Materials Journal, Vol. 91, No. 2, July-Aug. 1994, ACI, USA.

4. ZAKARIA, M. Tensile strength behaviour of structural concrete using crushed brick aggregate. Proc. 2nd East Asia Pacific Conf. on Struc. Engineering & Construction, Thailand, Vol. 2, 1989.

5. MAHER, A. Utilisation of waste brick ballast coarse aggregates for structural concrete. 'Building Materials for Low Income Housing - Asia and Pacific Region', 1987, E & FN SPON, London.

6. MOORE, W. M. Fired clay aggregates in flexible bases. Transportation Research Record No. 307, 1970, TRB, Washington DC, USA.

7. LEDBETTER, W. B. Synthetic aggregates from clay and shale : recommended criteria for evaluation. Road Res. Rec. No. 430, 1971, TRB, Washington D C, USA.

8. BONIFAY, D. W., et al. Rotary kiln-fired synthetic aggregates from Texas lignite fly ash. Highway Research Record No. 353, 1971, TRB, Washington DC, USA.

9. OWENS, P. L. Lightweight aggregates for structural concrete. Section 1, 'Structural Lightweight Aggregate Concrete', Blackie Academic & Professional, 1993, London.

10. CRIPWELL, J. B. What is PFA? Concrete, Vol. 26, No. 3, May/June, 1992.

11. NEWMAN, J. B. Properties of structural lightweight aggregate concrete. Section 2, 'Structural Lightweight Aggregate Concrete', Blackie Academic & Professional, 1993, London.

12. BRITISH STANDARD INSTITUTION. BS 812, BS 1881, BS 12, CP 110, CP 114, BSI, London, UK.

13. ZAKARIA, M. AND CABRERA, J. G. The performance and durability of demolition waste and artificial pfa-clay aggregates. International Symposium on "Inert Waste – An Opportunity to Use", Leeds, 1995, UK.

14. POPOVICS, S. Verification of relationships between mechanical properties of concrete like materials. Materials and Structures, Vol. 8, No.45, 1975, pp 183-91.

PROPERTIES AND PERFORMANCE OF RECYCLED CEMENTITIOUS MORTARS

T Rad

D G Bonner

University of Hertfordshire

United Kingdom

ABSTRACT. The study reported herein focused on the strength development of cementitious mortars containing young recycled mortar as a replacement for sand. Having used the same mass of cement in the samples, the analysis of the test results revealed that a higher strength of mortar utilising recycled material could be achieved compared to that of normal mortar. An attempt is made in this report using petrological observations, to link the microstructural characteristics of cement gel interactions with the compressive strength in the mortar mixes. This indicates that some activation in the cement gel of recycled mortar must have taken place.

Keywords: Recycled, Mortar, Compressive strength, Workability

Mr T Rad is currently a research student at the University of Hertfordshire, UK.

Professor D G Bonner is Head of Civil Engineering at the University of Hertfordshire, UK.

INTRODUCTION

The increasing prices and rising shortage of aggregates have been a major stimulus to develop new sources of aggregates for the concrete industry. Recycling of concrete at freshly set stages as well as old and deteriorating concrete is one obvious source for conserving aggregates.

It is generally accepted that mortar can be investigated in many cases as a model to simulate the properties of concrete. Mortar containing sand particles and cement-based material can be considered as a model concrete [1]. Therefore, the present study was conducted to investigate and evaluate properties of mortar incorporating fresh recycled mortar.

The study reported here focused on the production of recycled mortar mixtures made with different combinations of 1-day old crushed mortar, ordinary Portland cement (OPC), pulverised fly ash (PFA), superplasticiser and water to investigate the potential for a hydraulic reaction to be re-initiated in recycled mortar. This was explored by means of measuring strength development, and examining the chemical and physical microstructure using petrographical techniques.

EXPERIMENTAL PROGRAMME

Materials

Ordinary Portland cement conforming to BS12 [2] was used for the mixes in the investigation. River washed sand was used for production of control and recycled mortar mixes. Fly ash was obtained from the discharges of ex-West Burton power station. The superplasticising admixture employed in the mixes was Sikament 10, a sulphonated vinyl copolymer type.

Production of Recycled Mortar

The source mortar mix was made to prepare 50mm cubes for compressive strength testing and 400x200mm slabs for recycling purposes. This original mixture was designated as L1. The mix L1 contained OPC, oven-dried sand passing the 1.18mm sieve, and water. After 24 hours, cubes were demoulded and transferred to a water tank maintained at $25 \pm 2°C$. For each recycled mortar mix, two source mortar slabs were crushed at 24 hours using a laboratory jaw crusher and oven-dried. This material was then divided into two batches. One batch was pulverised manually using a pestel and mortar to pass a 1.18mm sieve. This was used for the recycled mixes designated LR.

The other batch was also pulverised manually, and then further ground using a mechanical pulveriser in 150 gram batches for 5 minutes each. This was used for the recycled mixes designated MR.

Most of the mixes investigated were prepared to achieve a constant workability as measured by the flow-table test [3]. Constant workability was chosen rather than constant water cement ratio, because the quantity of unhydrated cement contributed to the mixes from the recycled material could not be determined. However, some consideration of water cement ratio was made as will be discussed. The mix proportions were determined as the basis that the pulverised one-day mortar (POM) replaced the sand in a mortar mix. However, it was anticipated that the POM still had an hydraulic potential. Thus, the mixes containing PFA were proportioned so that the PFA was a fixed percentage of the POM or OPC.

Foe each recycled mortar mix (LR and MR) indicated in Table 1, 50mm cubes were prepared, as for the source mix, for subsequent testing and analysis.

The mixes LR1, MR1, LR2 and MR2 containing recycled mortar fines at different water contents were cast in an attempt to investigate the level of unhydrated cement grains in the mix. Mixes LR3 / MR3 and LR4 / MR4 were prepared to investigate whether the unhydrated cement and calcium hydroxide in the recycled mortar could be activated using PFA. Mixes LR5 / MR5 enabled the study of OPC alone with the recycled material. Mixes LR6 and LR7 incorporated both PFA and OPC. Mixes LR8 and LR9 / MR9 incorporated the superplasticiser in order to reduce the cohesion in the mix.

It was observed that the control mortar L1 had low cohesion for the first few hours which is a well-known characteristic of normal mortars. Conversely, the recycled mortars were very cohesive, sticky and stiffened in a short period of time. Various researchers [4, 5] also report this behaviour. In order to decrease the cohesion in the recycled mortar mixtures, either more free water or small dosage of superplasticiser was used depending on the mixture contents to achieve the same workability as the original mortar.

Mixes LR9 and MR9 contained superplasticiser at the manufacturer's maximum recommended dosage. The dosage of plasticiser in specimen LR8 was deliberately chosen to maintain constant workability as well as water cement ratio.

The recycled mortar was a product of OPC, sand passing 1.18mm sieve and water. By crushing and pulverising using pestel and mortar with / without mechanical pulveriser, the obtained recycled mortar contained more fine particles per unit volume than the coarser natural sand used in the source mix. This made the recycled mortar mix gradation too uniform. This provides the recycled mortar with a greater surface area and, consequently, a high water demand.

In addition to the mixes in Table 1, a cement paste mix CP was made from OPC and water. The hardened cement paste (hcp) was crushed and pulverised to pass a 1.18mm sieve. The crushed hcp was used to cast two recycled cement pastes. The mix CPR1 contained recycled hcp and water. The next mix CPR2 was cast using recycled hcp, OPC and water. These mixes were cast containing similar proportions of OPC and the ground recycled hcp to mixes LR1 and LR5. All the mixes were prepared on the basis of equal flowability. These mixes were only considered for thermal analysis in this report.

Table 1 Mortar mix proportions by weight for original and recycled mortar specimens

SPECIMEN DESIGNATION	OPC	POM	SAND	WATER	PFA	SUPER-PLASTICISER	FLOWABILITY SPREAD
L1	1.0		3.32	0.59			163
L2	1.0		3.32	0.79			-
LR1 / MR1		3.32		0.51			-
LR2 / MR2		3.32		0.66			-
LR3 / MR3		2.99		0.62	0.33		-
LR4 / MR4		2.66		0.62	0.66		-
LR5 / MR5	1.0		3.32	0.79			148
LR6	1.0		2.82	0.76	0.50		165
LR7	0.85		3.32	0.82	0.15		167
LR8	1.0		3.32	0.79		0.6%	163
LR9 / MR9	1.0		3.32	0.71		1.5%	182

Note: Superplasticiser was used as percentage weight of OPC.
Flow-table spreads were measured only for mixes pulverised manually.
POM = Pulverised one-day mortar

Compressive Strength

The compressive strengths of normal and recycled mortar were tested on 50mm cubes in accordance with BS1881: Part 116 [6] at ages 7, 28 and 90 days. The densities of the 28-day old cubes were also determined using the method outlined in BS1881: Part 114 [7].

Petrographical Examination

Thin sections of mortars L1, LR1, and LR5 were prepared from 28-day old specimens. The petrographical investigation was carried out using a polarising microscopic technique to outline useful observation. The mortar remnants from compressive strength tests were stored in liquid nitrogen at $-195°C$ for 24 hours to stop hydration [8]. The specimens were then oven-dried for 24 hours and impregnated with resin for the preparation of thin sections.

Calcium hydroxide determination

The quantitative analysis of calcium hydroxide (CH) was determined for specimens CP, CPR1 and CPR2 with a thermogravimetric analyser type 951 TGA on a null-balance principle. The analyser is used to determine the weight change of a substance at a constant rate of heating, while weight loss is simultaneously recorded. The rate of temperature increase adopted in this work was $0°C/min$ from room temperature to $845°C$. After the sample temperature was brought up equilibrium with atmosphere, it was held for a short period in order to drive any remained absorbed surface moisture.

RESULTS AND DISCUSSIONS

Strengths of Mortar

Figure 1 shows the 28-day compressive strength of the recycled mortars using manually and mechanically pulverised recycled material, along with the mean strength of the source mortar mix.

Test specimens LR1, MR1, LR2, and MR2, utilising fresh ground mortar fines alone as hydraulic cement with the addition of different water contents obtained a compressive strength of 9.0 MPa or less. This revealed that the recycled mortar on its own showed little hydraulic behaviour. However, a simple but important fact is that there was some activation in the mixture. The compressive strength of specimen MR1 is similar to that of specimen MR2. This shows the importance of the fineness of the recycled material in reactivating the hydration process. The rehydration in the finer material completely offset the effects of a higher water cement ratio, whereas in the coarser material, the higher water cement ratio reduced strength by 50%.

The additional effect of a pozzolanic reaction in test specimens LR3, MR3, LR4, and MR4 contributed much to the hydration mechanisms compared to the specimen LR1, demonstrating the availability of free calcium hydroxide in the recycled mortar. This is in agreement with the prediction reported by Hansen [9]. However, it showed lower strength than the mean value of compressive strength for normal mortar.

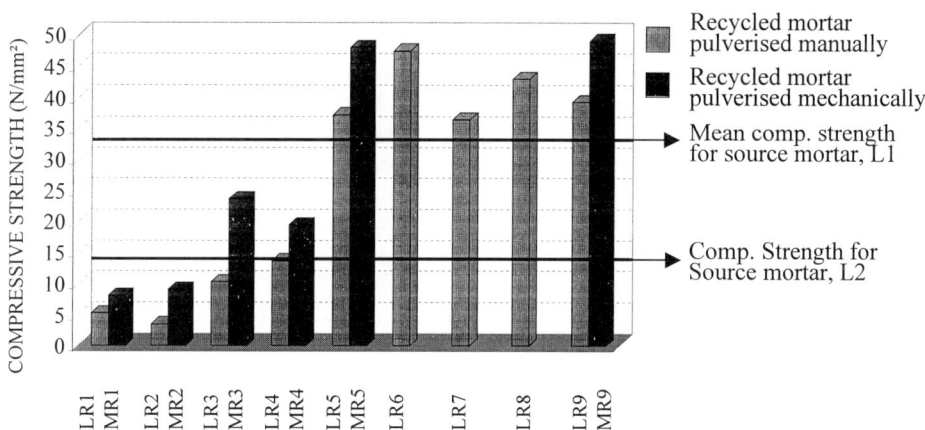

Figure 1 Bar chart for 28-day compressive strength of selected recycled mortars

The most important result illustrated in the bar chart is that the addition of OPC in all recycled mortar mixes LR5 / MR5 has increased the compressive strength significantly, and above that of the source mortar. Same amount of OPC and more water compared to source mortar, and yet the strength exceeds that of the source mix by an amount greater than the strength achieved from mix LR1 containing only recycled mortar. Therefore, this suggests that the reactions between the OPC and the recycled mortar which were not evident when the recycled mortar was mixed with water alone. The influence of the fineness of the recycled material (MR5) is again apparent; the more finely ground material achieved a 37% improvement in strength.

Since mixes LR5 / MR5 have a higher water content than the source mix L1, Mix L2 was prepared in order that mixes LR5 / MR5 could be compared with an equivalent normal mortar. The 28-day strength of mix L2 is significantly lower than that of mix L1 (Figure 2). This demonstrates even more clearly the strength enhancement achieved from the recycled material in mix LR5 / MR5.

The addition of PFA to the above mixes resulted in test specimens LR6 and LR7. As shown in Figure 1, the compressive strength of LR6 was the highest of all the recycled mortar mixes containing <u>manually</u> pulverised recycled material. Calcium hydroxide necessary for pozzolanic reactions was obtained from either existing portlandite in the recycled material or the hydrating added Portland cement. Mixes LR6 and LR7 contain both PFA and OPC. In mix LR6, the proportion of PFA is linked to the POM content, being 15% of the POM. In mix LR7, the proportion of PFA is linked to the OPC content, being 15% of the OPC. The enhanced strength achieved in mixes LR6 and LR7 is evidence of PFA reacting with both the portlandite from the POM and that from the hydrating added OPC.

Mix LR8 was prepared with addition of superplasticiser in order to reduce the water content as near as possible to that of the source mix without any change in the flowability of the mix. The mix LR8 obtained a higher strength compared to the source mortar L1 due to the addition of lubricant and water reduction. Conversely, mix LR8 was not high in strength compared to MR9, because the amount of water was lower in the mix. The compressive strength of mortar mix MR9 was about 50% more than the original mortar. Comparing two mixes MR9 and LR9, the more finely ground mortar mix MR9 was 28 percent stronger than the mortar mix LR9 which confirms the effect of the grinding procedure.

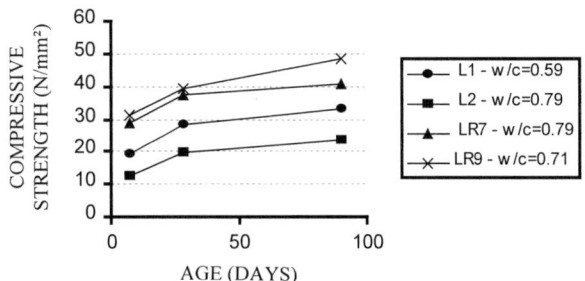

Figure 2 Graphs showing Compressive strength vs age

Petrographic Observations and Role of Calcium Hydroxide

Figures 3 and 4 shows a typical thin-section of the source and recycled mortars L1 and LR5 viewed in crossed polarised light. CH crystals can be clearly seen particularly at the boundaries of the sand particles, but also within the cement paste. It was anticipated that the CH would be present in greater quantities in LR5 than in the source mortar because of the added OPC. However, this appears not to be the case.

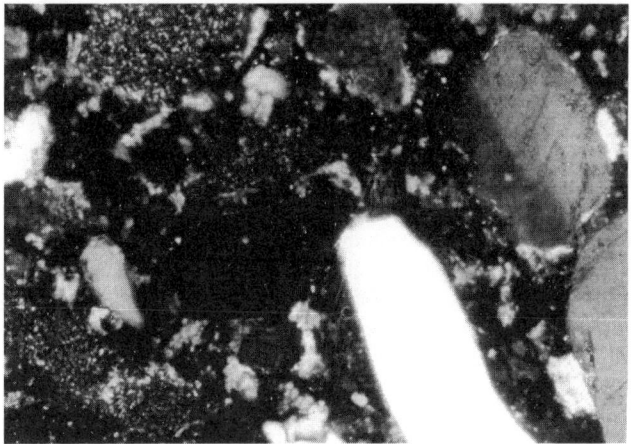

Figure 3 Structural arrangement in a recycled mortar mix L1

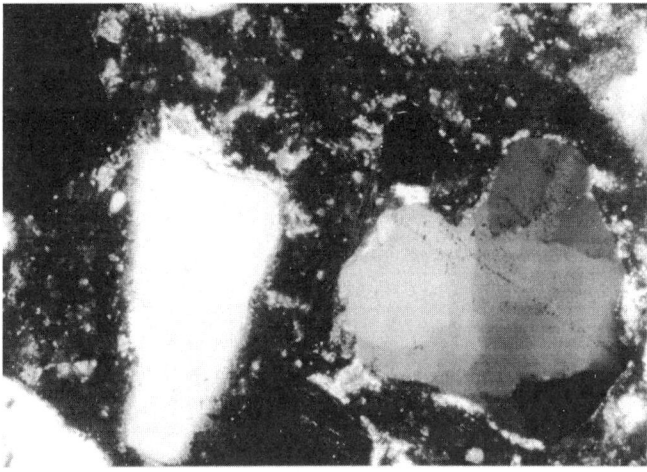

Figure 4 Structural arrangement in a recycled mortar mix LR5

In order to confirm this observation, the CH content in plain cement pastes (CP, CPR1 and CPR2) containing similar proportions of OPC and ground recycled hcp to mixes LR1 and LR5 were subjected to TG analysis. The results shown in Table 2 demonstrated that the CH content of all mixes at a particular age is approximately the same.

Table 2 Calcium hydroxide content of normal and recycled pastes (wt%)

SPECIMEN DESIGNATION	1 DAY	3 DAYS	7 DAYS
CP	8.5	12.5	13.0
CPR1	10.0	11.9	14.6
CPR2	9.2	11.9	13.2

In general, the appearance in thin sections of all the mixes containing ground recycled mortar was similar to the examples in Figures 3 and 4. However, the strength data shows clear differences between the mixes. This suggests that the crushed mortar being used as sand replacement did not remain inert. This is further supported by the fact that the ground sand particles from the recycled mortar can be clearly seen, but the ground hcp particles from same material cannot be distinguished, particularly in mixes containing added OPC. Therefore, the ground hcp particles must have reacted both through hydration, and reaction with components of the added OPC. X-ray analysis of the ground recycled material indicates the presence of a range of components including sulphates, iron oxide and alkalis which may have some contribution in these reactions, and may explain the increased amount of water needed for the recycled cementitious mixtures.

CONCLUSIONS

Based on the test data of this investigation, the following conclusions can be drawn:

1. There was clear evidence of rehydration of the remaining unhydrated cement in the fresh recycled mortar mixes. Furthermore, the petrological examination and thermal analysis of normal and recycled mortars and pastes suggest that there may be additional complex reactions between fresh recycled hcp and added OPC.

2. The fineness of the ground recycled mortar had a significant effect on the rate of rehydration. Finer, mechanically ground mortar always resulted in stronger recycled mix at a given age; in some cases the mix strength was more than twice that of mixes containing hand ground recycled mortar.

3. It has been demonstrated through strength measurements that the rehydration can be enhanced in a variety of ways. PFA has been found to react with portlandite in the POM. Additional OPC enhances strength through complex reactions with the POM. Also, the recycled mixes containing both OPC and PFA replacing a portion of the POM increased significantly the strength. PFA particles provided more surfaces for the precipitation of cementitious gels and combined with different constituents of the POM. These are being further investigated.

4. The mix producing the highest strength was prepared using mechanically pulverised recycled material, added OPC and superplasticiser. The compressive strength of this mix was very high compared to the conventional mortar mixture having the same water cement ratio or even the same workability with a small dosage of plasticiser. The role of superplasticisers needs to be investigated more fully.

REFERENCES

1. BANFILL, P F G. The rheology of fresh mortar. Magazine of Concrete Research, Vol. 43, No. 154, March, 1991, pp13-21.

2. BS12. Specification for Portland cement. British Standards Institution, 1996, 10pp.

3. BS4551: Part 1. Methods of testing mortars, screeds and plasters. Part 1: Physical testing. British Standards Institution, 1988, 32pp.

4. BUCK, A D. Recycled concrete, Highway Research Record, Highway (Transportation) Research Board, 1973, pp1-8.

5. RASHWAN, M S and ABOURIZK, S. The Properties of recycled concrete – Factors affecting strength and workability,' Concrete International, Design And Construction, Vol. 19, No. 7, July, 1997, pp56-60.

6. BS1881: Part 116. Testing concrete. Part 116: Methods of determination of compressive strength of concrete cubes. British Standards Institution, 1983 11pp.

7. BS1881: Part 114. Testing concrete. Part 114: Methods for determination of density. British Standards Institution, 1983, 11pp.

8. CHANDRA, S, HEDBERG, B and BERNTSSON, L. Freezing as a method of study of early cement paste hydration. Cement and Concrete Research, Vol. 10, 1980, pp467-469.

9. HANSEN, T C. Recycled concrete aggregate and fly ash produce concrete without Portland cement. Cement and Concrete Research, Vol. 20, 1990, pp355-356.

SEPARATION AS A REQUIREMENT FOR A HIGH LEVEL RECYCLING OF BUILDING RUBBLE

A Mueller
F Splittgerber
Bauhaus University Weimar
Germany

ABSTRACT. This paper examines the importance of the crushing process in the separation of components of building rubble for reuse. Specifically, the influence of the type and the value of stress on the particle size distribution, the content of cement paste and aggregate in the fractions and the generation of cement free secondary aggregates have been examined for several different concretes recovered from building rubble.

The results produced by the distinctive types of stresses (pressure, impact, shearing) imposed by using different crushers and grinders, make it clear that significant enrichment effects for the cement paste can only be obtained in the fine fractions (< 63 µm) of the crushing products. Such effects appear to be most favourable, if the is stressed by pressure or impact, *and* by shearing.

The influence of the value of the stress on the separation was tested in a special device, with which a cylindrical concrete specimen was accelerated to velocities of about 40 to 80 m/s before impacting on a steel slab and being destroyed. Two concrete samples were treated this way. The first one was made from aggregates from natural gravel, the other one from crushed gravel. Analysis of the crushed concrete from the destructive testing device shows a higher size reduction and a higher rate of liberation of the aggregates from the cement paste for the concrete made with natural, round gravel.

Keywords: Recycling, Crushing and grinding process, Particle size distribution, Enrichment, Separation, Concrete properties, Aggregates, Cement paste.

Prof Dr Ing habil Anette Mueller holds the Chair of Processing of Building Materials and Reuse at the Bauhaus-University in Weimar, Germany. Her special interests are the recycling of construction debris into building materials, the influence of processing on bulk materials and the chemistry of cements.

Dipl Ing Frank Splittgerber is an Employee of the Chair of Processing of Building Materials and Reuse at the Bauhaus-University in Weimar, Germany. His special interests are the recycling of concrete and the chemistry of cements. He is at work on his dissertation, which is entitled "Identification of Cements Used in Hardened Concrete and Mortars".

INTRODUCTION

The starting points for the investigations on the crushing of concrete described in this paper were the following three different aspects:

The Recycling Aspect

High level recycling of concrete can only be successful if there is a consequent separation of the different materials. The first step in this direction is the selective demolition of buildings. The second one is the mechanical liberation of the components which are connected in some building materials. The third step is the separation of the different components. The most usual process for the mechanical liberation of cement paste and aggregate in concrete debris is the crushing process. The material is stressed by pressure, impact or shearing, resulting in particle size reduction and liberation of the connected components. These latter can then be separated. The more complete the liberation of the aggregates in the concrete debris, the greater the potential for its reuse without any restrictions.

The Sample Preparation Aspect

An excellent separation of cement paste and aggregates means a high enrichment of cement paste in the fine fractions. This is important for the reuse of this material as binder component [1,2]. Also, when preparing samples for analysing the type of cement used in the initial concrete [3] a high degree of enrichment of the cement paste fraction is required for good results.

The Material Science Aspect

The separation of cement paste and aggregate is also a material science problem. Questions about the influence of specific properties of concrete on the crushing process have not been adequately explored. However, it should be expected that typical material properties are reflected in the crushed concrete. The boundary as the weakest bond in the concrete [4] should be the preferred location of the fracture during the crushing process. Rough or smooth surfaces of the aggregates should result in more or less coarse crushing products.The first of these aspects was investigated and reported on earlier [5]. The emphasis of this paper lies with aspects 2 and 3. These aspects were investigated in two series of tests with different devices, materials and methods.

EXPERIMENTAL DETAILS

Influence of Crushing and Grinding on Enrichment of Cement Paste Fine Fractions

The program for investigating the influence that the kind of stress has on the separation of cement paste from aggregates is shown in Figure 1. The used concrete specimens have a defined composition and were older than 1 year.

They were stressed in different crushers and mills. Stress by pressure was effectuated in a *jaw crusher*. A combination of pressure and shearing stress was realized in the *roll crusher*. Stresses by impact and shearing were applied in different *impact mills* and in a *hammer mill*. Stress by shearing is dominant in the *mortar grinder* and by *autogenous grinding*.

The crushers and mills used in this inquiry are laboratory scale because only in this scale are such a large variety of different types available to us. This scale dictated the small particle size (< 4 mm) of the feed.

CONCRETE SPECIMENS USED

- Cement: 360 kg/m^3 CEM I 32,5 R
- Water-cement-ratio: 0,5
- Aggregate: 1803 kg/m^3 sand+gravel A/B 32 mm
- Dry bulk density: 2,24 g/cm^3
- Strength after 28 d: 45 N/mm^2

Precrushing < 4 mm

Comminution in different crushers and mills

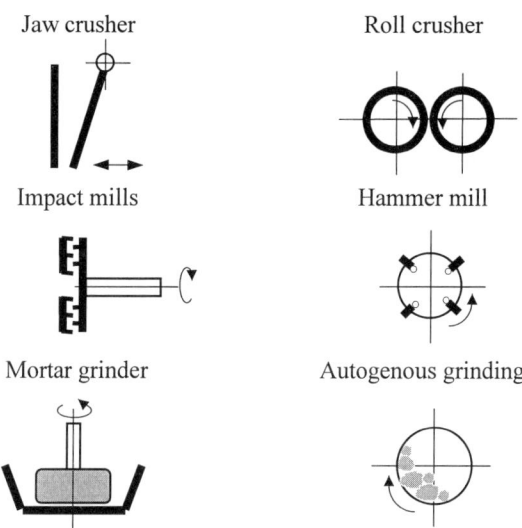

Determination of the particle size distribution
Analyses of the content of cement in the fraction < 63 µm

Figure 1 Program for investigating the influence of different kinds of stress on the comminution of concrete

The products of the crushing process were analysed for particle size distribution in a Coulter LS 230. The success of enrichment was determined by measuring the content of insoluble components in hydrochloric acid, which is representing the aggregates, and by the weight loss on ignition which divulges the hydration water of the cement paste. The content of cement referred to water free condition is computed by the equation:

$$C_{Cement} = \frac{1 - C_{Aggregat} - C_{HydrationWater}}{1 - C_{HydrationWater}} \; ; \; C_{Cement} + C_{Aggregat} = 1$$

The particle size distributions of the products from the jaw crusher, the roll crusher, the impact mills and the autogenous grinding are very similar. The example of the material from the jaw crusher, shown in Figure 2, is a typical monomodal distribution in the particle size range between x = 19,83 µm, for a cumulative undersize of 10 %, and x = 586,8 µm, for a cumulative undersize of 90 %. If the hammer mill or the mortar grinder are used the distribution is shifted to smaller particle sizes as shown in Figure 3. The range for the cumulative undersize of 10 and 90 % respectively, is between x = 3,9 µm and 443,4µm. Furthermore, polymodal distributions are generated by these grinders.

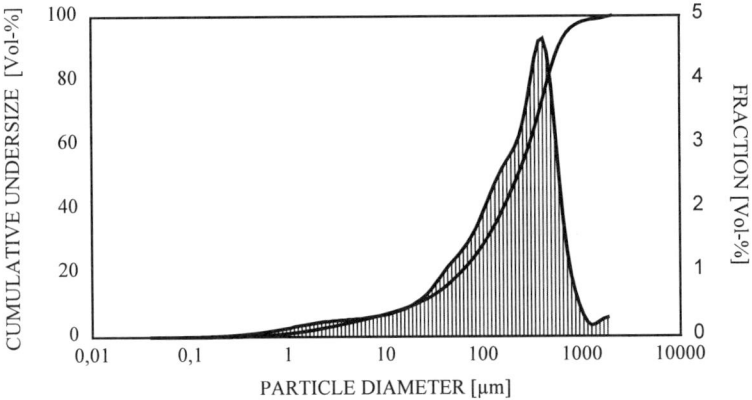

Figure 2 Log Table showing particle size distribution of material crushed in the jaw crusher

Enrichment of the cement paste in the fraction < 63 µm, compared with the original concrete, can be established for all the crushing and grinding devices employed in this study. As shown in Figure 4, the clearest, most consistent effect was obtained for the products of the hammer mill and the mortar grinder.

This effect may result from the "peeled" of cement paste from the surface of the aggregate grains when shearing stress is applied, with the cement paste passing into the fine fraction.

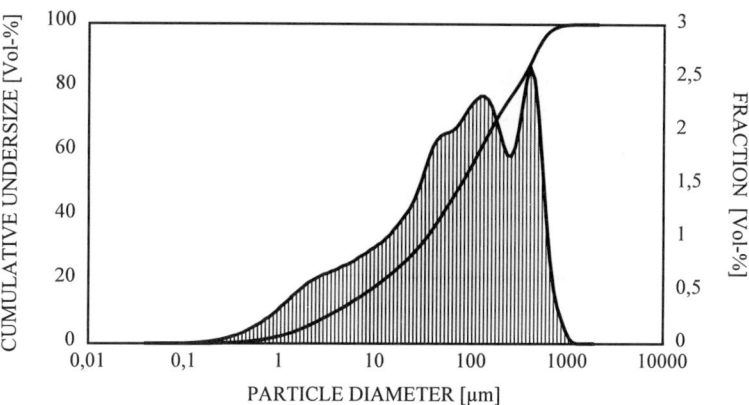

Figure 3 Log Table showing particle size distribution of material ground in the hammer mill

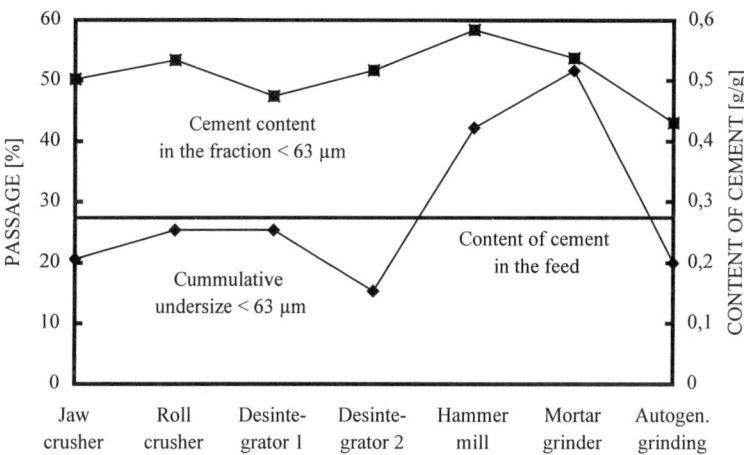

Figure 4 Content of cement in the fraction < 63 μm, for all the crushing and grinding devices

DISCUSSION

Influence of Stress and Properties of Concrete on Separation of Paste from Aggregates

The experiments which dealt with the interrelations between the crushing process and the nature of the concrete as composite material, were carried out in a special device, illustrted in Figure 5, which was developed and built at the Otto-Guericke-University in Magdeburg.

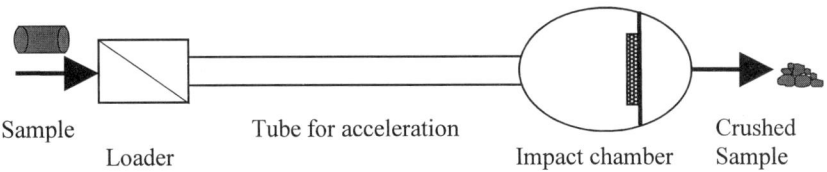

Figure 5 Main components of the equipment for impact tests with concrete specimen [6]

Cylindrical concrete specimen were accelerated to a certain velocity in a barrel. Upon hitting an impact plate they are destroyed. Figure 6 shows an overview of the program for these impact experiments [7]. Both the amount of the applied stress and the type of concrete were varied in these studies. Concrete A was made from natural sand, and gravel with a round shape. In concrete B, crushed gravel was used. The chemical and mineralogical compositions and the particle size distribution of both aggregates are approximately the same. As shown in Table 1, the velocities achieved by the specimens ranged from a minimum of 42,4 m/s, to a maximum of 65,2 m/s. This meant that the typical peripheral speed for impact crushers was exceeded to a small degree. Of course, in an impact crusher contact between the impact tool and the material occurs more than once.

Table 1 Program of interrelations between the crushing process and concrete properties

CONCRETE SPECIMEN	
Concrete A	Concrete B
Cement: 345,45 kg/m^3 CEM I 32,5 R Water-cement-ratio: 0,55 Aggregate: 1784,6 kg/m^3 sand + gravel A/B 16 mm	Cement: 345,45 kg/m^3 CEM I 32,5 R Water-cement-ratio: 0,55 Aggregate: 1796,2 kg/m^3 crushed gravel A/B 16 mm
PROPERTIES OF THE SPECIMEN	
Compressive strength of 15cm- cube after 28 days	
44,0 N/mm^2	49,9 N/mm^2
Tensile strength of 10cm- cube after 90 days	
3,50 N/mm^2	3,65 N/mm^2
Dry bulk density	
2,169 g/cm^3	2,223 g/cm^3
Preparing of cylindrical specimen 150 mm Ø x 150 mm for the impact tests	
CRUSHING WITH DIFFERENT VELOCITIES	
(42,4 ± 3,2) m/s (51,2 ± 3,0) m/s (65,2 ± 3,0) m/s	
Determination of particle size distribution Analysis of the fractions − Content of aggregate in the fractions − Content of cement in the fractions	

The particle size gradation curves of the two concrete samples are shown in Figures 6 and 7. The influence of sample velocity is not discernible. However, a small difference can be perceived between the compared concrete specimens. The secondary aggregates from concrete B (crushed gravel) are coarser than the secondary aggregates from concrete A (smooth gravel). Also, the cumulative undersize of 16 mm fraction is 45,2 % for concrete A and 28,9 % for concrete B.

This later observation points to a stronger bond between cement paste and aggregate in concrete B. The amount of the coarser grains in concrete B, without any cement paste retained on the surface, compared to the amount of paste free grains in concrete A, as shown in Table 2, confirms this fact.

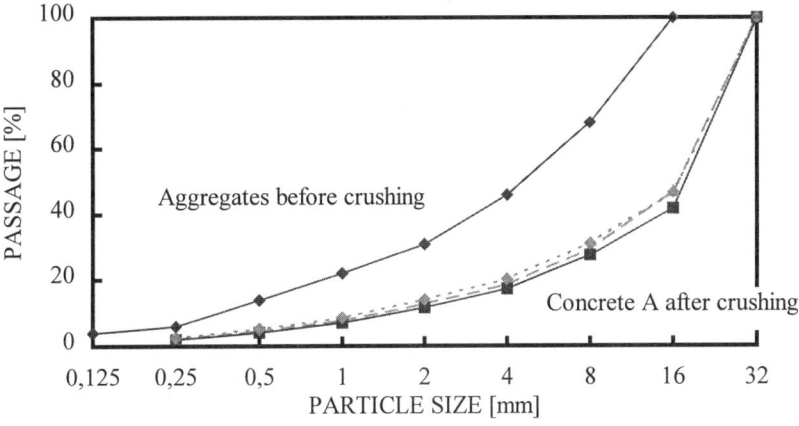

Figure 6 Grading curve of concrete A, after crushing

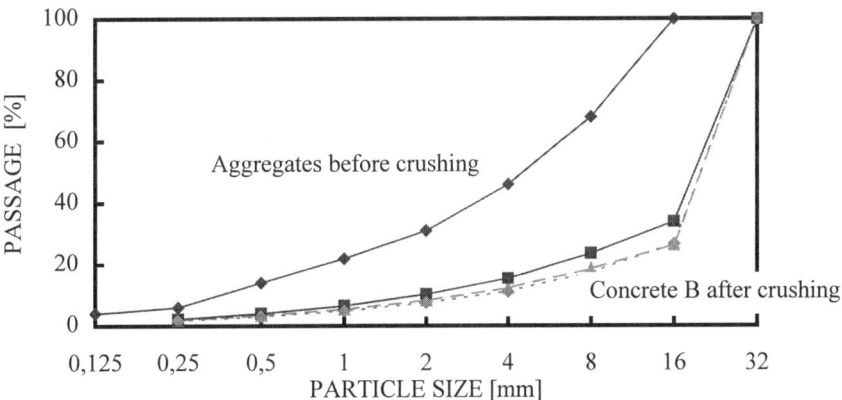

Figure 7 Grading curve of concrete B, after crushing

Table 2 Portion of grains without cement paste at surface after the impact crushing

PARTICLE SIZE (mm)	PORTION OF GRAINS WITHOUT CEMENT PASTE RELATED TO FRACTION MASS [%]	
	Concrete A	Concrete B
8/16	15,9	12,8
4/8	26,6	13,8
2/4	13,5	7,4

The measured contents of aggregates in the fractions of the crushed concrete, displayed in Figures 8 and 9, also reveal differences between concrete A and B. For concrete A, in the fractions 0/0,125 mm, 0,125/0,5 mm and 0,5/1,0 mm, a clear decrease of the content of aggregate, as compared to that of the original cement is recognisable. In the middle range of the particle sizes a small increase of the content of aggregate is revealed.

This typical effect also occurs when secondary aggregates are produced with a technical impact crusher. It results from the partial cracking of the cement paste from the surfaces of the aggregates. The composition of the coarse grains do not differ from the original concrete.

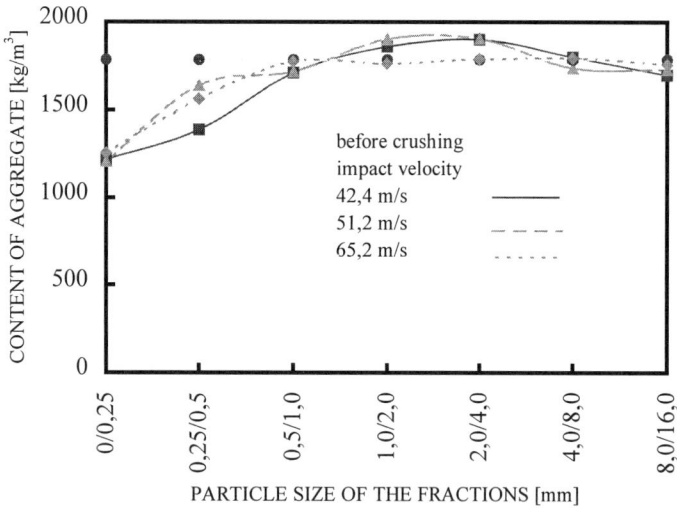

Figure 8 Content of aggregate in the fractions of concrete, after crushing

In case of concrete B only the factions 0/0,125 mm and 0,125/0,5 mm have a somewhat lower content of aggregates. All other fractions agree in their composition with the original concrete. Nevertheless, there is no clear idea of the dependence on particle size as in case of concrete A.

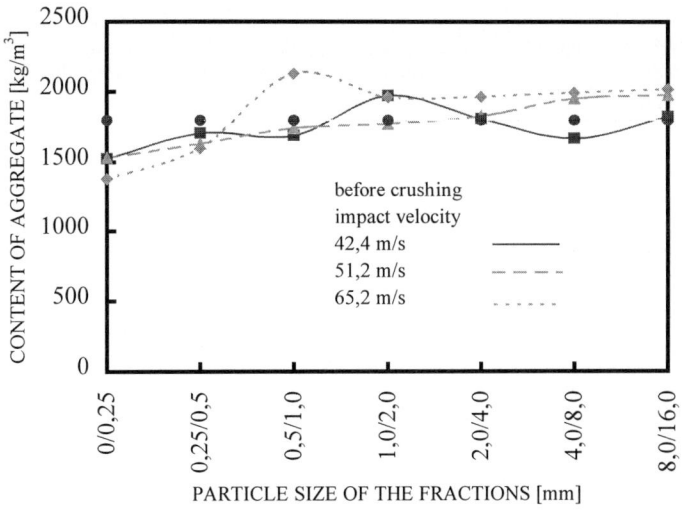

Figure 9 Content of aggregate in the fractions of concrete B, after crushing

CONCLUSIONS

The technology of the reuse of concrete debris as secondary aggregate requires a crushing process. At present, impact crushers are preferred for the comminution of this material. The results of the tests of different crushers and mills confirm this choice. Better effects may be achieved if the material is stressed not only by impact but also by shearing.

During the crushing process effects other than reduction of the particle size can be obtained, because concrete is a composite material. The composition of the fractions generated by crushing depends on the particle size. In the fine fractions the cement paste is dominant, as the more porous, weaker material is accumulated. In the middle fractions a small enrichment of the aggregates occurs. In the coarse fractions, the composition of the crushed concrete agrees with the composition of the original concrete.

The effects pointed out with concrete with round, smooth aggregates on the one hand, and crushed, angular aggregates on the other, verify that the typical material behaviour of concrete is also recognisable in the crushing products. The stronger bond between crushed aggregate and cement paste than between smooth aggregate and cement paste is reflected in the results of the particle size distribution, as well as in the appearance and the analysis of the crushing products. The effects of materials which improve the bond between cement paste and aggregates could be an interesting question for further research.

Summarising all results achieved with the different methods of mechanical crushing, it can be established, that the bond between the cement paste and the aggregate in concretes meeting today`s requirements is generally strong enough to withstand the stresses encountered during crushing. Thus, the separation of cement paste from aggregate is only possible up to a certain limit. For a more effective separation, modified or new methods of the crushing process must be developed.

One of these alternative ways may involve increasing the shearing stress during the comminution. Another one is used in further examinations. This method, developed at the Otto-von-Guericke-University in Magdeburg uses a kind of sonic pressure to stress the material. First results show much better separation effects compared with the examined conventional mechanical crushing.

REFERENCES

1. LOO, V W. Closing concrete Loop-From Reuse to Recycling. Sustainable Construction: Use of recycled Concrete Aggregate. Proceedings, London 1998

2. LANDER, S. Nacherhärtung von Zementstein aus aufbereitetem Betonbruch: Diplomarbeit 1997. Bauhaus-Universität Weimar, Fakultät Bauingenieurwesen, Professur Aufbereitung von Baustoffen und Wiederverwertung.

3. SPLITTGERBER, F. MUELLER, A. Identification of the Type of Cement in Hardened Concrete and Mortars Creating with Concrete, 1999

4. ZIMBELMANN, R. A method for strengthening the bond between cement stone and aggregates. Cement and Concrete Research, Vol. 17, 1987

5. MUELLER, A, WINKLER; A. Characteristics of Processed Concrete Rubble. Sustainable Construction: Use of recycled Concrete Aggregate. Proceedings, London 1998.

6. TOMAS, J, GRÖGER, T. Abtrennen von Wertstoffen aus dem Bauschutt. Proceedings, 6. Weimarer Tagung über Abfall- und Sekundärrohstoffwirtschaft, Bauhaus-Universität Weimar, Fakultät Bauingenieurwesen, Professur Aufbereitung von Baustoffen und Wiederverwertung, Oktober 1998

7. SCHRIMPF, M. Aufschlußzerkleinerung von Beton. Studienarbeit 1998. Bauhaus-Universität Weimar, Fakultät Bauingenieurwesen, Professur Aufbereitung von Baustoffen und Wiederverwertung.

THEME TWO:
ALTERNATIVE OPTIONS

Keynote Paper

SECONDARY MATERIALS: A CONTRIBUTION TO SUSTAINABILITY?

J M J M Bijen
INTRON BV
Netherlands

ABSTRACT. Increasing amounts of secondary raw materials are being used in concrete. It is generally assumed that replacing primary materials with secondary materials contributes to sustainable development of the environment.

Using the results of life cycle assessment studies, it has been shown that this is the case for current additions in concrete or in cements. Where primary aggregates are replaced the environmental benefits are less obvious. For some impacts they can even be negative. This is due to the amount of energy needed to upgrade wastes to aggregates that meet the requirements.

An important factor is durability. The durability of concrete should be in accordance with the design service life of the structure. Attention should be paid to this aspect when new secondary raw materials are used.

The usage of secondary raw materials increases the variety of raw materials used in concrete, which can affect its recyclability. The quality of granulates produced from demolished structures will show a larger variance.

Keywords: Sustainability, Additions, Aggregates, Environment, Durability, Wastes, Recyclability.

Professor Jan M J M Bijen is director of INTRON BV the Dutch quality assessment institute for the building industry. He is also part-time professor in materials science at the Faculty of Civil engineering and Geoscience of Delft University of Technology. His field of expertise is the durability and sustainability of building products.

INTRODUCTION

Secondary raw materials are widely used in concrete manufacturing, as fuel and raw material for Portland clinker production, as addition and as aggregate. The amounts are still growing.

It is generally assumed that usage of wastes as secondary raw materials is contributing to a sustainable development of the environment. Smaller quantities of primary materials are used and less waste has to be disposed of.This paper will discuss whether this assumption is, in general terms, correct.

A distinction will be made between the various types of secondary raw materials. They are:

- Fuels and raw materials for cement manufacturing.
- Additions (mineral admixtures) including their use in composite cements.
- Aggregates produced from waste concrete.
- Aggregates from other secondary resources.

As far as aggregates are concerned, the dominant factors with respect to environmental effects will be considered. Finally, the effects of using these secondary materials on the recyclability of concrete will be discussed

ASSESSMENT OF ENVIRONMENTAL EFFECTS

In determining the environmental effects of a product all immissions and emissions over the life cycle of the product have to be inventorized. In this paper use will be made of environmental effects calculated using a life cycle assessment method that is widely accepted in the Netherlands, the CML LCA method [1] and in a more aggregated form the so-called environmental measures [2]. These methods are in accordance with the draft ISO standards of the 14040 series, see Table 1. The method used does consider, in quantitative terms, the environmental effects at a global and regional level. Local effects such as change of landscape and nuisance are not taken into account.

Table 1 The ISO 14040 series, ISO standards on LCA

ISO STANDARD		STATE-OF-THE-ART
14040	Environmental Management – Life Cycle Assessment – Principles and Framework	Available (1998)
14041	Environmental Management – Life Cycle Assessment – Goal and scope definition and life cycle inventory analysis	Draft international standard (FDIS), 1997
14042	Environmental Management – Life Cycle Assessment – Life cycle impact assessment	Committee draft version 2, July ; expected to be upgraded to a draft international standard (DIS) in 1998
14043	Environmental Management – Life Cycle Assessment – Life Cycle Interpretation	Committee draft version 2, July 1997; expected to be upgraded to a draft international standard (DIS) in 1998

ADDITIONS

Additions are used to replace cement or to improve concrete properties or both. The most well known additions are fly ash, ground granulated blast furnace slag and silica fume. These additions are also used in cements. The European pre-standard prEN 197 defines a number of cements consisting of Portland clinker and additions (CEM II to CEM V). The use of these additions does reduce the environmental impact of the cement. Figure 1 shows the environmental measures of blast furnace slag cement (CEM III/B 42.5) and Portland cement (CEM I 42.5) [3].

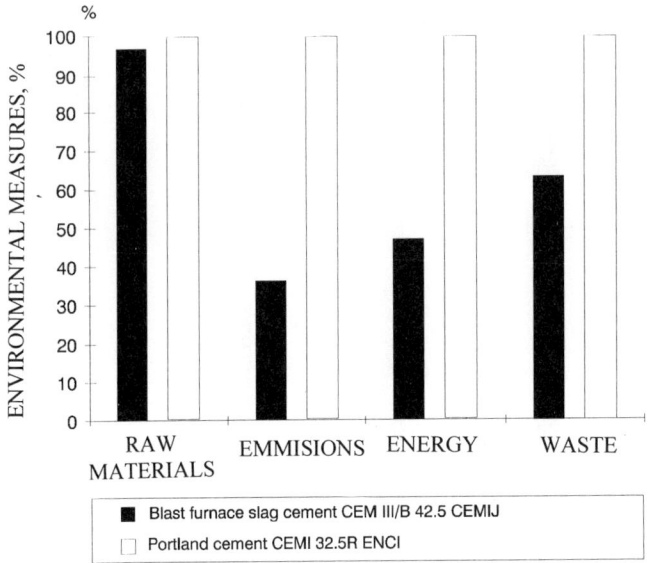

Figure 1 Environmental ratings (of a relative nature) of blast furnace slag cement, CEM III/B 42.5 (CEMIJ) and Portland cement, CEM I 32.5 R (ENCI) in a life cycle assessment from cradle to grave

If we use additions to improve properties of concrete such as durability, the effects of the environment load can be substantial. In 1981 and 1982 the author was involved as a consultant in the construction of the King Fahd Causeway between Saudi Arabia and Bahrain. The use of blast furnace slag cement was proposed instead of the specified type V sulphate resisting Portland cement. It was predicted that with the specified cement corrosion should occur within 10 years, probably within 5. With the blast furnace slag cement proposed, it was expected that it would last at least 100 years before penetrating chlorides would have reached the reinforcement. Now, 14 years after completion, it is obvious, we were conservative in estimating the performance of the blast furnace slag cement concrete [3].

The beneficial effects for both the environment and the economy are obvious. Durability is a major issue for the long cyclic products of the building industry. It should only be claimed that an alternative product contributes to sustainability where a durable performance over its service life has been demonstrated.

AGGREGATES FROM CONCRETE

The use of concrete granulates produced from waste concrete as a substitute for river gravel has a positive effect on the environmental profile in terms of raw materials and waste materials but a negative effect on emissions and energy consumption. The production of granulates requires more energy than the dredging of river gravel.

Figure 2 shows the changes in environmental measures of an increase from 20% to 40% substitution of river gravel by concrete granulate in a bridge construction project [4].

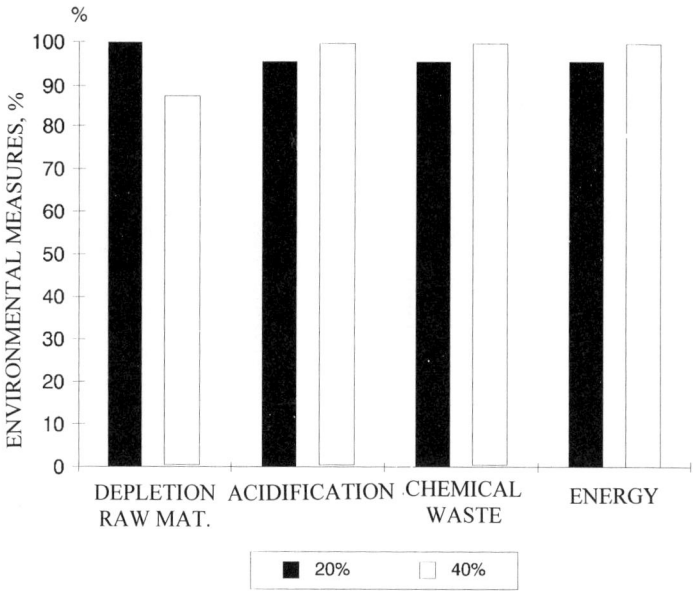

Figure 2 Effect on some aspects of the environmental profile when increasing the amount of concrete granulate from 20% to 40% in a concrete bridge [4]

The replacement of river gravel with up to 20% concrete granulates does not have a substantial effect on concrete properties. However, where larger percentages are replaced, the change in properties can be substantial. For instance shrinkage and creep will increase, while E-modulus decreases. This requires an adjustment of the design of structures with often an increase in the dimensions of the structure, which to some extent counteracts any beneficial environmental effects of the replacement.

AGGREGATES FROM OTHER SOURCES

Fine Aggregates

For the province of South Holland LCAs were made for primary and secondary mineral sand [5].

The materials considered are:

- Primary sand extracted from the embankments of rivers,
- Secondary sand processed from contaminated soil,
- Secondary sand separated from contaminated dredging mud,
- A secondary sand (sieve sand) from a selection plant for building and demolition wastes. For details see reference 5.

Figure 3 shows on a relative scale the environmental measures for the sands per kilogram. In this case, only part of the life cycle is taken into account, from cradle to gate e.g. to the place of delivery of the sand. Obviously the primary sand obtains the best scores in all environmental measures; see the next paragraph.

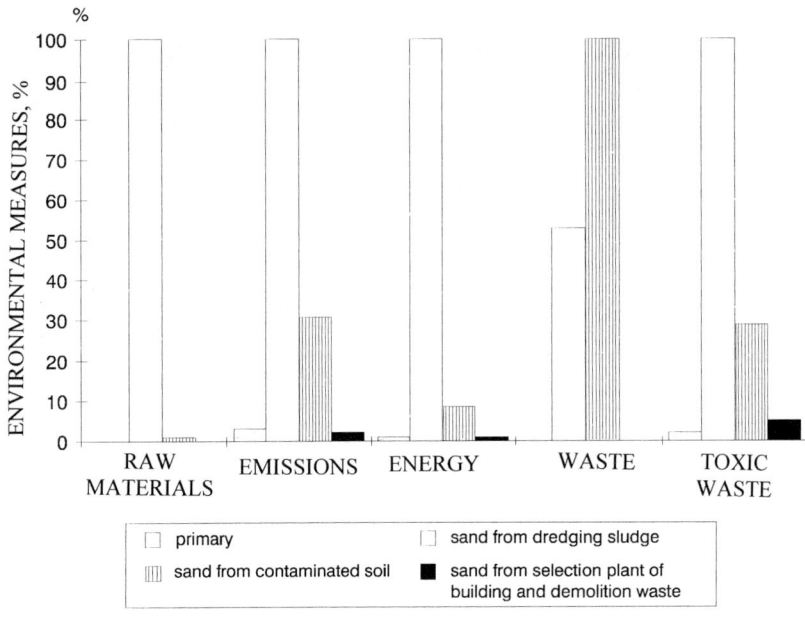

Figure 3 Environmental measures of 1 kg sand (fine aggregate), including transport, from cradle to gate on a relative scale [5]

Where it is not the fine aggregate, but the concrete of which the fine aggregate is a constituent, that is considered, the differences between the sands are less obvious. The contribution of other raw materials such as cement in concrete makes the image more diffuse, see for instance Figure 4 comparing one kilogram of concrete made of primary sand and one made with secondary sand.

Here again it must be stressed that local effects such as the influence of excavating primary raw materials on the life support system and biodiversity have not been taken into account.

Coarse Aggregates

A number of secondary materials from production loops other than that of concrete are used as a coarse aggregate in concrete. For instance air-cooled blast furnace slag, phosphorous slag, granulates from ceramic and calcium silicate masonry, lightweight aggregates made from fly ash etc.

If the environmental effects of upgrading these materials to a quality that meets the requirements for a coarse aggregate in concrete are wholly attributed to the life cycle loop from which the materials originate, the replacement of primary coarse aggregates may be advantageous. However, if the aggregate is of a lower quality than the raw material it replaced, the compensating measures to decrease the water/cement ratio, such as increasing the cement content, may counteract this.

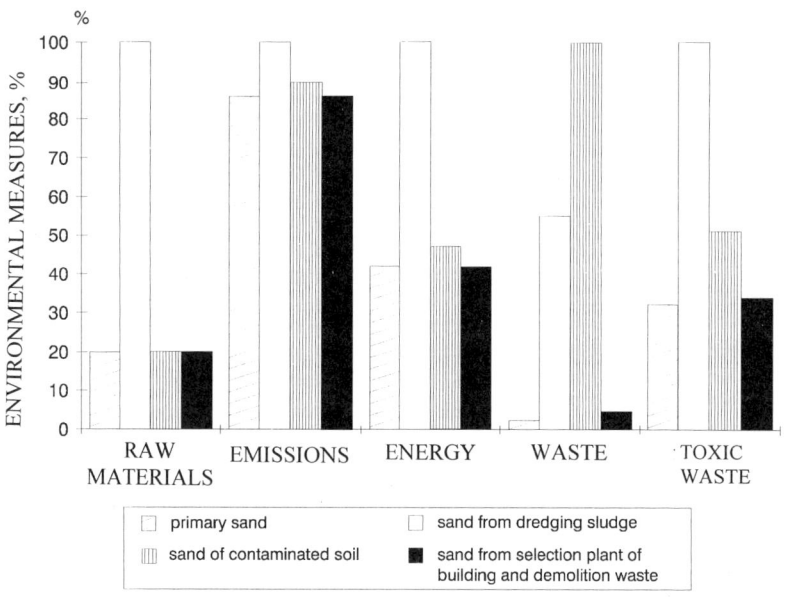

Figure 4 Environmental measures of 1 kg of ready mix concrete made with alternative sands on a relative scale (including transport) [5]

DOMINANT FACTORS

The dominant factor in terms of the environmental effects of aggregates is the use of energy. The amount of energy required to process secondary raw materials is often higher than for primary materials. Energy is not only reflected in the measure "energy" but also in "emissions" and "raw materials". The use of energy entails emissions, which contribute substantially to the aggregated and weighed emissions of the product cycle. As far as raw materials are concerned, energy sources are regarded as non-renewable, while the extraction of aggregates from primary natural sources is not taken into account because it is assumed they are abundantly available. On a world scale this is certainly the case.

Of course it can be argued that the energy required for processing secondary materials to a quality level required for usage should not be assigned to the secondary raw material but to the previous product cycle.

In that case the environmental effects of secondary materials will improve substantially. This is certainly true for a number of secondary materials such as the granulates produced from brickwork or fly ash. But in other cases, such as the sand manufactured from contaminated mud and dredging soil, as discussed in previous paragraphs, contamination has occurred in the past and the processing will be an environmental load now. It is this load that has to be taken into account in comparisons with alternatives such as disposal.

RECYCLABILITY

Concrete is a material with a great capacity to absorb secondary materials produced from wastes. As shown above, replacing primary raw materials with secondary materials often has positive effects on the environmental loading of concrete. However, in taking a general view, it is possible to envisage some negative aspects, such as the recyclability of concrete. A larger variety of raw materials means a larger variance in concrete compositions and in the quality of granulates made after demolition. This can affect the recyclability of the concrete in the longer term. Recyclability is likely to become a major feature of materials over the next few decades.

CONCLUSIONS

In general, the replacement of cement by additions contributes to sustainable development of the environment. It is beneficial for all environmental impacts. Improvement of additions on durability for concrete applied under aggressive natural conditions is another contribution to sustainability.

As far as the replacement of aggregates is concerned, the situation is less clear-cut. If secondary aggregates are used which need a lot of processing to meet requirements the effect on a number of environmental impacts can be negative. If the quality of the aggregate is lower than that of the substituted raw material compensating measures may cancel out the positive environmental effects of the replacement.

Attention should be focused on the recyclability of concrete. A greater variety of raw materials makes recycling less suitable, which can be regarded as disadvantageous in the longer term.

REFERENCES

1. GUINÉE, J B, et al. Environmental life cycle assessment of products. Part 1: manual, Part 2: backgrounds, (production: Centre of Environmental Science, Leiden University, TNO, B&G), NOH report, NOVEM, 9253/54,October 1992.

2. SCHUURMANS A., BIJEN J., "MBB milieumaten voor de Bouw, Fase 1 milieumaten voor de bouwproductgroepen niet dragende binnenwanden, dakgootsystemen en

bouwwerven" (environmental measures for the building industry), Intron report 94215, November 1994.

3. BIJEN, J, Blast Furnace Slag Cement for Durable marine Structures, Betonprisma, The Netherlands, 1996, ISBN 90-71806-24-3.

4. FRAANJE, J, Environmental effects of two approach bridges, IVAM, Amsterdam, May 1992.

5. BIJEN, J, VAN SELST, R, SCHUURMANS, A, Environmental impact of recycling, Mineral Planning in a European Context, Demand and Supply, Environment and Sustainability, ed. Van de Molen, Richardson, Voogd, Geo press 1998, pp 151-156.

SIMPLE TREATMENTS TO REDUCE THE SENSITIVITY TO WATER OF CLAYEY CONCRETES LIGHTENED BY WOOD AGGREGATES

A Ledham	R M Dheily
A Bouguerra	M Queneudec
University of Rennes	University of Picardie Jules Verne
France	France

ABSTRACT. Clayey concretes lightened by wood aggregates allow to utilise waste derived from two industrial ranges, aggregates (clay) and wood industries. However, one of the principal flaws put foward in the case of wooden concretes, in general, is their sensitivity to water giving significant size variations. In the case of clay-cement matrix, this flaw could be increased because of the reactivity of clay to moisture variations. This experimental study focuses on the development of a clay-cement-wood load-bearing insulating composite with small size variations. It consists in performing simplified treatments on the aggregates with the objective of reducing the retaking of water and size variations after hardening. The authors show the possibility to reduce the size variations at limit states down to 1mm/m.

Keywords: Clayey concretes, Wood aggregate treatments, Size variation reductions.

Dr A Ledhem is a post doctoral researcher in Civil Engineering, University of Rennes. His speciality is the design of building materials.

Dr A Bouguerra is a post doctoral researcher in Civil Engineering, University of Rennes. His main research interest concerns heat and mass transfer in building materials.

Dr R M Dheily is a lecturer in Process Engineering, University of Amiens. Her major centre of interest is the relationship between microstructure and reactivity.

Pr M Queneudec is Head of the « Laboratoire des Transferts et Réactivité dans les milieux condensés », University of Picardie Jules Verne, Amiens. She is specialised in the design and characterisation of building materials.

INTRODUCTION

An inventory of the aggregates available on the market was recently conducted by Pimienta [1].Chemical substances in the wood chips may affect the properties of the concrete made with the chips. As a result it was found that the aggregates in use generally undergo treatment intended to decrease water absorption during both mixing and service life. These various treatments [2 to 11] often rely therefore upon elaborate processes or require the consumption of considerable energy and hence are not very accessible to developing countries.

The work conducted herein lies within the scope of the development of local building materials. Simple solutions, which enable both limiting the water absorption of the chips and reducing the dimensional variations within a clayey concrete lightened by wood aggregates, will be proposed.

MATERIALS AND OPERATING TECHNIQUES

The mineralogical study has shown that the clayey fines used in this work were composed essentially of kaolinites, along with small percentages of quartz and mica [12]. The resultant material was not highly plastic (plasticity index: $I_p = 12\%$), and the upper granular limit value was close to 70 µm. Grains with a diameter of less than 2µm constituted 55% of the cumulative fines.

The cement used was a CPA CEMI 52.5. The lime was a hydraulic lime XHN100 (NF P15310). Fir was used as the wood for this study due both to its low density and to its widespread application in the building industry. The apparent density of the chips was around 0.05 while their real density was 0.44.

The schistous fines were residue of the crushing of Brioverian schist which had undergone a slight general metamorphism. The portion included in our investigation corresponded to what remained once the fraction of sand and fine gravel particles had been removed. The separation process was performed on dry material. The fines used in this work were characterized by a gap grading from 88% passing at 500µm to 100% passing at 1 mm. The proportion of grains less than 2µm was practically zero and the residue constituted a granular material that lacked both cohesion and plasticity. Its apparent density was 1.05 and its absolute density 2.70. The oil used was linseed oil with a density of 0.86.

The processes carried out were: coating by hydraulic binders, impregnation by oil, a boiling water treatment under atmospheric pressure. The oil treatment was performed by spraying, while the lime and cement were added in a grout form on the chips during a mixing at low speed (EN 196-1). The quality of the coating was assessed by scanning electronic microscopy; density was tested by weighing. After processing, the chips were stored in a damp room for 48 hours and then moved to an air-conditioned room (R.H. = 50%, T = 20°C) during 28 days before the concrete was made.

The clayey material, schist fines and cement were initially mixed at low speed in a standardized mixer for mortar (EN196-1). The chips were then added gradually while keeping the mixer at low speed for 2 minutes. The water, whose quantity is to be determined by use of an empirical formula [12], was then added gradually while continuing the mixing at low speed for another 2 minutes. Water added was calculated with the formula W = 0,35.Ce+0,7.Cl+0,2.S+K.W which takes into account the water being absorbed by wood aggregates and thereby enables operating at a constant workability. Ce, Cl, S, W are respective percentages in weight of cement, clay, schist and wood. K takes into account the absorption of water by chips. For non treated chips it is equal to 1,8. After mixing, the mixture was poured into molds. The filled molds were then held in a hygrometrically-controlled and temperature-controlled room (R.H. = 95%, T = 20°C) for 24 hours. Following removal from the molds, the test specimens were placed in a storage room (R.H = 50%, T = 20°C) for 28 days.

Table 1 Influence of the treatment on the characteristics of the composite

TREATMENT	WATER ABSORPTION (%)	DRY DENSITY	R_C (MPa)	EXTREMAL SIZE VARIATIONS (mm/m)	λ (W/M.K)
None	49	0.85	8.9	3.5	0.22
Boiling water (6 hours)	34	0.85	9.7	2.00	0.22
Lime (lime/wood = 1)	25	1.11	12.7	1.18	0.27
Cement (cement/wood =1)	25	1.19	12.4	1.28	0.28
Impregnation with oil (oil/wood = 0.5)	22	1.00	8.5	2.20	0.25
Boiling water (6 h) and lime (lime/wood = 1)	23	1.12	13.1	0.95	0.27
Boiling water (6h) and cement (cement/wood=1)	21	1.20	12.6	1.00	0.27
Target value		< 1.40	> 5.0	< 1.00	< 0.3

The composition and characteristics of the basic composite are presented in Table 1. The intensity of the process is to be determined by the material mass ratio used for the treatment / chips.

The dimensional variations were determined on 4x4x16 cm prisms using a retractometer (NF P 15-433) with balls able to detect variations on the order of 10^{-3} mm and equipped with a printer for entering the measurement data. The variations in mass were determined using a scale accurate to 1/100th of a gram. Determining the water content at saturation of the chips was a delicate process. The method which consists of weighing the aggregates in a dry state, then placing them in a sieve and immersing the entire set-up until saturation leads to a significant dispersion of the results; a random quantity of water remains bonded to the chips. This water could be eliminated by means of centrifugation (3 cycles of drying to 120 revolutions/min for 30 seconds).

EXPERIMENTAL RESULTS AND ANALYSIS

Influence of Treatment on the Water Absorption of the Chips

Figure 1 presents the evolution in water absorption as a function of the intensity of the process applied to decrease the water absorption of the wood aggregates. It can be noted that for all of the processes, water absorption decreases more significantly as the process becomes more intense. In the case of coating with a hydraulic binder, it appears that the treatment with cement is more effective than that with lime. The oil-based treatment considerably limits the chips' water absorption. Moreover, it is more effective than the coating with a hydraulic binder. For an equal mass ratio (0.75), the cement-based treatment reduces water absorption by 42% and reduces oil absorption by 87%. The result of the boiling water treatment is obvious. However, at the end of 4 hours, the decrease in water absorption is indeed insignificant. It can thus be concluded that this duration is sufficient to eliminate almost all of the substances capable of hydrolysis under these specific temperature and pressure conditions.

Influence of the Treatment on the Water Absorption of the Composite

Figure 2 shows the influence of the various processes on the water absorption of wood concrete. The treatment of the chips, in whatever form, yields an improvement in the composite's behavior. For a cement / wood or lime / wood mass ratio of 0.5, the water absorption of the composite is already lowered by 40%. The thickness of the chips' coating layer exerts an undeniable influence, yet one which remains relatively insignificant in comparison with that of the chips themselves on the water absorption. In the case of the oil-based treatment, water absorption decreases as the intensity of the process decreases until reaching an oil / wood ratio of 0.5. An increase can nonetheless be observed for a mass ratio of 0.75. This phenomenon is to be attributed to the presence of oil droplets within the matrix.

This finding has been confirmed by the color of the test cubes, which undergoes a change when the ratio stands at 0.75. An optimal process which corresponds with the chips' absorptive capacity does in fact exist. In the event of excess absorption, a proportion of the oil no longer seeps into the structure of the wood, but rather remains on the surface and then migrates into the matrix mixture during mixing.

In this manner, the oil drops do occupy a certain volume in the fresh state. During aging, these drops retract, thereby generating additional porosity, which could account for the increase in water absorption.

This phenomenon can be highlighted by means of scanning electronic microscopy (see Figure 3). With respect to the boiling water treatment, the same trend as observed for the chips can be noted. Those substances capable of hydrolysis that get eliminated during the first few hours are thus partially responsible for the composite's water absorption.

Concretes Lightened by Wood Aggregate 221

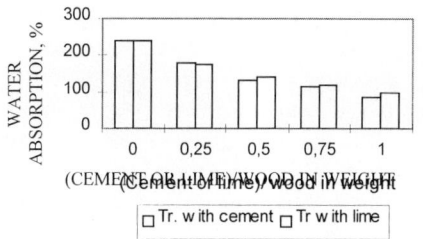

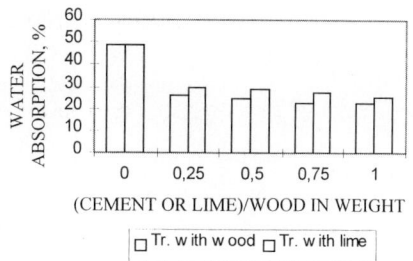

Figure 1a Treatment with hydraulic binders

Figure 2a Treatment with hydraulic binders

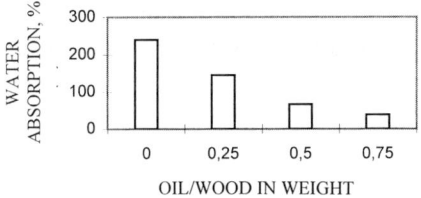

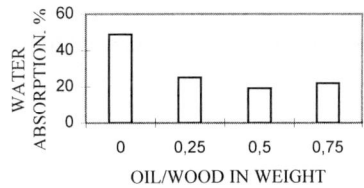

Figure 1b Treatment with oil

Figure 2b Treatment with oil

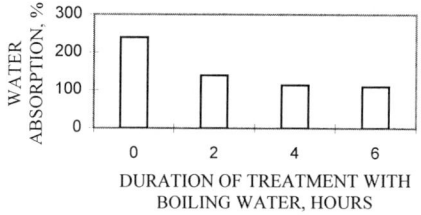

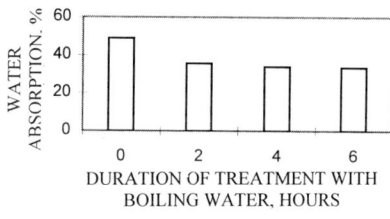

Figure 1c Treament with boiling water

Figure 2c Treatment with boiling water

Figure 1 Influence of the treatment of the wood chips on variations on weight

Figure 2 Influence of the treatment of wood chips on variations on weight in cement-wood-fines composite

Influence of the Treatment on the Composite's Dimensional Variations

Figure 6 displays the influence of the various processes on the dimensional variations of wood concrete. The dimensional variations, i.e. the difference between the saturated and the dry states, observed herein do tend to be extreme.

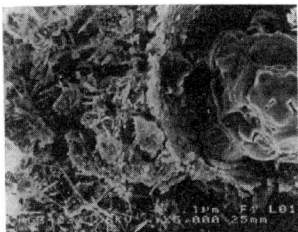  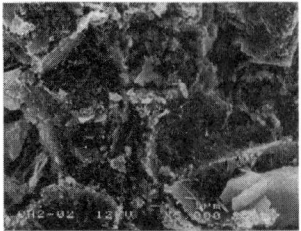

Figure 3 Fines-cement composite with wood chips treated with linseed oil (oil/wood = 0.75) G = 5000

Figure 4 Fines-cement composite with wood chips treated with lime (lime/wood = 0.25) G = 5000

Figure 5 Fines-cement composite with wood chips treated with lime, (lime/wood = 0.5) G = 5000

In the case of a process using a hydraulic binder, the thickness of the coating layer is of critical importance. The lime-based treatment turns out to be more effective.

The decrease in the dimensional variations of the composite is more significant than the decrease in the chips' water absorption. The dimensional instability is therefore not solely due to the absorption of water by the chips. This finding corroborates the work carried out by E. Mougel [13]. Figures 4 and 5 reveal that the compactness of the matrix increases as the intensity of the process becomes greater. The decrease in the microporosity of the intergranular medium would then appear to be an essential element of dimensional stabilization. In the case of oil-based treatment, the dimensional variations have not been significantly reduced. The treatment with boiling water serves to enhance dimensional stability. The majority of this decrease in microporosity is obtained after 2 hours of treatment.

Combined Treatments and Synthesis

The aim of the work conducted herein is to lower the sensitivity to water. The European program "FOREST" has set a dimensional variation limit of ≤ 1 mm / m as its target. Figure 7a presents the dimensional variations of the composites for the optimal values of the various processes. Among the simpler processes, those involving the use of coatings prove to be the most effective. The lime-based treatment is more powerful than the treatment using cement. The advantage of the oil-impregnation process, with respect to both the chips' and the composites' water absorption levels, is not necessarily apparent from the decrease in the composite's dimensional variations.

The study using scanning electronic microscopy has shown that the oil-based treatment leads to additional porosity in the matrix, whereas the coating process induced a decrease in the porosity of the composite's matrix. It therefore must be emphasized that the interaction of the processing of chips with the matrix's porosity is essential in these dimensional variations.

These various processes, when employed alone, do not allow achieving the target of 1 mm / m. A coupling of the most powerful treatments was thus studied.

Concretes Lightened by Wood Aggregate

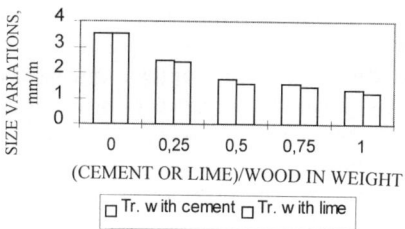

Figure 6a Treatment with hydraulic binders

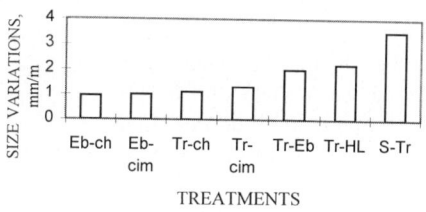

Figure 7a

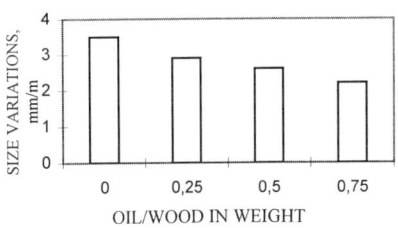

Figure 6b Treatment with oil

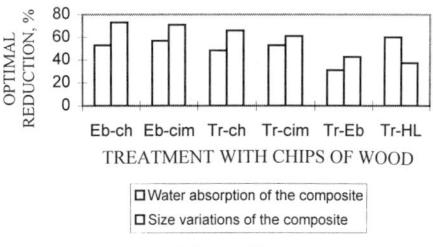

Figure 7b

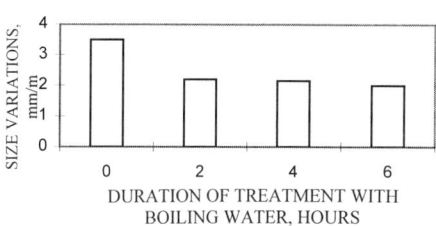

Figure 6c Treatment with boiling oil

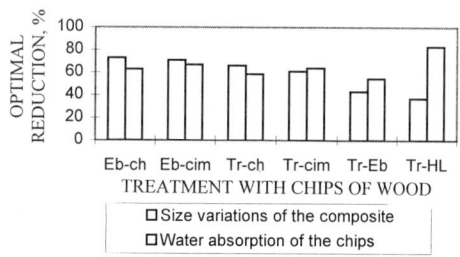

Figure 7c

Figure 6 : Influence of the various treatments of the wood chips on the dimensional variations of the cement-wood-fines composite.

Figure 7 Comparison of the effects of the various treatments

Tr-cim : treatment with cement
Tr-ch : treatment with lime
Tr-HL : treatment with linseed oil
Tr-Eb : treatment with boiling water

Eb-cim : treatment with boiling water and cement
Eb-ch : treatment with boiling water and lime
S-Tr : none treatment

The lime- or cement-based treatment was carried out on chips which had been pretreated with boiling water for 6 hours. It can be observed that the "FOREST" program's target value is in fact attained for the following combined treatment: boiling water and lime.

Figure 7b presents the decreases in the dimensional variations in comparison with the water absorption of the composite as a function of the nature of the process. Two types of behavior can be distinguished. The coating processes with hydraulic binders and/or the treatment with boiling water yield dimensional variations which are smaller than the composite's water absorption. The oil-impregnation processes give opposite results. This same phenomenon can be observed when the decreases in the dimensional variations of the composite and in the water absorption level of the chips are compared (see Figure 7c). However, we can note that in the case of a treatment with boiling water, the decrease in water absorption by the chips is higher than that observed in the dimensional variations.

This finding demonstrates that a strong decrease in the chips' water absorption coupled with a strong decrease in the composite's water absorption does not necessarily induce a similar decrease in the dimensional variations. The mechanism of dimensional variation is directly influenced by the nature of the transformations induced by the chips in the matrix following the particular process carried out. As the matrix becomes more compact and rigid, the dimensional variations decrease. A summary of these results is presented in Table 1.

CONCLUSIONS

The purpose of this study has been to lower, by means of simplified treatments, the water sensitivity of clayey concretes lightened by wood aggregates. The relationship between the water absorption of the chips and that of the composite, as well as between the extreme dimensional variations, was investigated. The target set for the extreme dimensional variations of less than or equal to 1 mm / m was reached with a combined treatment which consisted of: hydrolysis with boiling water - hydraulic lime coating. This treatment did not induce any decrease whatsoever in the material's thermal performance. This result allows us to envisage the use of local materials in insulating materials manufacture.

REFERENCES

1. PIMIENTA, P, CHANDELLIER, J, RUBAUD, M, DUTRUEL, F AND NICOLE, H. Etude de faisabilité des procédés de construction à base de béton de bois. Cahiers du CSTB, Janv-Fev 1994, No 346.

2. BROKER SIMATUPANG, Dimensionstabilisierung zementgeburdener Holwerkstoffe, Holz als Roh 32, 1974, pp 188-193.

3. DARYANTO Etude des caractéristiques physico-mécaniques du bois imprégné et du bois densifié par imprégnation suivi d'une compression plastique, Thèse 3ème Cycle, Institut National Polytechnique de Lorraine, Nancy, 1988.

4. YOUNGQUIST, KRZYSIK AND ROWELL, Dimensionnal stability of acetylated aspen flakeboard. Wood and Fiber Science,1986, Vol. 18, No 1, pp 90-98.

5. LECLERC AND de BUSSY, Procédé et dispositf pour le grillage suivi de la torrefaction de fragments de bois humides. Brevet d'invention n° 86 01969, 1987.

6. ROWELL, TILLMAN AND ZHENGTIAN, Dimensional stabilization of flakeboard by chemical modification. Wood Science Technology, 1986, No 20, pp 83-95.

7. GUEVARA AND MOSLEMI, The effect of alkylene oxides furan resin and vinylpyrrolidinone on wood dimensional stability. Wood Science Technology, 1984, No 18, pp 225-240.

8. GUEVARA AND MOSLEMI, Hygroexpansive and sportive behavior of wood modified with propylene oxide and oligomeric discocyanate. Journal of Wood Chemistry and Technology, 1983, Vol. 3, No 1, pp 95-114.

9. ROWELL, Chemical modification of wood. Forest Products Abstracts, 1983, Vol 6, No 12.

10. RAMIO, Procédé pour la fabrication de granulés végétaux minéralisés. Brevet d'invention n°1283 272, 1962.

11. BLACK, MRAZ, Inorganic surface treatments for weather resistant natural finishes. USA Forest Service Research Paper FPL 232, 1974.

12. AL RIM, K. Etude de l'influence de différents facteurs d'allégement des matériaux argileux : le béton argileux lager. Généralisation à d'autres fines de roches et applications à la conception d'éléments de construction préfabriqués. Thèse de l'Université de Rennes, 1995.

13. MOUGEL, E, BIEGALKE, C, BERTHIER, G, FRIMAT, A AND ZOULALIAN, A. Mise en œuvre de divers conditionnements du bois-ciment en présence d'eau liquide ou vapeur. Actes du 3ème colloque Science et Industrie du bois, Bordeaux, 1990. pp 14-15

THE USE OF FLY ASH IN THE COMPOUND WOOD-CEMENT

F Z Mimoune
University of Sétif
M Mimoune
University of Constantine
Algeria
M Laquerbe
INSA of Rennes
France

ABSTRACT. The use of wood in the field of construction materials is not a new practice, it has already been exploited for many years in various ways. From an economical point of view, such material creates a particular interest with the use of fly ash. In optic, a programme of tests has been made to study a compound of cement-wood containing fly ash. The insulating properties of wood, combined to fly ash as filler have lead to the possibility of making a construction material by compression cheaper than existing materials in addition to its principals characteristics (lightness; insulation; etc) which enable it to be adopted in many ways. The investigation is mainly based on the addition of fly ash to a wood-cement mixture, as well as the impact of the cement type on the physical and mechanical characteristics of the product. The results show a diminution of mechanical strengths with increased dosage but they remain acceptable regarding specific weight obtained. The same tendency has been observed with the dimensional variation of lengths. The best results are obtained with CPA55. Observations with a scanning interaction electron microscope show a high porosity of product and a poor wood-matrix liaison.

Keywords: Cement, Compound, Fly ash, Liaison, Sawdust, Strength, Wood.

Dr F Z Mimoune is a lecturer in Civil Engineering, University of Sétif Algeria. Her main research focuses on the ways of shaping materials and damping problems.

Dr M Mimoune is a lecturer in Civil Engineering, University of Constantine Algeria. He specialises in the use of fibrous materials. He was the head of department of Civil Engineering.

Professor M Laquerbe is Director of Laboratory GTMa , INSA Rennes France.

INTRODUCTION

The use of wood waste in the production of materials gives a particular interest specially that this procedure improve the environment protection; helps the development of cheaper construction materials [1] in the one hand to achieve products of weak density which are used in several fields on the other hand[2] [3].

The applications of this materials are multiple, such as fields of filling elements. They achieve a good compromise between mechanical strengths and thermal performances; and gives a solution of constructions lightening. In this article, we intend to study physical and mechanical characteristics of cement-wood compound (sawdust), and the effect of fly ash additions, cement nature and to show the state of cement wood links by using a sweeping microscope.

MATERIALS USED

Binders

To study the effect of cement nature, two types of cement were used CPJ45 and CPA55 which are used currently.

Fly ash

Ashes used are silicate-alume supplied from the thermal of Cordemais (Loire atlantique-France). Their physical properties are given in reference[4]. The interest of adding fly ash in the compound cement-wood (CCW), may give an effect of uncoumbrance or filler and minimise the poisoning of wood by cement.

The presence of ash improved also the overlay of fly ash and gives to the product a better surface state.

OPERATING USE AND TESTS

Mixes Design

Before the compression setting in place, mixtures undergo a dry and humid blending. The average pressure is 0.3 MPa. During the manipulation we have observed that the compound looks like a pulverulent material which makes it insensitive to any vibration.

It's why we adopt the setting in place by compression, which gives some cohesion to the product.

Curing Conditions

After casting in moulds 4x4x16 cm specimens are cured in a humid room (20° C and 100% H.R) during 24 hours before demoulding. After demoulding, specimens are stocked in a room at 20° C and 50% H.R.

Test Method

During curing in the regulated room, specimens undergo the measurement of ponderable and dimensional variations as well as the test resistance (bending and compression) at different ages. Visual observations with SEM show the state of liaison between wood and cement paste.

RESULTS AND DISCUSSION

Effect of Sawdust Dosage

The compounds used and studied are reported in Table 1.

Table 1 Mixes tested

	V1	v2	v3	v4
Cement/Sawdust	1.0	2.0	3.0	4.0
Density(kg/m^3)	512	790	1024	1094

Ponderable and Dimensional Variations

We observe that ponderable and dimensional variations are inversely proportional to the cement/sawdust ratio. The results show a resumption beyond the ratio of 3. The hygroscopic nature of wood is without any doubt the cause of these variations. The latter, gives rise to intense stresses which makes the internal structure of the material very weak by microcracking, Figure1.

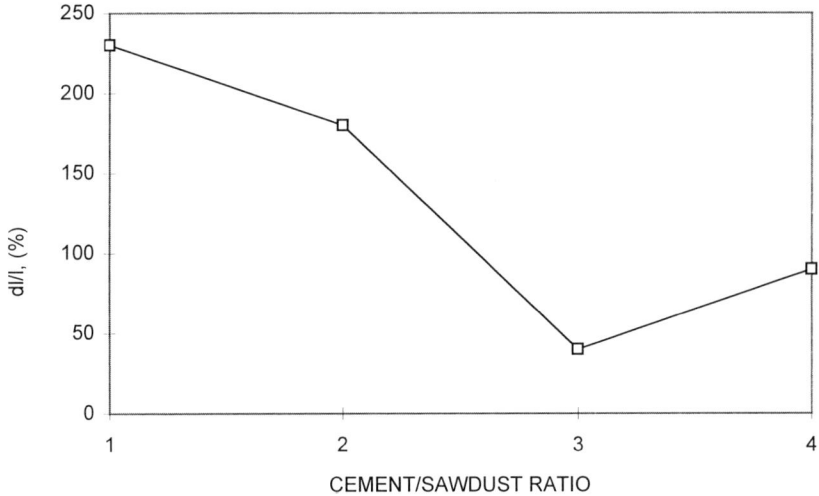

Figure 1 Length variations as function of sawdust dosage

Mechanical Strengths

The evolution of mechanical strengths is proportional to the increasing ration of cement/sawdust. The weak strengths obtained from higher dosage of sawdust are attributed to the mixes quality indeed, during cement crystallisation, we observe that the latter, cannot make the binder. Moreover than the void made by wood, we ride a high porosity which is expressed by weak weight density.

The Effect of Fly Ash Additions

Testing results don't show any important amelioration in dimensional variations with the presence of ash. The same tendency seems to be confirmed with mechanical strengths; as the filling effect has not been obtained. On the contrary, we observe falling strengths; the ratio between strengths of mixes with ash and mixes without ash may reach 1.68.

The Effect of Cement Nature

The cement nature seems to effect dimensional variations and mechanical strengths. Indeed, we observe on Figures 2 and 3 that CPA55 cement gives better results in comparison to CPJ45 cement in presence of fly ash. In comparison the gain in strength fluctuate between 21% and 43% the most important information given by these sets of tests is that indispensable to work with high cement dosage.

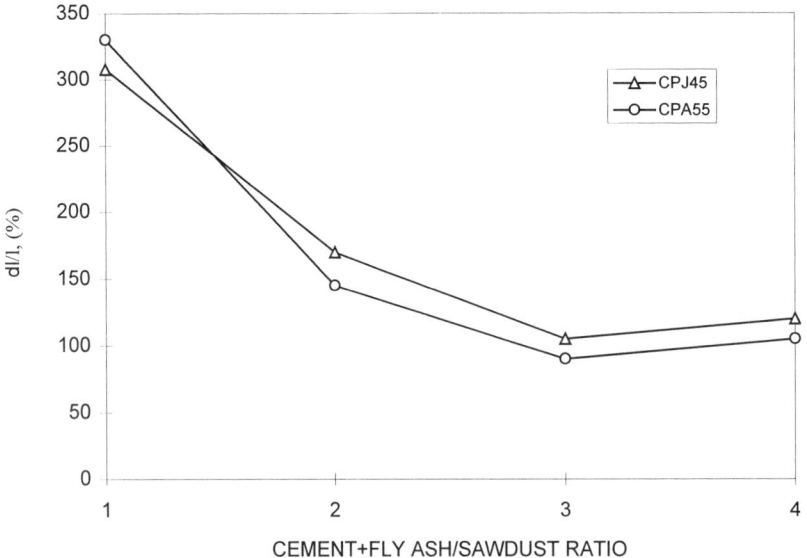

Figure 2 Effect of cement nature on length variations

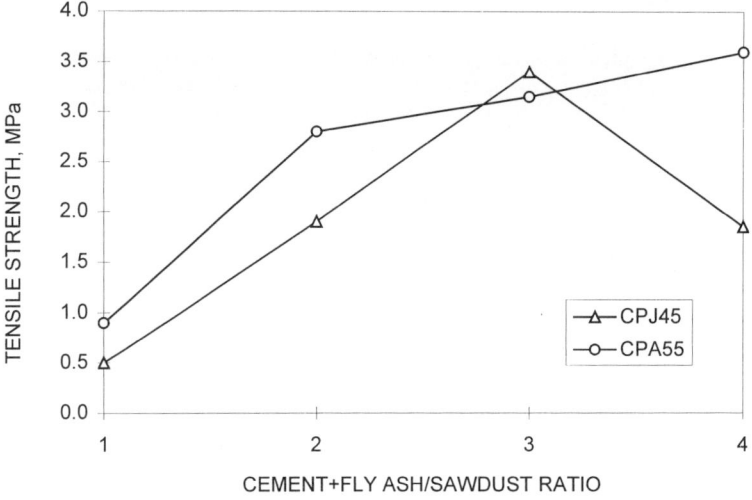

Figure 3 Effect of cement nature on the tension by flexural strength

STATE OF CEMENT-WOOD LIAISON

The liaison cement-wood may be mechanical or chemical [5]. The observations made with SEM have shown that the adhesion wood-cement is obtained by the penetration of the paste cementry in the wood channels, Figure 4 and 5. Other testing results, which we have, dare in humid conservation, show that hydrogen liaisons may be produced between wood particles and cement paste.

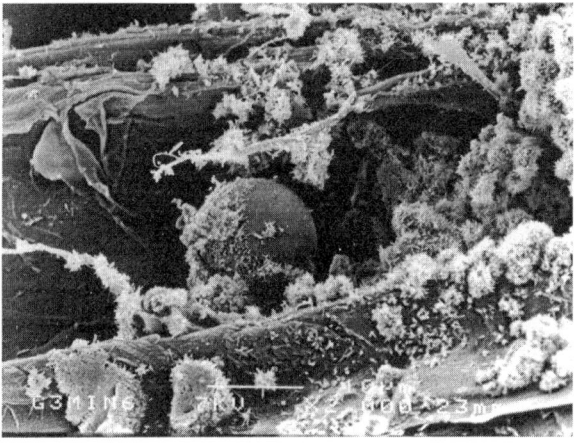

Figure 4 Penetration of cement and fly ash in the wood channels

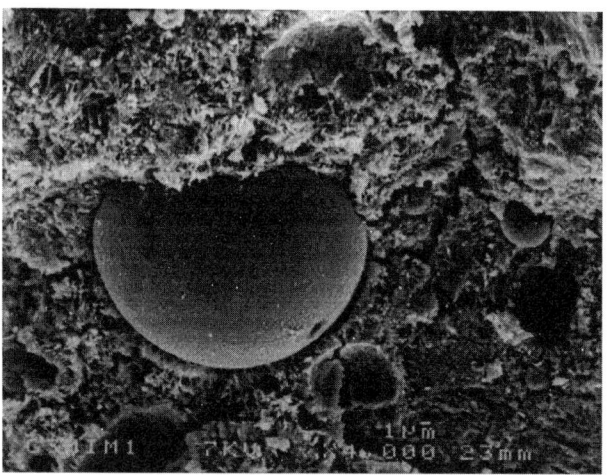

Figure 5 Cement-fly ash liaison

CONCLUSIONS

Compounds cement-wood give acceptable mechanical strengths in comparison to the weight-density obtained.

The addition of fly ashes do not give any amelioration and do not resolve elasticity problems during decompression. The best results are obtained with CPA55.

REFERENCES

1. MIMOUNE, M, AOUADJA, F Z, DOUSTENS, A and LAQUERBE, M. Etude de la faisabilité de mise en forme par extrusion du composite argile-ciment-fibres. Annales de l'ITBTP n°538, février 1995. pp.25-40.

2. MIMOUNE, M and MIMOUNE, F Z. Influence of sawdust on physical and mechanical characteristics of stabilized clay when cold. International congress proceedings. University of Dundee Scotland 24-28 June 1996, pp 645-650.

3. AOUADJA F.Z, MIMOUNE M et LAQUERBE M. Etude experimentale sur les bétons à base de résidus de bois. Revue Algérie-EQUIPEMENT n°18 mars-avril 1995. pp.24-27.

4. TEMIMI M. Utilisation des cendres volantes dans lélaboration des matériaux argileux stabilisés à froid à l'aide de différents liants et mis en forme par extrusion. Thése de Doctorat. INSA de Rennes France. 1993.

5. COUTTS R.S.P, KIGHTLY P. Bonding in wood fibre-cement composite J. of materials science 19(1984) 3355-3359.

SHRINKAGE AND CREEP OF RECYCLED PAPER WASTE CONCRETE

R P West

C A Ryan

A Thompson

Trinity College Dublin

Eire

ABSTRACT. The established properties of dry lean concrete which incorporates waste from the paper recycling industry are reviewed. The effects of such an addition on the shrinkage and creep characteristics are concentrated on in this paper. It will be shown that the increase in voids gives rise to an increase in water loss on drying and, hence, given higher elasticity, the shrinkage increases significantly with increasing waste content. In contrast, the creep strain reduces with increasing waste content because the fibrous nature of the waste contributes to restraining movement under load. The results of this pilot study will be used to plan a more comprehensive investigation using field trials.

Keywords: Creep, Recycled paper, Recycled concrete, Roller-compacted concrete, Shrinkage.

Dr Roger P West is a Senior Lecturer and Director of the Structural Laboratory at Trinity College Dublin. His research interests in concrete lie in durability, rheology and new materials. He is Secretary of the Irish Concrete Durability Committee and is the current Vice-Chairman of the Council of the Irish Concrete Society. He is also a former Chairman of the Structures and Construction Section of the Institution of Engineers of Ireland, and is a Fellow of that Institution.

Cathy A Ryan having graduated with a first class honours Civil Engineering degree in 1997, is currently a postgraduate research student at Trinity College where she is studying the confinement effects in RC members, employing wireball reinforcement, tested under cyclic loading.

Andrew W Thomson graduated from Trinity College with an honours degree in Civil Engineering in 1997 and has continued his studies at Trinity as a postgraduate student. He is researching the inelastic behaviour of steel connections using the cyclic and pseudo-dynamic testing methods.

INTRODUCTION

Recycled Paper Waste

It is becoming increasingly evident that, due to the large quantities of aggregate being used by the construction industry world wide, alternative materials for use in concrete are being sought and some have already been successfully used. For example, the relatively recent changes in environmental legislation ensure that the recyclability of concrete as an aggregate in future projects is considered at an early stage in the design process. Other work on alternative materials, such as wood pulp[1], recycled plastic[2] and recycled paper waste[3], have been reported elsewhere.

This matter is all the more relevant given the stringent accountability now being enforced in Europe in the disposal of domestic and commercial waste. In Ireland alone, only 7.4% of the 1.7 million tonnes of waste generated in 1993 was recycled[4], despite a cost of approximately £7 per tonne for dumping to landfill. In the paper recycling industry, about 80,000 tonnes of paper were recycled in that year, accounting for less than 20% of the total amount of paper waste. Of this raw material, about 90% is recyclable, composed mainly of old corrugated packaging, magazines and newspapers.

During the pulping and filtering processes, the non-recyclable materials are separated out and dumped to landfill at present. This, currently unusable, recycled paper waste (RPW) is mostly composed of a variety of soft plastics, namely polyethylene, filled polycarbonates and rubber composites [3], all of which are shredded into thin fibres of length 3 to 4 cm by the pulping process (Figure 1). As the separation process inevitably involves large quantities of water, the waste material has a high moisture content when dumped, averaging about 50% by weight where the density of the dry waste is about $400 kg/m^3$.

Established Properties of RPW Concrete

In a three year pilot study, the fundamental properties of RPW and its inclusion in concrete were established [3]. From the outset it was clear that, from the consequent strength and

Figure 1 Photograph of waste from the paper recycling mill, composed mainly of shredded plastics

impermeability losses in the RPW concrete arising from the inclusion of the waste, the most suitable application for this material is in low strength dry-lean concrete, such as that used for a road base material. The moisture content of any concrete mix was, therefore, critical to achieve maximum compaction and, hence, maximum strength from the specified compactive effort.

Mix constituents were varied to establish strength characteristics for samples with 120 (Mix A), 150 (Mix B) and 180kg/m^3 (Mix C) of Normal Portland Cement with RPW contents from 0 to 20% by volume. It was concluded [3] that, typically, with a 150 kg/m^3 cement content, a minimum cube strength of 10N/mm^2 [5] with up to 10% waste could be achieved (Figure 2).

The densities ranged from 2400 to 2100kg/m^3 for 0 to 20% waste content respectively. Further, the 10 % waste concrete was shown to have high permeability with no better than poor to fair classification from Figg air and water tests.

The stress-strain characteristics under monotonic loading revealed a highly elastic material (Figure 3) which, from an inspection of the failed specimens, would be suitable for absorbing energy where they exhibited an unusually good re-loading capacity. Further, significant elastic recovery of the "failed" cube samples was observed in high waste content samples.

Additional tests to establish the degree of compaction of a RPW concrete slab under passes of a conventional roller highlighted the "sponginess" of the material. Under cyclic loading at 7.5Hz, a considerable amount of energy absorption and creep were evident in the response of a typical cylindrical specimen extracted from the test slab (Figure 4).

In this context, it was felt appropriate, as the final part of the pilot study, to investigate the creep characteristics of the material and the not-unrelated phenomenon of shrinkage is also reported here as it was studied simultaneously.

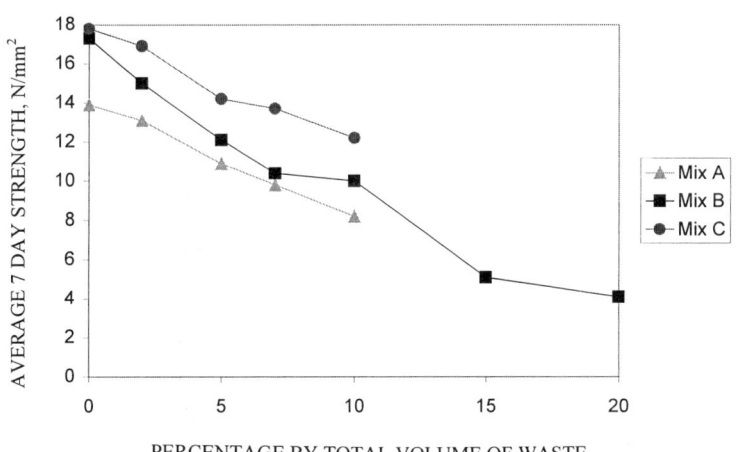

Figure 2 Average 7-day compressive strengths of cubes versus percentage of waste present for Mix A, B and C [3]

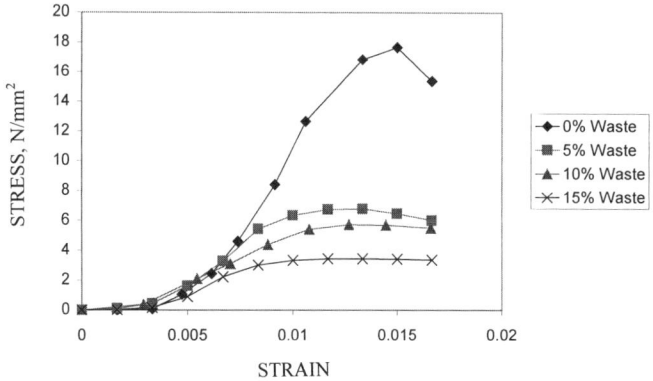

Figure 3 Typical stress-strain curves for Mix B (150kg/m^3 of cement) under monotonic loading for 0, 5, 10 and 15% waste [3]

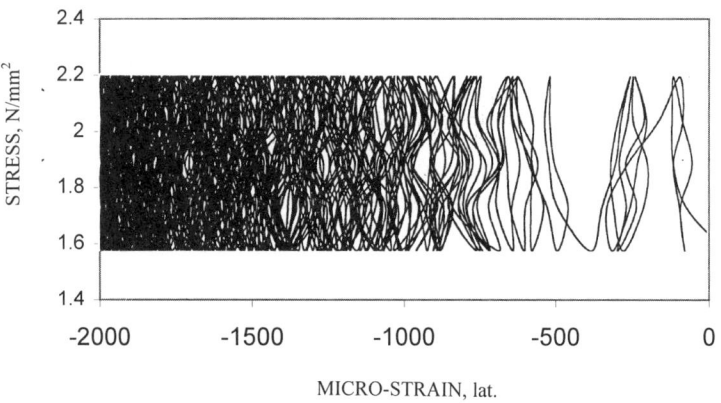

Figure 4 Typical data logged results of load versus deflection for a cored sample from a 200mm deep roller-compacted slab of recycled paper waste concrete with 10% waste and 150kg/m^3 of cement [3]

SHRINKAGE OF RPW CONCRETE

Experimental Work

The drying shrinkage of normal roller compacted concrete has been investigated elsewhere [6], but in this application it is anticipated that the fibrous nature of the waste will have a significant bearing on the behaviour.

The drying shrinkage and wetting expansion of the RPW concrete was determined by monitoring the length changes of test specimens on drying and wetting [7]. The test specimens were prisms (300mm long and 75mm square) and contained 0%, 5%, 10% and 15% paper waste by volume. The apparatus used to measure their length change consisted of a simple retort stand. A dial gauge, mounted in a vertical direction, was rigidly attached to the stand. At the base of the stand, in vertical alignment with the dial gauge tip, was a conical spigot. At the base of the prisms, in a central position, a conical recess was provided to slot into the conical spigot. A ball bearing at the top of the prism, also in a central position, fitted into the dial gauge tip. This allowed accurate centring of the prisms to be achieved, thus ensuring that the prisms were in the same position for each length measurement. An invar rod, 300mm in length, which, like the prisms, slotted over the conical spigot and into the dial gauge tip, was used as a standard of length against which the readings of the gauge could be compared.

To determine the drying shrinkage and wetting expansion of the RPW concrete, a test schedule was put into place whereby the length and weight changes of the prisms were monitored on a drying, wetting and re-drying cycle. The prisms were removed from the curing tank after 28 days and their lengths and weights measured. This first length measurement is referred to as the original 'length' of the prism. The prisms were placed in a self-desiccating oven for an initial drying period of 21 days. Their length change after this initial drying is the initial drying shrinkage. After the initial drying period, the prisms were immersed in a tank of water (at a constant temperature of 20°C) for 4 days in order to determine their wetting expansion. The wetting expansion, or swelling, of a prism is the difference between the dry measurement and the final wet measurement. Finally, the prisms were replaced in the oven for 7 days and their length and weight changes with re-drying recorded. Each length change, for drying, wetting and re-drying, was calculated as a percentage of the original length of the prism and hence the shrinkage strains were determined.

Discussion

Figure 5 illustrates the weight change of each percentage RPW concrete on initial drying, wetting and re-drying. It is evident from the figure that, for each percentage RPW, the same trends, with respect to weight change, are observed. There is rapid weight loss for the first 7 days and then a slower weight loss until there is almost none between 15 and 21 days. From 21 days to 25 days all the graphs change to a downward slope as the weights increase due to the absorption of water. During the re-drying period (from 25 days to 32 days) almost all the absorbed water evaporates from the concrete.

The drying shrinkage and swelling strains developed upon drying and wetting of the RPW concrete are shown in Figure 6. The shrinkage strains for each percentage RPW show similar trends with the level of initial drying shrinkage being directly proportional to the percentage RPW in the concrete. This is mirrored by the weight loss where the higher the percentage RPW the more weight loss there is. On immersion in water the prisms develop swelling strains which reverse the shrinkage strains. However, it is evident from the figure that not enough swelling occurs in the 4 day immersion period to completely counteract the shrinkage.

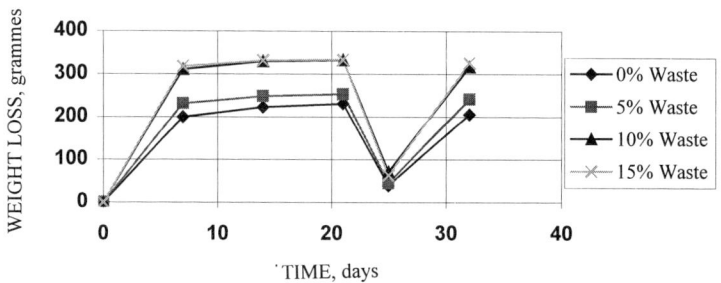

Figure 5 Average weight change of 0, 5, 10 and 15% paper waste concrete on initial drying, wetting and re-drying

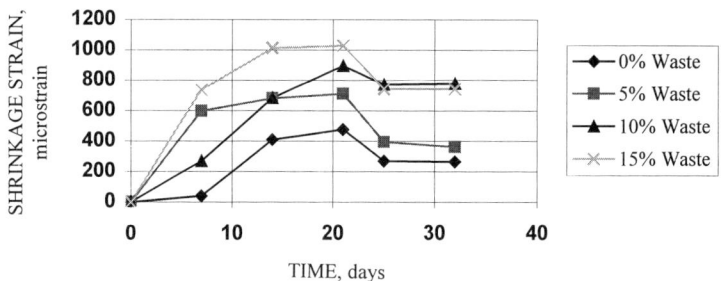

Figure 6 Average shrinkage strains of 0, 5, 10 and 15% paper waste concrete on initial drying, wetting and re-wetting

After being immersed in water for 4 days the prisms were put back in the oven and re-dried. What is unusual here is that while there is a rapid decline in weight on re-drying no corresponding shrinkage develops. Looking at the behaviour of the shrinkage strains on initial drying, which show a rapid development of strain during the first seven days, one might expect a similar trend on re-drying for 7 days, especially since there is a rapid weight loss on re-drying and weight loss and shrinkage are normally directly related. A reason for this might be that that the concrete is older and stiffer than it was when it first displayed such a rapid development in strain on initial drying. Also conventional concrete exhibits initial shrinkage which is greater than the shrinkage that occurs on wetting/re-drying cycles [6].

What can be concluded from the shrinkage test carried out is that the RPW concrete shrinks more than concrete without RPW, and that the shrinkage strains induced on initial drying are directly proportional to the percentage RPW included. The RPW fibres do provide restraint against shrinkage but may not offer enough resistance to overcome the lack of restraint against shrinkage introduced by voids in the concrete. The RPW concrete is difficult to compact and inadequate compaction leads to voids.

The concrete with the higher percentages RPW also displays more weight loss than those concretes containing lower percentages. This is again due to the high moisture content of the waste and the reduced density. Obviously the more waste included the more moisture available to evaporate off and to cause weight loss. A lot of weight loss is recoverable in each percentage RPW concrete on wetting and the shrinkage strains are diminished on wetting. However, neither the weight loss nor the shrinkage strain is entirely reversible on wetting, as expected.

CREEP OF RPW CONCRETE

In order to assess the recycled waste paper concrete as a possible road base material [5] its behaviour under constant loads was investigated. It had been seen in previous work [3, 8] that the waste paper concrete had a high elastic modulus and that creep played a large part in the behaviour of the waste paper concrete under cyclic loading, and would have to be further investigated.

Experimental Work

The specimens used for the test were standard 100mm diameter cylinders consisting of 0, 5, 10 and 15% waste content which were tested in a specially designed apparatus. In order to accurately assess the creep characteristics of the waste paper concrete, this apparatus had to be capable of applying a load evenly over the top of the cylinders and sustaining this load for long periods of time. The maximum load that the waste paper concrete would be capable of sustaining was determined using compacted cubes tested at 28 days, and a simple arrangement of three steel plates with tightening bolts was decided on (Figure 7) [9]. The apparatus was instrumented with strain gauges on the four legs and calibrated to apply a maximum load of 75 kN, as it was determined from previous work that this would be the maximum load that the concrete would be capable of sustaining. To accurately assess the creep of the waste paper concrete, two phases of testing were followed. In the first phase, six out of every ten cylinders were tested to failure to determine the ultimate load of the concrete. The other four cylinders were then placed in the testing apparatus and the load set to predetermined percentages of the ultimate failure load based on the first phase of the testing procedure. The load was applied to the cylinders, and was monitored to ensure the load remained constant throughout the test. When the load was observed to drop during the test, the original load was regained by tightening the bolts. Each test ran for approximately 72 hours, and the creep tests were carried out on a total of seventy samples. The cylinders were all tested to 20% of the ultimate failure load.

Results

The results were plotted in a graph of creep versus time (Figure 8) [9] and compared to discover the overall impact of the addition of recycled waste paper concrete to a dry lean concrete mix.

Figure 7 Creep Testing Apparatus

The cylinders containing no waste paper material were used as a control. When these cylinders were tested it was noted that a very high increase in strain occurred in the first 200 minutes after loading (Figure 8). This increase was measured at 40% of the initial strain in the cylinders. As the test continued, the rate of increase of creep in the cylinders began to slow. After 1000 minutes the creep in the control cylinders had increased by only another 17% of the initial strain.

When the cylinders containing 5% waste paper material were tested, it was noted that there was a significant decrease in the total creep of the specimen over the same time period. The initial rate of creep in the cylinders was approximately two-thirds of that of the control cylinders within the first 200 minutes with a total creep of 26% of initial strain (Figure 8) being measured. This slowly rose to 39% after 1000 minutes. It was also noted that the change in the rate of creep with respect to time was significantly slower than that of the control cylinders.

Again, it may be seen from Figure 8, that the initial rate of creep in the cylinders containing 10% waste paper material has been further reduced. It may be seen that the cylinders have a total creep of 18% in first 200 minutes which then rises to 29% after 1000 minutes. The rate of creep in the 10% waste paper concrete appears to be approaching equilibrium at a much faster rate than the control cylinders or the 5% waste paper concrete cylinders.

Finally, the concrete cylinders containing 15% waste paper material had a creep behaviour that was very similar to that of the 10% waste paper concrete. Again the majority of the creep occurs in the first 200 minutes with a total creep of 18 % of initial strain. This value is identical to the value achieved for the 10% waste paper concrete. As the test proceeded the creep began to approach equilibrium slightly faster than the other specimens, although, it still remained very close to the values measured for the 10% waste paper concrete with a creep of 27.5% of initial strain being recorded after 1000 minutes.

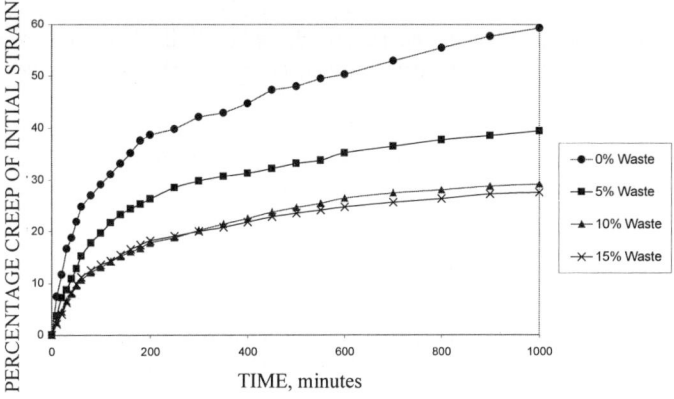

Figure 8 Graph of Creep versus Time

Discussion

It was observed from the experimental work that, as the percentage volume of waste paper material is increased, the creep in the specimens decreases significantly. However, this relationship does not appear to be linear in nature. The tests showed a 40% drop in the creep for the 5% waste paper concrete and a 60% drop in creep for the 10% waste paper concrete when compared to the control specimens. However, the creep in the 15% waste paper concrete dropped by 62% compared to the control concrete cylinders. This reduction in the rate of creep in the recycled waste paper concrete is caused by the bonding of the waste paper material with the concrete. This bonding increases the tensile strength of the concrete and restrains any movement in the concrete and, hence, reduces the overall creep of the recycled waste paper concrete. It is also anticipated that the creep in each case would reach equilibrium eventually, although this process could take some time. However, it can be seen from Figure 8 that the concrete which contains waste paper material is approaching this state faster that the concrete with no waste paper material.

CONCLUSIONS

This paper describes the work carried out as the final part of a pilot study into the fundamental properties of lean concrete incorporating fibrous waste from the paper recycling industry. It concentrates on the shrinkage and creep characteristics of the so-called recycled paper waste concrete.

The conclusions reached suggest that the amount of shrinkage increases approximately proportionately with initial drying and that, as expected, on further drying and wetting cycles it exhibits non-reversible shrinkage. The higher shrinkage is due in part to the more elastic material and to the initially higher degree of water content due to the more permeable and porous structure for higher waste contents.

The results from the large number of samples in the creep study indicate that there is a considerable reduction in the amount of creep when waste is added (as compared with the initial strain under load), although the benefit decreases with increasing waste content. Samples with higher waste content seem to stabilise faster than plain lean concrete samples.

These preliminary results from the pilot study are currently being assessed with a view to planning a more comprehensive research program, including full-scale field trials, into the characteristics and applications of recycled paper waste concrete.

ACKNOWLEDGEMENTS

The Authors wish to acknowledge with gratitude the assistance given by Smurfit Paper Mills in providing the recycled raw materials for this research. The assistance of Sarah Prichard in preparing some of the diagrams is also gratefully acknowledged.

REFERENCES

1. COUTTS, R S AND MITCHELL, A S. Wood pulp fibre-cement composite. J. Appl. Polymer. Sc., Vol. 37, 1983, pp 829-844.

2. REBEIZ, K S AND FOWLER, D W. Flexural strength of reinforced polymer concrete made with recycled plastic waste. ACI Struct. J., Vol. 93, 5, 1996, pp 524-530.

3. WEST, R P. Recycled Paper Waste Concrete. Proc. of FIP Symp. On The Concrete Way to Development, Johannesburg, 1997, pp 699-703.

4. DEPARTMENT OF ENVIRONMENT. Recycling for Ireland: a strategy for recycling domestic and commercial waste, DOE, Dublin, 1994.

5. DEPARTMENT OF TRANSPORT. Specification for roads and bridgeworks. HMSO, London, 1988.

6. PITTMANN, D W AND RAGAN, S A. Drying shrinkage of roller-compacted concrete for pavement applications. ACI Mat. J., Vol. 95, 1, 1988, pp 19-26.

7. RYAN, C A. Drying shrinkage of paper waste dry-lean concrete. BAI Disstn., Trinity College Dublin, 1997, pp 87.

8. SOROUSHIAN, P, SHAH, Z AND WAN, J P. Ageing effects on the structure and properties of recycled waste paper fibre cement composites. Mat. And Struct., Vol. 29, 1996, pp 312-317.

9. THOMSON, A W. Waste Paper Concrete and the Property of Creep. BAI Disstn., Trinity College Dublin, 1997, pp 168.

THE EFFECT OF MOISTURE CONTENT AND TEMPERATURE ON THERMAL CONDUCTIVITY OF LIGHTWEIGHT ENVIRONMENTAL CONCRETE

M L Benmalek
University of Guelma

M S Goual
University of Laghouat

A Bali
Polytechnic School of Algiers
Algeria

A Bouguerra
University of Rennes
France

M Queneudec
University of Amiens
France

ABSTRACT. Results of an experimental work on thermal conductivity evaluation of lightweight concrete using industrial waste and wood aggregates are presented in this paper. The measurements have been performed on five mixtures having the same formulation and matrices different in their mineralogical nature. Thermal conductivity of each mixture has been measured in the range 0-60°C from a fully saturated water content to an oven dry condition. The various states of water content have been obtained by progressive drying in a microwave oven while thermal conductivity measurements were made using a new transient probe developed recently by J.P.Laurent.

Generally an increase of thermal conductivity was observed with increase in temperature at every moisture ratio. At the usual temperature of use (~20°C), the schistose lightweight concrete presents the highest thermal conductivity, probably because of its raw material density.

Keywords: Industrial waste, Lightweight concrete, Thermal conductivity, Moisture content, Temperature.

Mr Mohamed Larbi Benmalek is a senior lecturer at Civil Engineering Department, University of Guelma, Algeria. His main research interest concerns the valorisation of local materials and their thermophysical characterisation.

Dr Ahmed Bouguerra is a senior lecturer at Civil Engineering Department, University of Rennes, France. He is specialised in heat and mass transfer in building materials.

Mr Mohamed Sayeh Goual is a senior lecturer at Civil Engineering Department, University of Laghouat, Algeria. His main research interest concerns building materials and structures.

Professor Abderrahim Bali is a head of research team at the Polytechnic School of Algiers, Algeria. He is specialised in building materials.

Professor Michele Queneudec is a head of research team at the University of Amiens, France. She specialises on the design of building materials and their mechanical and thermal properties.

INTRODUCTION

It is well known that moisture within building materials modifies considerably their performances, in particularly their thermal and mechanical characteristics. Hence their durability is affected owing to their porous nature. When these porous materials are unsaturated of water and exposed to varying temperatures, the mechanism which governs the heat and mass transfer becomes extremely complicated. Several researches have been carried out to master the thermohygrometric behaviour of building materials and the authors have pointed out this difficulty [1,2]. The parameters to be taken into consideration are numerous. The dry density of material, the nature and geometry of its porosity, the type of used aggregates and the mixing of constituents have been the principal cited parameters.

This paper presents some experimental results concerning the effect of the water content and temperature on the thermal conductivity of five different concretes elaborated from solid industrial wastes and lightened by wood aggregates. This study enters into the framework of a program which aims at the valorisation of local resources for promoting insulating building materials with low environmental impact. It is devoted to the hardened materials to be used principally in the perfecting of partition wall units.

EXPERIMENTAL DETAILS

Raw Materials

The raw materials used to elaborate the studied concretes are described in table 1. They have been dried at 105°C, reduced to a ground form and maintained in a dry cell before use.

Table 1 Raw materials

RAW MATERIAL	ORIGIN	DRY DENSITY (KG/M^3)	MINERALOGICAL QUALITATIVE COMPOSITION
GR	Mud resulting from sawing operations on massive granite rock.	2610	Quartz Sillimanite Orthoclase
SA	Mud resulting from the exploitation of sandstone quarry by washing aggregates obtained by crushing.	2640	Quartz Illite (very little)
SC	Powder resulting from extracted crushing aggregates by dry process.	2800	Quartz Pyrophillite Kaolinite (little)
C1	Mud resulting from the exploitation of alluvial aggregates by washing.	2600	Kaolinite (principally) Quartz (little)
C2	Mud resulting from the exploitation of alluvial aggregates by washing.	2620	Quartz (~ 50%) Kaolinite, Halloysite Illite (little)

Elaboration of The Lightweight Concretes

The lightweight concretes studied have been elaborated with the same dry-weighted mix. The proportions used were, by weight:

- Solid waste material: 50% - Cement CPA CEM I 52.5 (European standard EN 196 1): 25%
- Treated wood aggregates, 3 – 8 mm (French patent N° 8102941): 25%.

The quantity of workable mixing water is variable according to the solid waste material employed. It has been determined experimentally such that the mixture exhibits with water a normal consistency [3]. The procedure followed has been widely described in a former paper [3]. The elaborated concretes have a low dry density ranged between 700 kg/m^3 for the C2 concrete to 920 kg/m^3 for the SC concrete. The intermediate values are respectively for GR, SA and C1 concretes: 770, 730 and 740 kg/m^3. They are considered as lightweight environmental concretes in view of their densities and the nature of their matrices. Their mean compressive strengths measured on dry cubic specimens of 1 dm^3 volume are ranged from 2 MPa (C1 concrete) to 4.8 MPa (SC concrete). The other values are: 2.2 MPa (C2 concrete), 3.3 MPa (GR concrete) and 4.6 MPa (SA concrete).

Thermal Conductivity Measurements

The measurements were carried out by a transient method, the thermal shock probe method [4] developed in a recent past by JP Laurent [5]. Using this method, the water migration effects in moist lightweight concretes is reduced during the thermal tests. The schematic diagram of the figure 1 shows the experimental set up. It is composed of a thermal cylindrical probe, a power supply, a data acquisition system which enables measurements to be taken automatically at regular intervals, a computer for treatment of measures and a climatic cell.

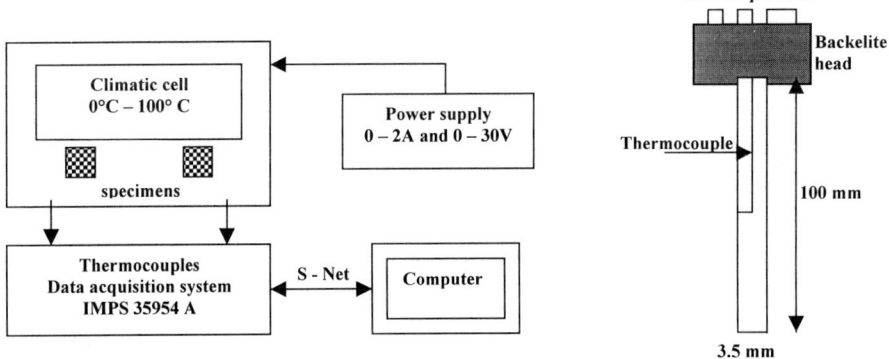

Figure 1 Schematic diagram of the experimental set up (a) Measurement instrumentation and (b) Thermal probe

The thermal conductivity was computed from the equation above using least squares approach to the data acquired between 220 and 400 seconds.

$$\Delta T = \frac{Q}{4\pi\lambda}[Ln(t) + C]$$

Where Q = Power per unit length supplied to the thermal probe (W/m), ΔT = Temperature rise measured at the time t (°C) and C = constant.

The measurements have been performed on samples of 100 mm x 100 mm x 100 mm dimensions in a range of temperature from 0 to 60°C. They were prealably drilled in their middles ten centimetres deep hole corresponding to the probe length, saturated of water until constant weight, placed into plastic bags and sealed by welding. They were then packed inside the climatic cell until temperature equilibrium has been reached.

To carry out measurements at series of water contents from saturated to dry states, the samples have been progressively dried in a microwave oven. This technique insures a homogeneous distribution of the water inside the material [6].

RESULTS AND DISCUSSION

Figure 2 gives the totality of the experimental data for the five lightweight concretes. The similar behaviour can be observed for all materials. The increase in temperature appears to reduce the thermal conductivity and vice-versa. Its increase is almost linear but a trend to become constant at 40 to 60°C above 50% of moisture (by weight) is however noticeable.

This phenomenon might be explained by the effects of evaporation-condensation pointed out by other authors [7,8] concerning several buildings materials. The total porosity of the five concretes, in the process of determination by mercury intrusion porosimetry, will permit calculation of water saturation degree and will reveal more preciseness on the phenomenon.

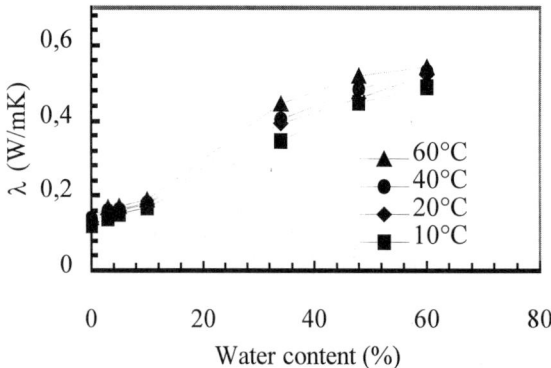

Figure 2 (a) Granite lightweight concrete

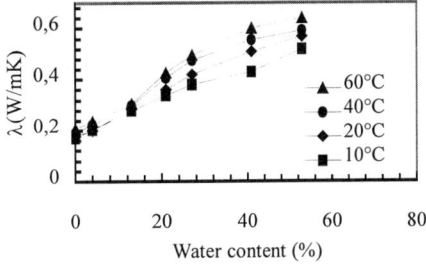
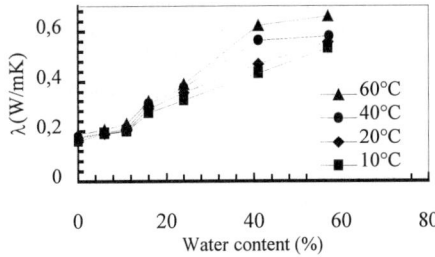

Figure 2 (b) Schist lightweight concrete

Figure 2 (c) Sandstone lightweight concrete

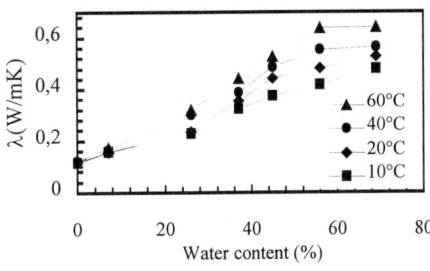
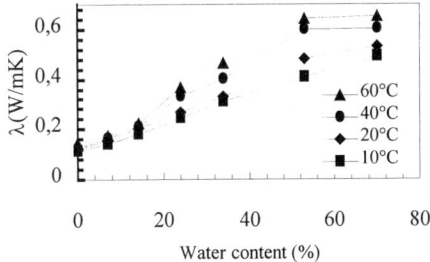

Figure 2 (d) Clay 1 lightweight concrete

Figure 2 (e) Clay 2 lightweight concrete

Figure 2   Evolution of thermal conductivity of lightweight concretes with water content at various temperatures

Figure 3 illustrates the dependence of thermal conductivity on the water content for the five concretes. It has been elaborated for the usual temperature of 20°C to appreciate the effect of the solid waste raw material.

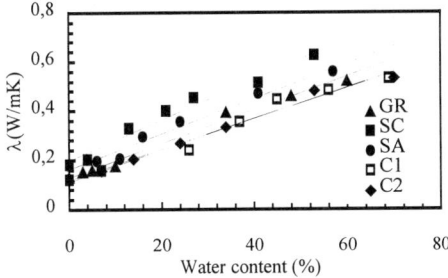

Figure 3   Dependence of thermal conductivity on the water content of the various composites

It can be seen that the thermal conductivity follows a quasi-linear evolution with the increase of water content for all the lightweight concretes. This linearity indicates the non appearance of the evaporation-condensation above phenomenon. On the other hand, the clayey concretes elaborated with both C1 and C2 clays present at this temperature the same behaviour; the lines curves describing their behaviour are practically confused. The mineralogical specific nature of the clays seems to have a little effect. However, the schistose concrete is the more heat conductor at every state of moisture contents due probably to the schist raw material density.

CONCLUSIONS

Thermal conductivity of lightweight concretes made with different solid wastes have been measured at various temperatures and moisture contents. The measurements have been performed by a transient method. The totality of experimental results has been presented. It has been shown that the increase in temperature reduces the thermal conductivity at every moisture content. However above 40°C, the phenomenon of evaporation-condensation described by many authors appears and the thermal conductivity becomes constant. On the other hand, the comparative evolution of the thermal conductivity with moisture contents of the various elaborated concretes at 20°C has shown the linearity of the evolution.

Furthermore, the relation of thermal conductivity with dry density of raw materials of the concretes has not been well established. The schistose concrete has nevertheless the high value of λ and its raw material is the densest.

REFERENCES

1. PERRIN, B, AND JAVELAS, R. Transferts couplés de chaleur et de masse dans des matériaux consolidés utilisés en génie-civil. International journal of heat and mass transfer. 30(2), 1987, pp 297-309.

2. GROUHELL, M C AND GIAT, M. Conductivité thermique apparente de la terre cuite humide non saturée. Revue générale de thermique N° 323, Nov 1988, pp 591-596.

3. BENMALEK, M L, BOUGUERRA, A, LEDHEM, A, DHEUILLY, R M AND QUENEUDEC, M. The effects of mineral fines on thermal properties of wood composites. Proceedings of CIB Congress (on CD-ROM), Gävle, Sweden, 7-12 June 1998.

4. HLADIK, J. Fil et cylindre chauds en régime instationnaire. Chap XIII de "Métrologie des propriétés thermophysiques des matériaux". Ed Masson, Paris, 1990, pp 141-160.

5. LAURENT, J P. Sonde à choc thermique monotige. Communication privée. LTHE, Grenoble, France, 1992.

6. HORTON, R, WIERENCE, J P AND NIELSON, D R. A rapid technique for obtaining uniform water content distributions in unsaturated soil column. Soil Science, 133(6), 1982, pp 397-399.

7. PHILIP, J R, AND DE VRIES. Moisture movement in porous material under temperature gradients. Transactions of the American Geophysical Union, D.A, 1957, pp 222-232.

8. AZIZI, S, MOYNE, C AND DEGIOVANNI, A. Approche expérimentale et théorique de la conductivité thermique des milieux poreux humides. I, expérimentation. Journal of heat and mass transfer, 1988, pp 2305-2317.

FINE GRAINED CONCRETE CONTAINING AGGREGATE FROM SPOIL OPEN CUTS

S I Pavlenko
Siberian State University of Industry
Russia

ABSTRACT. The properties of concrete containing both fine-grained material, from spoils from open-cut mining of coal and ash from the hydrodumps of a thermal power generating station at Ekibastus (Kasakhstan), have been studied. Technology has been developed for the production of higher density concrete for load bearing applications and for lower-density, insulating, applications. Both are appropriate for use in low-rise buildings. The burnt spoils from open cuts are heterogeneous in their mineralogical, chemical and granulometric compositions, about 30% of them being decomposed under the influence of the atmosphere. They disturb the ecological balance in the region. As shown by the investigation, they should be ground to a particle size of 0 to 5 mm so that they can be used in concrete. Three thermal power plants which are in action in Ekibastuz deliver annually to the so-called hydrodumps up to 20 million tons of ash as the coal mined by an open cast method contains up to 50% rock. 20 million tons of spoils and 20 million tons of dumped ash per year is an ecological disaster for a small town. That is why, our Academy under a contract to the Etekugol association studied the utilization of these by-products for the production of various kinds of building materials (brick, small wall blocks, various types of concrete).The present paper gives the results of the study of the use of fine-grained M 50 to 150 (5 to 15 MPa) concrete and M 35 to 75 (3.5 to 7.5 MPa) concrete for load-bearing and enclosing structures, respectively, in construction of single, two-storey cast in-situ cottages. Secondary sodium alkyl sulphate (the detergent called "Progress") was used in a lightweight concrete for enclosing structures as an air-entraining admixture. The compositions of the concretes are in accordance with the requirements of Building Regulation and State Standards for fine-grained concretes. In Ekibastuz, the construction of cottages of the above concretes has begun.

Keywords: Burnt spoils, Processing and grinding of rocks, Hydro-removed ash, Fine-grained concrete, Air-entrained fine-grained concrete, Load-bearing and enclosing structures.

S I Pavlenko is a Professor of the Department of Civil Engineering, Siberian State University of Industry, Novokuznetsk, Russia.

INTRODUCTION

In Ekibastuz, in mining by the open cast method up to 20 million tons of stripping burnt rocks are dumped polluting land, air and reservoirs. Meanwhile, the town has problems with aggregates such as gravel, crushed stone, sand for heavy concretes and claydite, aggloporite, foam polystyrene for lightweight concretes in the construction of dwellings and other objects. According to the contract made with the coal association, we decided to solve this problem by studying the possibilities of utilization of burnt mucks from open cuts of Ekibastuz and of ash from the hydrodumps of the Ekibastuzskaya thermal power plant in concretes for load-bearing and enclosing structures. As shown by the research, we have succeeded in solving the above problem.

MATERIALS

Burnt Mucks

Mineral composition

The initial materials for the burnt mucks formation are metamorphise coal-measure rocks consisting of natural blends of argillites, aleuralites and sand rocks. The rocks are inhomogenous incorporating clay states (about 48%), sand shales (about 27%), sand rocks (about 20%), shales (about 3%) and carbonate rocks (about 2%).The colour of the burnt rocks may vary from deep purple to black in which the particles of coal are contained.

Grading

To determine the grading, 100 kg of rocks were taken from the dumps by the quartering method and were sifted through the screens having 5, 10 and 40 mm meshes. The results are given in Table 1.

Table 1 Grading

PARTICLE SIZE, mm	RESIDUE ON THE SIEVE, %
0 - 5	36.3
5 - 10	13.5
10 - 40	24.2
above 40	26.0

The presence of more than 30% small grading fraction shows that a part of the rocks has been destroyed under the impact of wind, moisture, sun and frost. These are weak rocks which may be decomposed in concrete as well, if their structure is not destroyed in good time. It has been found that more than 30% of mucks are not decomposed and have stable granulometric composition. These are particles of 40 mm in size and larger which may be used as coarse aggregates for heavy concrete.

Testing of Specimens

Rocks consisting of particles of 3 sizes (0 to 10, 10 to 40 mm and larger) were tested for density, frost resistance, water demand, resistance to ferriferous and silicate decompositions. The data on density of the muck are given in Table 2.

Table 2 Density of mucks

PARTICLE SIZE, mm	DENSITY, kg/m^3	
	Bulk	True
0 - 10	1100	1870
10 - 40	1180	1920
above 40	1410	2250

Specimens of each grading fraction were tested for frost resistance in a climatic chamber by alternate freezing and thawing during 4 hours performing 15 cycles. The results are given in Table 3.

Table 3 Weight loss of specimens after freezing and thawing tests

PARTICLE SIZE, mm	WEIGHT LOSS, %	
	After 10 cycles	After 15 cycles
0 - 10	13.0	16.4
10 - 40	16.8	23.2
above 40	0	0

As can be seen from the above table, the rock having particle size above 40 mm passed 15 cycles, the grading fraction of 0 to 10 mm passed 10 cycles which is in accordance with the requirements of GOST while the grading fraction of 10 to 40 mm had the weight loss more than 15% after 10 cycles which does not meet the State Standard requirements. Thus, the latter grading fraction cannot be used for concretes subjected to freezing and thawing without the processing required.

To determine the water absorption of the rock specimens they were oven dried then submerged in water at room temperature for 48 hours, then weighed. The average water demands were 4.8, 16.2 and 7.3% with the grading fractions above 40 mm, 10 to 40 mm and up to 10 mm, respectively. For defining resistance of the rocks to ferriferous decomposition the dry weighed portion of the muck was kept in a vessel with the distilled water, for 30 days. The data are given in Table 4.

Table 4 Ferriferous decomposition of the muck

PARTICLE SIZE, mm	WEIGHT LOSS AFTER 30 DAYS IN WATER, %
5	7.5
10	1.5
20	3.5
40	0

The results indicate that the weight loss of the 5 mm grade fraction alone with its high contents of iron oxides exceeds the State Standard requirements (up to 5%). The other grading fractions are resistant to ferriferous decomposition. This also confirms the necessity (in the case of complete utilization) of grading the rocks making their grain composition uniform.

Silicate decomposition was determined by the alternate steam curing and saturation with water of the specimens. The test results are shown in Table 5.

Table 5 Silicate decomposition of the muck

PARTICLE SIZE, mm	WEIGHT LOSS AFTER STEAMING & WATER SATURATION, %
5	6.3
10	2.8
20	1.2
40	0

As seen from the Table 5, only 5 mm grading fraction had the weight loss exceeding the norm (5%). The 40 mm grading fraction is resistant to silicate decomposition.

Chemical analysis of the rock are given in Table 6.

As can be seen from the analysis, the rocks are acidic containing large amounts of silicon oxides and aluminium oxides. The contents of coaly particles in the rock is normal (up to 5% for concretes) but they are irregularly distributed in the muck blends, the greatest portion of them being concentrated in the 10 to 40 mm grading fraction.

So, from the chemical point of view, grinding the rocks and making their grain composition uniform are necessary as well.

Table 6 Chemical analysis of the muck

OXIDES	QUANTITIES, %
SiO_2	61.71
Fe_2O_3	5.58
FeO	0.43
Al_2O_3	13.72
MgO	0.41
CaO	7.29
SO_3	0.49
MnO	0.17
TiO_2	0.77
Na_2O	4.40
K_2O	5.56
SO_2	0.29
LOI	3.72

Thus, investigations into physical properties, chemical and mineral compositions of the burnt mucks from the open cuts of Ekibastuz show the following:

1. The rocks cannot be used in concretes in their natural state without previous processing as they will not ensure their durability.

2. Only 40 mm and larger grading fractions meet all the requirements of State Standard for concrete aggregates. But as they constitute 30% it is necessary to separate them from the fine mass which does not completely solve the problem of utilizing all the muck.

There are two ways of solving this problem:

1. Screening the particles of more than 40 mm in size from the mucks and use them as coarse aggregates for heavy concretes of different types (including roads). If it is necessary, crush the particles to the size required. In this case, the rest mass (70% and more) should be studied for its use in the production of brick and other calcinated materials. This is possible but would require large capital investments.

2. Making 100% grinding of the whole muck into sand with a particle size distribution of 0 to 5 mm and then utilize it in different types of fine-grained concrete both for load-bearing (classes 50 to 150, 5 to 15 MPa, respectively) and enclosing structures (classes 35 to 75, 3.5 to 7.5 MPa) using air-entraining admixtures.

The second way requires much less capital investments, makes the grain composition of the material uniform, breaks its unstable structure and above all allows one to utilize all waste products for the construction of one and two-storey buildings. We have chosen the second way for our further research.

Ash from the Hydrodumps of the Ekibastuzskaya Thermal Power Plant

Ash from the hydrodumps of the Ekibastuzskaya TPP was studied in order to improve the properties of concrete already containing sand from the dumps as an aggregate, to save cement and to utilize waste from the TPP located near the construction site of the family cottages.

Physical Characteristics of Ash and Chemical Composition of Ash from the ETPP

The moisture content of the ashes from hydrodumps of the thermal power plant is 25 to 35%, the bulk density of fly ash is 600 kg/m^3 and its true density is 2100 kg/m^3. The specific surface is 2600 cm^2/g. Ash is dark grey in colour. The chemical analysis of the hydraulically removed ash is presented in Table 7.

By its physical characteristics and chemical composition, the ash is in accordance with the State Standard requirements for use as an admixture in concrete.

Table 7 Chemical analysis of the ash from the ETPP

OXIDES	QUANTITIES, %
SiO_2	50.7
Al_2O_3	21.55
Fe_2O_3	4.76
CaO	3.13
MgO	1.62
Na_2O	3.27
K_2O	0.49
TiO_2	0.82
MnO	0.11
SO_3	2.97
P_2O_5	0.58
LOI	10.2

Admixtures

Technical grade lignosulphonate (TGL), waste product of the pulp and paper industry is a hydrophilic plasticizing admixture. Secondary sodium alkyl sulphate. The synthetic detergent called "Progress" is an air-entraining admixture. It is used in the production of lightweight concrete for enclosing structures.

DISCUSSION

The objective of the study was to completely exclude Ekibastuz natural aggregates (crushed stone, sand) for heavy concretes as well as artificial ones (claydite, aggloporite and others) for lightweight low strength concretes in the construction of dwellings and other objects. The research was based on the method of rational planning the experiment [1,2]. The variable values in the concrete components were M 400 portland cement, ground muck in the form of sand with a particle size distribution of 0 to 5 mm, ash from the hydrodumps of the ETPP and water. The amount of plasticizing admixture TGL added to the concrete mix was constant during all series of tests constituting 0.3% by cement weight in a dry state. The amount of portland cement used was in accordance with the building code 5.01.23-83 for fine-grained concrete [3].

In the process of concrete design, the main control factors include mobility of concrete according to the slump, average density of fresh concrete, bleeding, average density of hardened concrete, compressive strength of concrete immediately after heat treatment and at 28 days after it. After each test cycle, the results were analysed and the best combinations of the effecting factors were chosen. Then the second cycle of tests was carried out with little deviations from the initially chosen values with the results analysed. The tests were performed until their results corresponded to the optimum combination of all the factors.

Concrete for Load-Bearing Structures

The concrete was mixed in the laboratory counter-current mixer. The sequence of charging the materials was as follows: sand from the muck, ash from the hydrodumps, cement, water containing TGL. The mixing time was 3 to 5 min. The specimens, 10 x 10 x 10 cm in size were moulded and vibrated on the platform vibrator. 3 specimens of each mixture hardened in the natural conditions and the rest were placed into the laboratory steam-curing chamber at the temperature of 90°C using 3+10+3 hours cycle. The compositions of concretes and their primary characteristics are shown in Table 8.

The analysis of the data obtained indicates that sand with a particle size of 0 to 5 mm received from the muck of the open cuts of Ekibastuz may be successfully used as an aggregate in fine-grained classes 50 to 150 (5 to 15 MPa) concretes for the construction of low-rise dwellings, that is, it may replace natural aggregates (crushed stone and sand) extracted for this purpose.

Concrete for Enclosing Structures

In order to obtain lightweight concrete and to correspondingly reduce its heat conductivity, the air-entraining (foaming) admixture, a detergent called "Progress", the secondary sodium alkyl sulphate, was introduced into the optimum concrete mixture for load-bearing structures in the quantity of I to 3% by cement weight. The design of the optimum compositions was performed, as in the previous test series, by the method of rational experiment planning where the amounts of admixture and cement varied. As a result of the investigations, 6 optimum compositions (3 necessary classes) of air-entrained fine-grained concrete for enclosing structures including 3 compositions with the ash and 3 compositions without the ash from ETPP have been obtained. The data are summarized in Table 9.

Table 8 Compositions and characteristics of fine-grained concrete incorporating sand from the mucks for load-bearing structures

MIX	QUANTITIES, kg/m³					SLUMP cm	AVERAGE DENSITY, kg/m³	COMPRESSIVE STRENGTH, MPa		
	M 400 PC	Ash from dumps	Sand from mucks	Water	TGL, % by cement weight			A	B	C
50	200	300	1250	226	0.3	6-8	2000	3.9	4.2	4.1
50	235	275	1270	235	0.3	6-8	2030	4.3	4.8	4.2
50	250	250	1312	239	0.3	6-8	2080	4.9	5.6	4.6
50	250	-	1536	194	0.3	6-8	2180	3.9	4.5	4.9
75	250	250	1288	236	0.3	6-8	2060	5.2	5.9	5.0
75	275	225	1310	244	0.3	6-8	2090	6.0	6.8	6.2
75	300	200	1320	258	0.3	6-8	2115	7.2	7.8	7.3
75	300	-	1602	200	0.3	6-8	2130	5.8	7.1	7.4
100	300	200	1324	252	0.3	6-8	2100	7.1	9.4	9.0
100	325	175	1372	257	0.3	6-8	2140	7.5	10.2	9.5
100	350	150	1398	260	0.3	6-8	2170	8.0	11.3	10.2
100	350	-	1647	215	0.3	6-8	2230	6.8	8.9	9.3
150	350	150	1397	258	0.3	6-8	2160	12.0	15.1	14.2
150	375	125	1422	266	0.3	6-8	2190	13.8	16.0	14.7
150	400	100	1439	275	0.3	6-8	2215	14.5	16.6	15.2
150	400	-	1646	232	0.3	6-8	2290	10.1	13.8	14.6

A - After heat treatment, B - At 28 days after heat treatment, C - At 28 days, no treatment

Table 9 Compositions and characteristics of air-entrained fine-grained concrete containing sand from mucks and ash from ETPP

MIX	QUANTITIES, kg/m³				Admixtures, %		SLUMP cm	AVERAGE DENSITY, kg/m³	COMPRESSIVE STRENGTH, MPa		
	M400 PC	Sand from mucks	Ash from ETPP	Water	TGL	Progress			A	B	C
35	300	700	200	360	0.3	2.0	10-12	1400	3.1	3.7	3.3
35	350	900	-	300	0.3	3.0	10-12	1430	2.9	3.4	3.6
50	340	730	200	380	0.3	2.2	10-12	1480	4.5	5.6	4.8
50	370	940	-	320	0.3	3.0	10-12	1520	4.2	5.3	5.0
75	375	745	200	390	0.3	2.4	10-12	1525	5.9	7.8	7.5
75	400	900	-	330	0.3	3.0	10-12	1580	5.7	7.4	7.7

A - After heat treatment, B - At 28 days after heat treatment, C - At 28 days, no treatment

CONCLUSIONS

1. Burnt mucks from open cuts of Ekibastuz cannot be used in concrete in their natural state without previous treatment as they provide durability of concrete.

2. In order to achieve 100% utilization of the above rocks in concretes, they should be processed in sand with a particle size of 0 to 5 mm and used in classes 50 to 150 (5 to 15 MPa) and 35 to 75 (3.5 to 7.5 MPa) finegrained concretes for load-bearing and enclosing structures, respectively.

3. Fine-grained concretes for load-bearing structures are 200 to 400 kg/m^3 lighter than ordinary heavy concretes of the same strength.

4. The best results are obtained by introducing 100 to 250 kg/m^3 of ash from the hydrodumps of the TPP into the fine-grained concretes from the mucks depending on concrete brand. Heat-treatment of the concrete is recommended.

5. Preliminary 6 month investigations into physico-mechanical and deformation properties of the concretes showed that they corresponded to the Russian standards for fine-grained concretes.

6. To draw final conclusions, the detailed study of durability of these concretes (deformation properties, corrosion of reinforcement, frost resistance, water permeability, resistance to weather, heat conductivity and others) is being carried out.

REFERENCES

1. PROTODYAKONOV, M M, AND TEDDER, R I. Method of rational experiment planning, Nauka Publishing House, M., 1970, 70 pp.

2. VOZNESENSKY, V A. Statistical methods of experiment planning in technico-economical investigations, Izdatelstvo Statistiki, M.,1974, 192 pp.

3. BUILDING CODE 5,01.23-83, Typical norms of cement quantities in preparing concretes for precast, cast in-situ concrete, reinforced concrete products and structures, Stroyizdat, M., 1985, pp. 22-23.

WASTE CLAY BRICK - A EUROPEAN STUDY OF ITS EFFECTIVENESS AS A CEMENT REPLACEMENT MATERIAL

S Wild
University of Glamorgan
United Kingdom

A Gailius
Vilnius Gediminas Technical University
Lithuania

J Szwabowski
Silesian Technical University
Poland

H Hansen
Danish Technical Institute
Denmark

ABSTRACT. This paper comprises a review of some of the results of a three-year research project funded under the European Copernicus Research Programme. The partners in the programme were the University of Glamorgan, U.K.; the Vilnius Gediminas Technical University, Lithuania; the Silesian Technical University, Poland; and the Danish Technical Institute, Denmark. The research concerned the potential use of ground brick as a partial cement replacement material. The results reported here are confined to mortars. The waste clay bricks were obtained from each of the four partner countries and the mortar properties examined were workability, strength, porosity and pore size distribution, sorptivity, water absorption, frost resistance, sulphate and seawater resistance, and resistance to alkali silica reaction. The results obtained are very positive and suggest definite potential applications for this waste.

Keywords: Clay brick, Waste, Pozzolans, Cement replacement, Mortar, Durability, Strength.

Professor Stan Wild is Head of the Building Materials Research Unit in the School of the Built Environment at the University of Glamorgan. His research currently concentrates on the production of novel cementitious materials with enhanced durability and reduced environmental impact. He has published widely in the areas of durability and performance of cement, mortar and concrete, and of stabilised soils.

Professor Albinas Gailius is Head of the Department of Building Materials at the Vilnius Gediminas Technical University. His research interests are in building materials production technology, properties of building materials and durability.

Professor Janusz Szwabowski is Professor of Civil Engineering at the Silesian Technical University. His particular interests are in the field of concrete technology with particular emphasis on the rheological properties of fresh mortar and concrete.

Helge Hansen is Head of the Chemical, Process and Environmental Department of the Danish Masonry Centre at the Danish Technical Institute. He has wide knowledge of the properties of clay bricks with respect to their chemistry and mineralogy.

INTRODUCTION

The environmental, economic and technical advantages of using pozzolanic materials as binders to partly replace cement, particularly if these are waste materials, are well established [1,2]. In the present work the pozzolanic waste product is clay brick obtained from four different countries, Britain, Denmark, Lithuania and Poland. The output of clay brick is related approximately to population size[3]. Thus the UK and Poland have respective annual outputs of 5-7 million tonnes and 4 million tonnes, whereas Lithuania and Denmark have respective annual outputs of 0.57 million tonnes and 0.75-0.9 million tonnes.

The amount of waste produced varies, depending on the technical sophistication of the firing process and the extent of recycling of any waste. Denmark produces the smallest amount of waste (2%), and Poland and Lithuania the greatest (5%), whereas the UK produces about 3.2% waste. However the amounts vary substantially between different plants and over different time periods, and the values quoted are broad approximations. Clearly utilisation of the waste as a cement replacement, with respect to quality control and economy of scale, would not be viable for small brick making plants and only the large scale plants should be considered as potential suppliers of waste brick for this purpose.

Modern clay bricks are fired at about 1,000°C, a temperature at which the clay has completely broken down to form much more refractory minerals such as mullite, crystobalite and feldspar. In addition liquid phase, formed at the firing temperature, cools to produce glass phase and it is this that provides the pozzolanically active material.

All the ground brick materials investigated (8 brick types, two from each country) have been shown, both from chemical pozzolanicity tests and mortar bar strength tests, to be pozzolanic [3]. This paper reports on the properties which four of these brick types (denoted B, D, L, and P) impart, when ground to cement fineness, to the properties of mortar.

EXPERIMENTAL DETAILS

Materials

Details of the method of grinding the brick and also the techniques for determining the mineralogical and chemical compositions are given in [3]. The specific surface is in the range 320-350m^2/kg and the energy required to grind the material in a ball mill was 12-13kWh/tonne which compares very favourably with that to grind cement clinker to similar fineness which is 37kWh/tonne. The oxide and phase composition of the four ground brick types are given in Table 1. The glass content was determined in a semi-empirical manner from the area of the diffuse band on the X-ray diffractograms [4].

Because of the constraints imposed by the various tests employed and the different countries in which the tests were performed, the types of cement and aggregate employed varied with respect to the particular test being carried out. Some details are given in Table 2.

Table 1 Analytical data for ground brick

PHASE	BRICK TYPE (wt.%)				OXIDE	BRICK TYPE (wt.%)			
	B	D	L	P		B	D	L	P
Quartz	35	53	35	70	SiO_2	54.83	69.99	68.79	72.75
Feldsp	4	20	19	2	TiO_2	0.97	0.55	0.85	0.84
Haemat	7	5	16	4	Al_2O_3	19.05	10.62	15.23	15.89
Cristob	9	11	11	3	Fe_2O_3	6.00	4.02	6.28	4.97
Spinel	+	+	+	-	MnO	0.06	0.08	0.07	0.02
Gypsum	12	-	-	-	MgO	1.77	1.39	2.02	1.20
Anhydri	4	+	+	-	CaO	9.39	8.86	1.79	0.87
Glass	28	12	19	21	Na_2O	0.50	1.02	0.26	0.27
					K_2O	3.15	2.61	3.71	2.17
+ indicates trace					BaO	0.04	0.05	0.04	0.05
- none detected					P_2O_5	0.2	0.11	0.07	0.10
feldsp - feldspar					Cr_2O_3	0.03	0.01	0.02	0.02
haemat - haematite					SrO	0.05	0.03	0.01	0.01
cristob - cristobalite					SO_3	2.90	0.04	0.13	0.07
Anhydr - anhydrite					LOI	1.48	0.25	0.19	0.36

Experimental Procedures

The principal tests carried out on the ground brick mortars were compressive strength, porosity and pore size distribution, water absorption, sulphate resistance, frost resistance and resistance to ASR. Details of the tests are given in Table 2.

Table 2 Test details

SPECIFICATION	STRENGTH	SULPHATE RESISTANCE	RESISTANCE TO ASR	FROST RESISTANCE
aggr/bind	3:1	3:1	2.25:1	2.5:1
aggr type	Standard sand to DIN EN 196-1	Standard sand to DIN EN 196-1	Standard sandto ASTM-227 + 5% opal	Standard sand to DIN EN 196-1
cem type	OPC to BS 12	High C_3A (11.7%)	High alkali 1.25% Na_2Oeq.	CP 45 N to DIN 1164
w/b ratio	0.5	0.5	0.45	0.4 and 0.55
cem repl levels	B,D,L & P, 10%, 20% & 30%.	B,D,L & P, 10% 20% & 30%	B,D,L & P, 10% 20% & 30%	B & P, 10%, 20% & 30%
mortar mould	Cubes 100mm	Prisms 20x20x160mm	Prisms 25x25x285mm	Prisms 40x40x160
curing regime (water)	20±2°C for up to 1yr	28 dys then exposed to test solution	Sealed vessels, samples in air 100%rh, 38°C.	20±2°C 28dys freeze/thaw for 25 cycles.
test	BS1881:1983	Europ.Prestan.. ENV 196-X		PN-85/B-04500

Porosity and pore size distribution measurements were made on the specimens produced for strength measurements. Samples were taken from the centres of the crushed mortar cubes (which had been water cured at 7, 28, 90 and 365 days), dried, and subjected to mercury intrusion porosimetry. Also water absorption measurements were made, according to PN-85/B-04500, on specimens prepared for frost resistance measurements. Full details of all the testing procedures are given in [4,5].

RESULTS AND DISCUSSION

Strength and Pore Structure

In general increase in ground brick content of mortar, as a cement replacement, increases the total pore volume of the mortar (Table 3). However the rate of pore refinement in the initial stages of curing (0-28 days) is greater than it is for the control mortar without ground brick, particularly at the highest replacement level (30%). Thus by 365 days the proportion of micropores with radius < 0.05μm is approximately the same as that of the control mortar, particularly in mortars with high ground brick (GB) content. Although all ground brick types investigated showed similar behaviour the British brick type B did produce a somewhat greater rate of pore refinement than did the other three brick types.

Table 3 Cumulative pore volume (mm^3/g) and pore fineness (%) for B mortars

SPECIMEN*	PORE VOLUME mm^3/g			
	7 days	28 days	90 days	365 days
Control	63.9	61.4	49.1	46.3
B10	75.9	58.6	42.4	41.3
B20	79.4	77.5	71.0	46.7
B30	87.0	74.6	71.0	68.5
SPECIMEN*	% PORE VOLUME OCCUPIED BY PORES < 0.05μm			
	7 days	28 days	90 days	365 days
Control	56.8	62.6	69.4	68.0
B10	51.5	64.8	71.0	70.2
B20	47.4	60.5	63.6	66.6
B30	44.3	58.8	69.9	69.9

* 10, 20 and 30 denote the % replacement of cement with ground brick

Increasing substitution of cement with ground brick in mortar results, at early ages (up to 28 days), in progressively decreasing compressive strengths relative to the control (Table 4). However, between 0 and 90 days, relative strengths (i.e. strength of ground brick mortar relative to control mortar without ground brick) increase substantially with increase in curing time. Thus beyond 90 days the strengths of some ground brick mortars, particularly at the lower replacement levels, exceed those of the control. This is particularly the case for mortars containing brick type B.

Table 4 Compressive strengths for control and GB mortars with w/b ratio 0.5

SPECIMEN	COMPRESSIVE STRENGTH (N/mm^2)			
	7 days	28 days	90 days	365 days
control	52.2	65.3	67.0	78.8
B10	43.8	61.2	74.1	87.3
B20	39.4	55.4	66.2	82.7
B30	31.3	45.2	55.7	69.2
D10	40.3	55.0	65.2	78.6
D20	30.2	45.0	55.0	66.0
D30	29.0	38.6	49.1	59.8
L10	44.7	59.4	69.1	81.3
L20	33.5	48.5	61.5	72.5
L30	27.8	40.5	55.1	65.2
P10	39.8	55.6	65.5	78.9
P20	34.5	50.1	62.7	74.2
P30	28.3	39.8	57.5	67.8

This increase in relative strength corresponds with increasing pore refinement and is attributed to the additional C-S-H gel produced from the pozzolanic reaction between the ground brick and the CH from the hydrating cement. In fact, from the strength and porosity data, clear empirical relationships are established between compressive strength and pore fineness and between compressive strength and total pore volume (see Figure 1). These demonstrate that as pore fineness increases, strength increases, and as pore volume decreases strength increases. The key factor, which influences the amount of cementitious gel formed from pozzolanic activity, and hence the strength contribution from this process, is the amount of glass phase present in the ground brick. This increases in the order D<L< P<B (see Table1). Inspection of the strength values of mortar in Table 4, shows that after 365 days curing (when the influence of pozzolanic activity on strength will be close to completion), the strengths attained are in the order D<L<P<B. This provides convincing evidence that the principal factor, which determines the effectiveness of GB as a pozzolan, is the glass content of theGB.

Sulphate and Seawater Resistance

The partial replacement of cement by ground brick in mortar (0 to 30%) greatly influences the resistance of that mortar to sodium sulphate solution and to synthetic seawater solution. In most cases the resistance is increased but in some cases it is reduced. In particular brick type D has a wholly detrimental effect on the resistance of mortar to both sodium sulphate solution (e.g. Figure 2b) and seawater. It accelerates expansion and failure, and the greater the level of replacement the more rapidly that failure occurs. Brick types B (e.g. Figure 2a), L and P however retard expansion significantly. The amount of expansion produced systematically decreases, as the amount of replacement of cement by ground brick in the mortar increases.

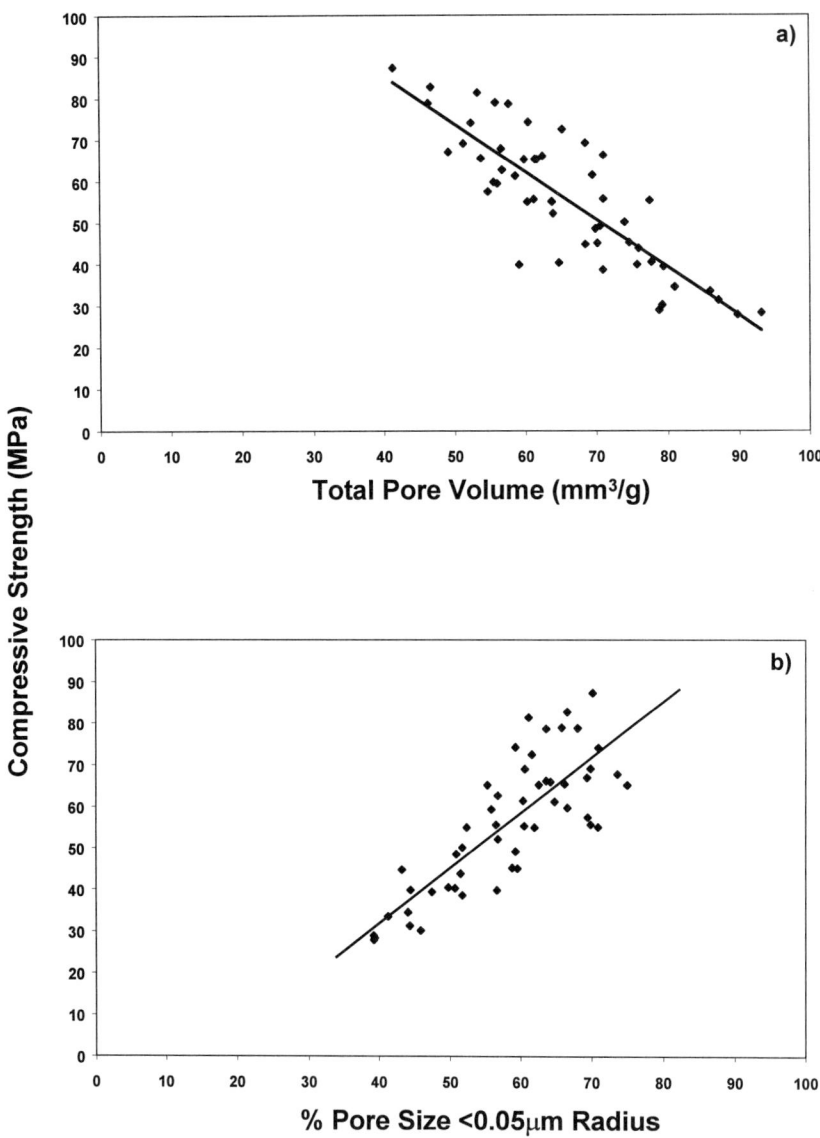

Figure 1 Compressive strength versus a) total pore volume and b) percentage porosity with pore size <0.05μm, for ground brick mortar type B, D, L and P at 10%, 20% and 30% replacement levels

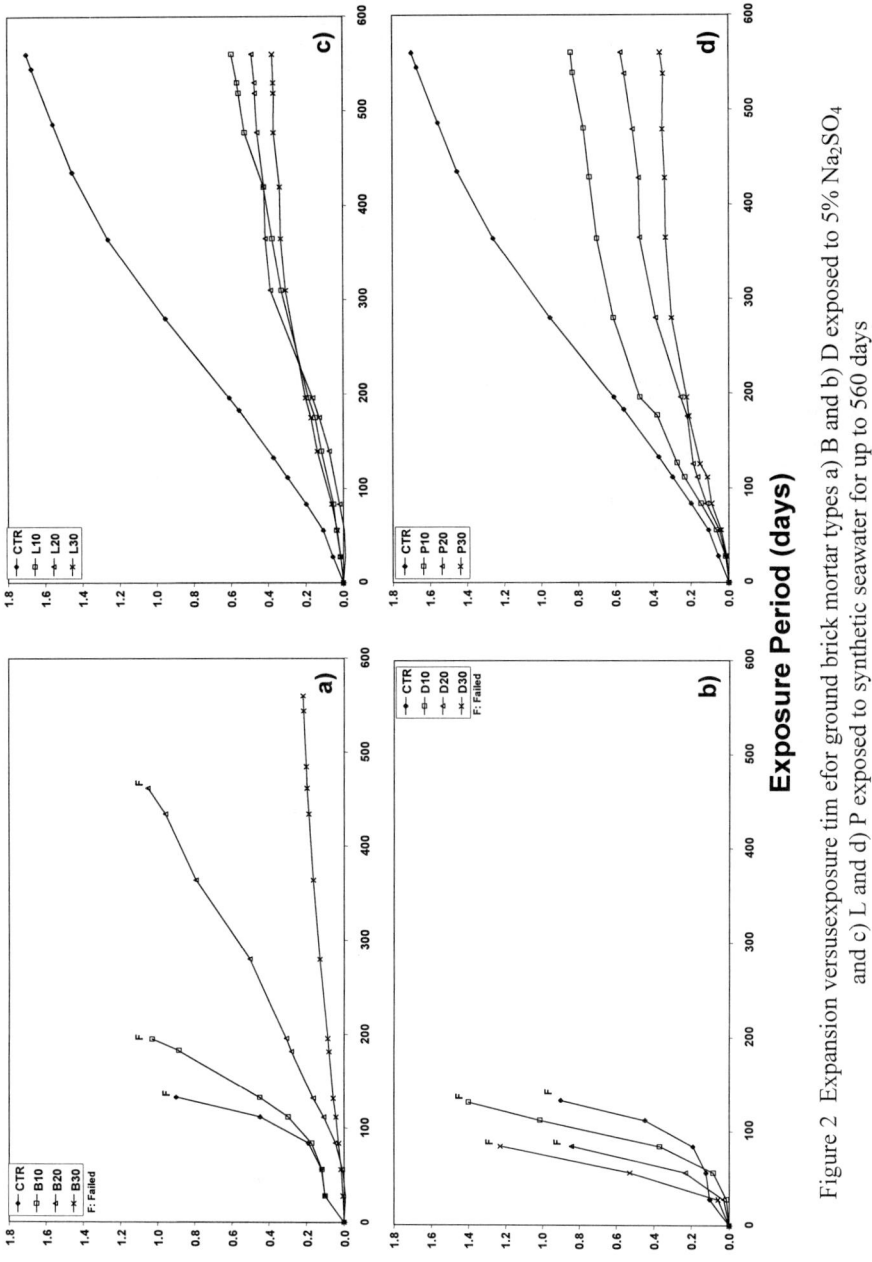

Figure 2 Expansion versus exposure time e for ground brick mortar types a) B and b) D exposed to 5% Na_2SO_4 and c) L and d) P exposed to synthetic seawater for up to 560 days

The effect is particularly pronounced for mortar exposed to synthetic seawater (e.g. Figure 2c-d). For mortar exposed to sodium sulphate solution there is a direct linear relationship between expansion and weight increase during exposure (Figure 3). This indirectly confirms that the component producing the weight increase, which is produced during exposure, is the product responsible for the expansion. Exposure to seawater produces a substantial drop in mortar compressive strength relative to that for mortar exposed to pure water, although this strength loss is progressively reduced by replacement of cement with increasing amounts of ground brick (see Table 5). This again emphasises the advantage of using ground brick to improve durability.

Table 5 Percentage loss in strength* of 28 day cured mortar prisms exposed to synthetic seawater for 560 days

BRICK	B10	B20	B30	L10	L20	L30	P10	P20	P30	CTR
STRENGTH LOSS	65%	44%	34%	63%	57%	39%	66%	51%	40%	76%

*expressed as a percentage of strength of water cured prisms at same age

The increased resistance to sulphate attack, of mortar in which the cement is partially replaced with ground brick (B, L and P), is attributable in part to the dilution effect (i.e. less cement and hence less C_3A), and in part to the pozzolanic reaction between the ground brick and CH. The latter removes CH from the system, which is then no longer available to react to form expansive calcium sulpho-aluminate products, with the available sulphate and alumina. In addition the increased pore volume produced in the mortar by the ground brick will provide additional space into which the products can expand. This is evident in Figure 4, in that specimens do not begin to expand until their weight has increased by about 1%. The formation of pozzolanic reaction products also results in pore refinement and reduced permeability but the results on mortar sorptivity and water absorption [6] suggest that this is not a significant factor in the improvement in sulphate resistance.

The very marked reduction in sulphate resistance of mortar containing brick type D, is attributable to the chemical and mineral phase composition of the brick. Comparison of the chemical compositions of the four ground brick types, (Table 1) shows that the greatest difference is in CaO content. Bricks L and P have very low CaO contents and B and D very high CaO contents. It is apparent, from the chemical and mineralogical analysis, that for brick B part of the CaO is fixed as gypsum and anhydrite. This is not the case in D. Some CaO will go into solution in crystalline plagioclase feldspar, and some will be incorporated in the glassy vitreous alumino-silicate component of the brick. The glass content of D is however less than half that of B, thus the CaO concentration in the glass phase in D would be expected to be very much greater than in B. It is suggested that the pozzolanic reaction with the high CaO content glass releases calcium aluminates, which are readily available to form expansive calcium sulpho-aluminates in the presence of sulphates. A similar effect has been observed in high calcium fly ashes [7].

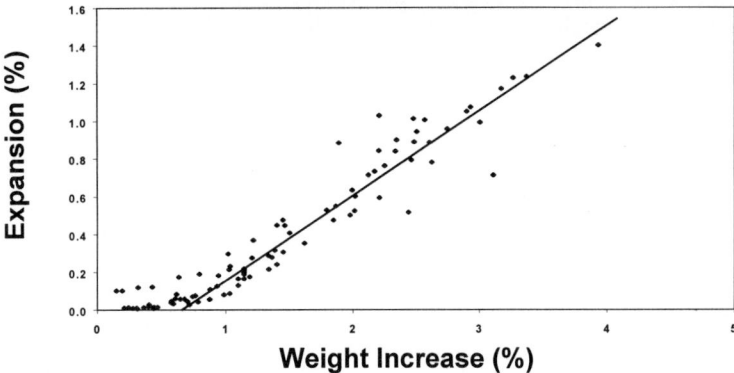

Figure 3 Expansion versus weight increases for ground brick mortar types B, D, L and P at 10, 20 and 30% replacement levels exposed to 5% Na_2SO_4 solution for up to 560 days

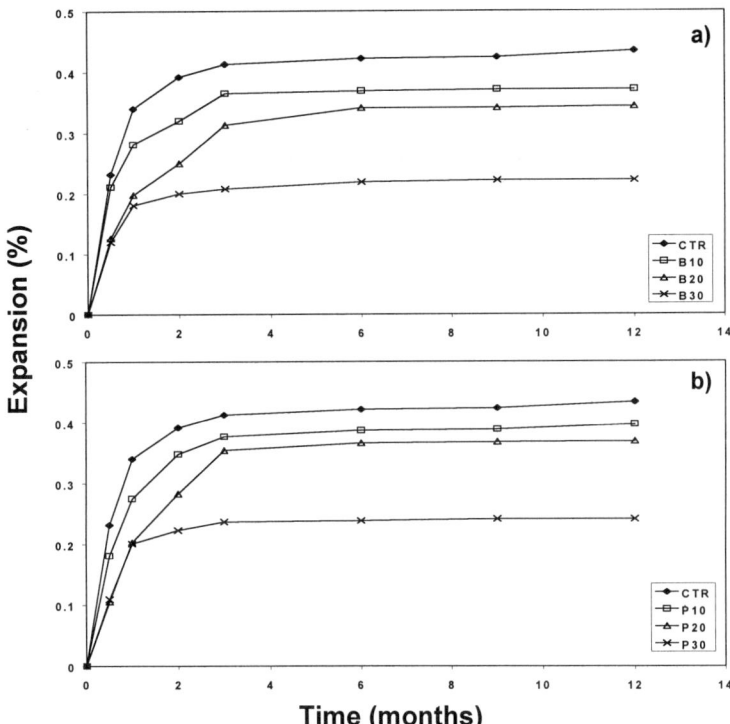

Figure 4 Expansion versus age for ground brick mortar type a) B and b) P containing 5% opaline agggregate and high alkali cement (1.25% eq Na_2O)

Not only will the CaO content of the glass for brick B be much less than in D, but the high gypsum (and anhydrite) content will enable the normally damaging expansive reactions to take place at a very early stage (i.e. in the period during and immediately after mixing when the mortar is in a highly plastic state). Thus the small levels of SO_3 in the brick may be advantageous in both accelerating strength development and enhancing sulphate resistance.

Resistance to ASR

Mortar made with high alkali cement (1.25% Na_2O equivalent) and aggregate containing reactive silica (5% local Lithuanian opal sand), was found in general to show progressively reduced expansion, as the cement was partially replaced with increasing amounts of ground brick (e.g. Figure 4). The type of ground brick had only a marginal effect on the amount of expansion produced. There are a number of factors, which determine the effect that the ground brick has on the amount of expansion produced. These include dilution of total alkali as the cement content is reduced, absorption of alkali by the additional gel produced by the pozzolanic reaction, and increase in available pore space which can be occupied by any expanding product. It is suggested, from the evidence available, that in this case dilution and increased pore volume make the greatest contributions to reduced expansion.

Frost Resistance

The frost resistance was determined on 28 day cured mortar bars, for two different w/b ratios (0.4 and 0.55). The cement had been partially replaced by brick types B and P. Susceptibility to frost attack was indicated by the magnitude of the fall in compressive strength, after 25 freeze thaw cycles, relative to equivalent mortar which had been cured for the same period in water at 20°C (Table 6).

Table 6 Percent loss in compressive strength after 25 freeze thaw cycles, and mortar water absorption values for 28 day cured GB mortars with ground bricks B and P

Specimen	w/b RATIO 0.4		w/b RATIO 0.55	
	% Strength Loss	% Water Absorption	% Strength Loss	% Water Absorption
Control	1.9	6.3	-	7.4
B10	2.6	6.4	5.1	8.4
B20	9.9	7.4	14.0	9.6
B30	5.7	7.7	21.5	10.1
P10	4.1	6.8	4.0	8.6
P20	-1.0	7.1	0.7	8.8
P30	0.9	7.2	13.9	9.6

At the low w/b ratio the fall in compressive strength was in most cases insignificant particularly for the P mortars. However at the high w/b ratio the reduction in compressive strength, although very small at low cement replacement levels, was substantial at high cement replacement levels where strength reductions of 20% were observed (i.e. 30% replacement of cement with B). This behaviour is closely related to the water absorption of the mortar, which is significantly higher at the higher w/b ratio particularly at high cement replacement levels. Therefore if ground brick is to be utilised as a partial cement replacement in mortar, in an environment in which the mortar is subject to freezing and thawing, mixes should have a low w/b ratio and large replacement levels should not be used. If this is not possible then additional precautions such as the use of air entraining agents should be considered.

CONCLUSIONS

The project has established that it is technically feasible to partially replace cement with ground brick (GB) in mortar, to produce a more durable, and less environmentally damaging material than that produced without cement replacement. However there are a number of limitations to the use of ground brick as a cement replacement, depending on the particular material properties required. The following set of guidelines are designed to take account of these limitations.

1. **Brick composition**. Bricks which have a high calcium content (CaO>5%), particularly if they also have a low glass content (<20%) and a low sulphate content (SO_3<1%) should not be used as pozzolans because they impair chemical durability. The damaging component is probably high calcium glass. Thus if there is any doubt about the viability of the material detailed analysis of the glass phase should be undertaken along with long term durability testing.

2. **Strength**. It should be noted that partial replacement of cement with GB results in a significant loss of strength in the early stages of curing (up to 90 days), particularly when cement replacement levels are high (30%). Therefore this loss in strength should be taken into account in the mix design. At long ages (>90 days) strengths of GB mortar may approach or even exceed those of standard mortar, depending on the glass content of the brick and the replacement levels employed.

3. **Frost resistance**. If GB at high cement replacement levels (30% plus) is being incorporated in mortars with high w/b ratios (>0.5), the frost resistance of the mortar will be significantly reduced. Thus under these conditions it would be necessary to take additional measures, such as the inclusion of air entraining agents, in order to ensure adequate protection against frost.

4. **Sulphate resistance and resistance to seawater**. As long as the brick composition criteria cited above are adhered to, partial replacement of cement by GB in mortar can be used to provide enhanced resistance to sulphate and seawater attack. To obtain substantial resistance at least 30% replacement is recommended.

5. **Resistance to ASR**. The replacement of cement with GB in mortar reduces the expansion produced by ASR. Substantial reduction in expansion is achieved only at high replacement levels (30%). It is therefore recommended that if GB is to be employed to reduce the effects of ASR replacement, levels of 30% and above should be considered.

In addition to the advantages of improved durability, widespread use of GB in mortar would reduce cement consumption, materials costs, and environmental damage associated with cement production. If the brick is a waste product then as the energy of grinding, to produce ground brick of equivalent fineness to cement, is only one third that of cement, the energy costs per unit weight of binder should be reduced. However the process should be considered only for large-scale brick producers, as collecting small amounts of waste from diverse locations would be inefficient.

ACKNOWLEDGEMENTS

The authors are particularly grateful to M.O'Farrell who did the work on strength and sulphate/seawater resistance, J.Golaszewski who did the work on frost resistance, I.Gerniene who did the work on ASR and L.Pederson who did most of the analytical work on the brick. They should receive due credit for their major contribution to this project. In addition the authors wish to thank the European Commission for funding the project, without which the work would not have been possible.

REFERENCES

1. MEHTA, P K. Role of pozzolanic and cementitious material in sustainable development of the concrete industry. Proc.6th CANMET/ACI Int. Conf. on Fly Ash, Silica Fume, Slag and Natural Pozzolans in Concrete, Bangkok, 1998, Ed. V M Malhotra, Vol 1, pp 1-20.

2. WILD, S, GAILIUS, A, HANSEN, H, PEDERSON, L AND SZWABOWSKI, J. Pozzolanic properties of a variety of European clay bricks. Building Research and Information, Vol 25, No.3, 1997, pp170-175.

3. WILD, S, TAYLOR, J, SZWABOWSKI, J, GAILIUS, A AND HANSEN H. Recycling of waste clay brick and tile material for the partial replacement of cement in concrete. Copernicus Research Project Contract No. CIPA-CT94-0211, First Annual Report Feb. 1st 1995 - Jan. 31st 1996.

4. WILD, S, TAYLOR, J, SZWABOWSKI, J, GAILIUS, A AND HANSEN, H. Recycling of waste clay brick and tile material for the partial replacement of cement in concrete. Copernicus Research Project Contract No. CIPA-CT94-0211, Second Annual Report Feb. 1st 1996 - Jan. 31st 1997.

5. WILD, S, TAYLOR, J, SZWABOWSKI, J, GAILIUS, A. AND HANSEN, H. Recycling of waste clay brick and tile material for the partial replacement of cement in concrete. Copernicus Research Project Contract No. CIPA-CT94-0211, Third Annual and Final Report Feb. 1st 1997 - Jan. 31st 1998.

6. SABIR, B,B, WILD, S AND O'FARRELL, M. A water sorptivity test for mortar and concrete. Materials and Structures, Vol 31, 1998, pp1-7.

7. TIKALSKY, P,J AND CARRASQUILLO, R,L. Fly ash evaluation and selection for use in sulphate-resistant concrete. ACI Mat. J., Vol 90, part 6, 1993,. pp 545-551.

ALKALI SILICA RESISTANCE GROUND BRICK MORTARS

A Gailius
Vilnius Gediminas Technical University
I Girniene
Building Construction College
Lithuania

ABSTRACT. The research is directed towards the utilisation of waste product as a partial cement replacement material in mortar and concrete. The waste material being investigated is from clay brick and tile manufactures although subsequently it can be extended to clay bricks and tiles from old buildings.

This paper, which forms part of a major international research programme investigates the resistance to alkali silicia reactions (ASR) of mortars, in which cement is partially replaced by a variety of European ground brick. For ASR test the mixes were designed with high-alkali cement and 10, 20 and 30 % cement replacement levels of British (B), Danish (D), Lithuania (L) and Polish (P) ground brick. A natural sand to which was added 5% local Lithuanian opal sand. Mortar bars (25x25x285 mm) produced from the mixes were stored at 38°C and 100% relative humidity and their lengths monitored with time. Currently are available up to 365 days. These are encouraging in the increasing replacement levels of all four ground brick types results in decreasing levels of expansion.

Keywords: Alkali-silica reaction, ASTM C277 test, Brickdust, Cement, Concrete, Expansion, Mortar, Replacement.

Professor Albinas Gailius is working at Department of Building Materials, Vilnius Gediminas Technical University, Lithuania. He specialises in the investigations dealing with concrete technology, especially on the optimization of the Material Structure, durability of concrete. A.Gailius author over 100 scientific publications, participant of over 60 scientific conferences.

MSc Ingrida Girniene is a Lecturer in Building Materials, Building Construction College, Vilnius, Lithuania. Research interests: Optimization of the Material Structure, composition and durability of cement based Composites.

INTRODUCTION

Expansive alkali - silica reactions (ASR) are a frequent and world-wide cause of poor durability in Portland cement concrete. Durability failure of Portland cement concrete may result from the expansion and cracking caused by reactions between some types of aggregates (opal, glassy, volcanic rocks, and other natural, manufactured, or waste materials) and strong alkalis derived mainly from the cement. Expansion and cracking may develop within a few weeks or months or may not appear for a number of years after concrete has been prepared. The time for the expansion to become excessive depends on the types of reactive aggregate, concentration of alkalis in the pore solutions availability of water, admixtures, and curing conditions [1, 3].

In this test, the retardation of expansion due to alkali silica reaction (by partial cement replacement with ground brick) is to be investigated. Mortar bars each containing the pessimum amount of reactive aggregate and various replacement levels of the cement by ground brick pozzolan have been exposed to the standard conditions recommended for ASR expansion (ASTM C227) and the expansion monitored. The length measurement and visual observations of the specimens were monitored after demoulding, after 14 days, and 28 days exposure, and subsequently at monthly (28 day) intervals up to one year. If necessary monitoring will continue at least every 6 months thereafter. Currently monitoring has been carried out for 365 days.

EXPERIMENTAL DETAILS

Materials

The materials employed in the current work are listed below.

Cement:	High alkali cement, 1.25% Na_2O equivalent.
Sand:	Sand graded to ASTM C227 recommendations.
Reactive Aggregate:	Local, Lithuania opal.
Water:	Distilled.
Admixtures:	Ground brick types.

Mix Proportions

The mix proportions are summarised in Table 1.

Procedure

The mix proportions used for the mortar were 1:2.25:0.45 (binder : sand : water), the binder consisting of high alkali cement with ground brick replacement levels of 0, 10, 20 and 30% by weight. The sand used was standard sand with 5% replacement with Lithuanian opal as reactive aggregate. The mix proportions are given in Table 1.

Four mortar bars (25 x 25 x 285mm), with stainless steel inserts at each end for length monitoring, were prepared from each mix. Moulds containing the bars were placed in a moist room for 24 hours before demoulding and their lengths were then measured (initial length) to the nearest 0.002 mm. The bars were then stored in sealed containers in 100% relative humidity at 38^0C. Before length measurement the bars were allowed to cool down to ambient temperature in their container for 24 hours. Length measurements were made at 14 days, 28 days and then every month up to three months, and subsequently at least every 6 months.

Table 1 Mix proportions of the mortar mixes

MATERIAL	CONTROL SPECIMEN	10% REPLACEMENT	20% REPLACEMENT	30% REPLACEMENT
Cement	300g	270g	240g	210g
Sand	641.25g	641.25g	641.25g	641.25g
Reactive Aggregate	33.75g	33.75g	33.75g	33.75g
B/dust	-	30g	60g	90g
Water	135g	135g	135g	135g

The length change is expressed as the percentage of the initial length and the value reported is the average of four separate measurements. At each monitoring stage the bars were examined for surface appearance, crack formation, surface deposits and exudations and the nature of these noted.

Table 1 gives values are based on a mix ratio of cement:sand:water of 1 : 2.25 : 0.45 and making two bars per batch (2 batches per mix). The pessimum level of reactive aggregate to be used being 5%.

RESULTS AND DISCUSSION

Table 2 and Table 3 gives the percentage expansions for monitoring periods of up to 365 days and Figures 1 to 4 plot the changes in expansion with time for the mortars containing the four different ground brick types: (a) - British (B), (b) - Danish (D), (c) - Lithuanian (L) and (d) - Polish (P) ground brick with 10, 20 and 30 % cement replacement levels.

There is a clear reduction in expansion with increasing ground brick content relative to the control. This can of course be attributed partly to the diluting effect of the ground brick on the total alkali content of the binder. However, the ground brick types also contribute alkali, and the amounts (equivalent Na_2O : B 0.71, D 0.48, L 1.04, and P 0.46) have previously been determined as part of pozzolanicity tests [4; 5]. Malhotra and Kumar [6] claim that there are three principal mechanisms whereby mineral admixture reduce expansion and control alkali-silica reaction.

Table 2 Expansion of mortar containing brickdust

TIME, DAYS	CEMENT+SAND +OPAL	LENGTH CHANGE %					
		B-10	B-20	B-30	D-10	D-20	D-30
14	0.232	0.211	0.126	0.120	0.191	0.119	0.106
28	0.340	0.281	0.198	0.181	0.237	0.205	0.183
56	0.392	0.320	0.250	0.200	0.277	0.308	0.207
84	0.413	0.365	0.313	0.208	0.303	0.378	0.228
168	0.422	0.369	0.341	0.219	0.316	0.379	0.234
252	0.424	0.371	0.341	0.222	0.319	0.382	0.238
365	0.434	0.371	0.343	0.222	0.321	0.381	0.241

Table 3 Expansion of mortar containing brickdust

TIME, DAYS	CEMENT+SAND +OPAL	LENGTH CHANGE %					
		L-10	L-20	L-30	P-10	P-20	P-30
14	0.232	0.182	0.127	0.115	0.181	0.105	0.109
28	0.340	0.284	0.239	0.180	0.275	0.203	0.201
56	0.392	0.358	0.356	0.210	0.348	0.283	0.223
84	0.413	0.380	0.385	0.236	0.377	0.354	0.237
168	0.422	0.383	0.388	0.244	0.387	0.366	0.239
252	0.424	0.388	0.389	0.246	0.389	0.368	0.242
365	0.434	0.391	0.397	0.249	0.397	0.369	0.242

These are:

- alkali dilution caused by the partial replacement of OPC by the mineral admixture,
- pozzolanic reaction producing a gel with a lower C/S ratio which can absorb greater amounts of Na_2O and K_2O,
- a less permeable microstructure which results in lower water absorption.

Clearly there will be a dilution effect in the current work although the ground bricks also release sodium and potassium alkali in an alkaline environment which will counteract the dilution effect. Evidence of pozzolanic activity has been clearly established from the pozzolanicity tests, strenght development tests, and measurement of changes in pore size distribution [4; 5]. The sorptivity tests suggest that depending on the curing period, permeability will not be reduced significantly by replacement of cement with ground brick and at short curing periods will almost certainly be increased. Therefore the reduction in expansion of the mortar with increasing replacement levels of cement by ground brick is attributed to a combination of the effects of dilution and pozzolanic activity.

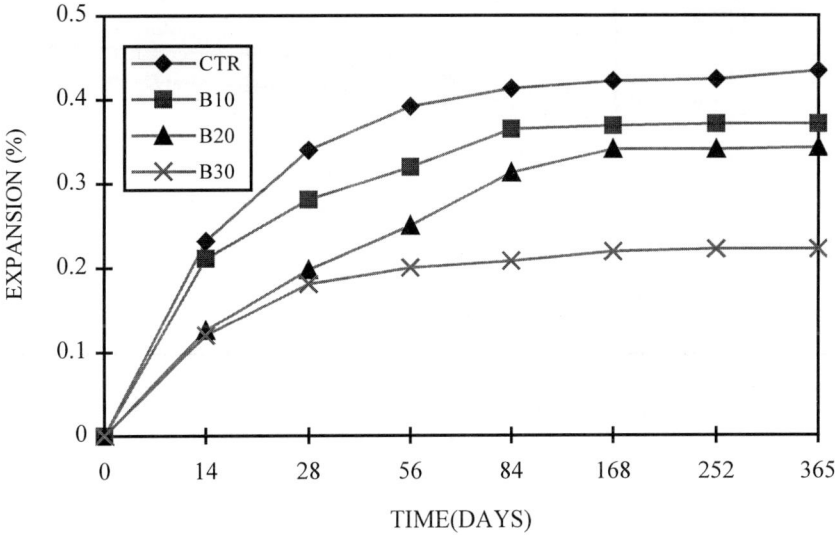

Figure 1 Effect of GB pozzolans on the reactive expansion of mortar made with high alkali cement and standard sand containing 5% Lithuanian opal sand and British Ground Brick

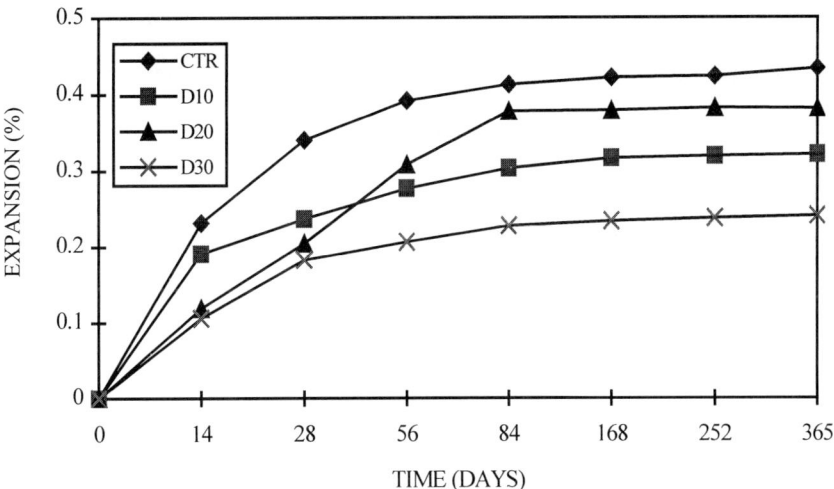

Figure 2 Effect of GB pozzolans on the reactive expansion of mortar made with high alkali cement and standard sand containing 5% Lithuanian opal sand and Danish Ground Brick

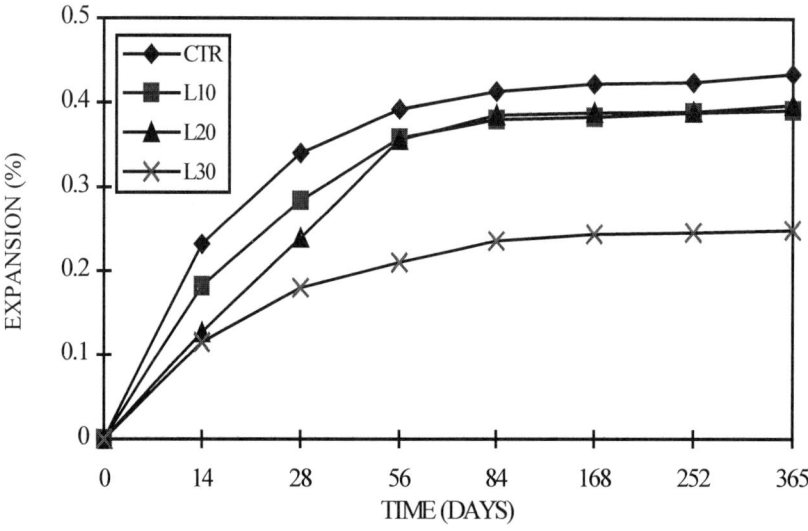

Figure 3 Effect of GB pozzolans on the reactive expansion of mortar made with high alkali cement and standard sand containing 5% Lithuanian opal sand and Lithuanian Ground Brick

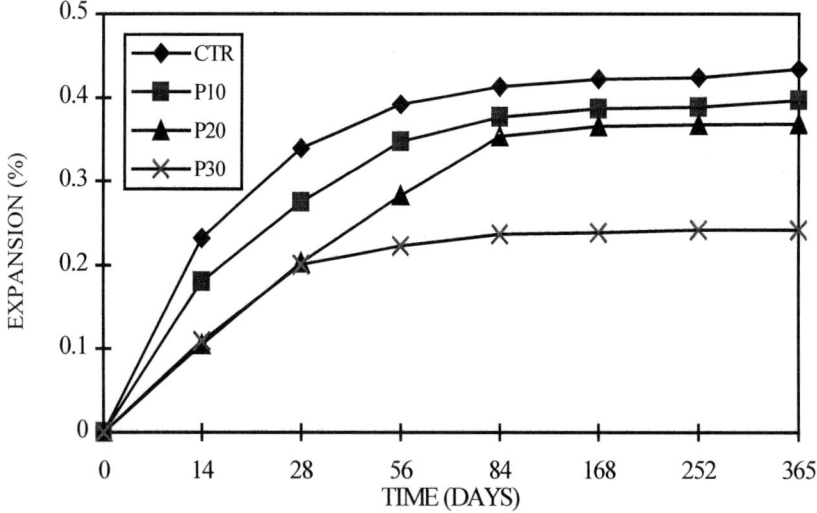

Figure 4 Effect of GB pozzolans on the reactive expansion of mortar made with high alkali cement and standard sand containing 5% Lithuanian opal sand and Polish Ground Brick

ACKNOWLEDGEMENTS

The authors would like to thank the European Commission for funding the research, which forms part of the research programme of the utilisation of waste clay brick and tile material for the partial replacement of cement in concrete.

REFERENCES

1. NEVILLE, A M. Properties of concrete. Longman Group Limited, England. 1995 (6th Edition).

2. TAYLOR, H F W. Cement chemistry. Academic Press, New York, 1990, p.p. 61-78.

3. TAYLOR, H F W. Sulfate reactions in concrete - microstructural and chemical aspects. In: Gartner, E M and Uchikawa, H (Eds): Cement Technology.

4. WILD, S, GAILIUS, A., HANSEN, H., PEDERSON, L. AND SZWABOWSKI, I. Pozzolanic properties of a variety of European clay bricks. Building Research and Information. Vol 25, No 3, 1997.

5. COPERNICUS RESEARCH PROJECT. Recycling of waste clay brick and tile material for the partial replacement in concrete. January 1997, Second Annual Report, January 1997, No CIPA - CT94211.

6. MALHOTRA, V.M. AND MEHTA, P.K. Pozzolanic and Cementitious Materials, Advances in Concrete Technology, Gordon and Breach, Vol.1, 1996, pp. 149-152.

DURABILITY OF CONCRETE CONTAINING RICE HUSK ASH AS AN ADDITIVE

P R S Speare

K Eleftheriou

S Siludom

City University London

United Kingdom

ABSTRACT. The study deals with the performance of concrete with finely ground rice husk ash added at up to 10% of the cement content. A moderately high performance basis was adopted for all mixes with a fixed cement content of 400 kg m^{-3}. The emphasis of the testing was on durability related parameters such as absorption characteristics and oxygen permeability, resistance to freeze-thaw surface scaling in the presence of de-icing salts and prolonged exposure to chlorides.

Keywords: Rice husk ash, Durability, Permeability, Absorption, Freeze-thaw resistance.

Dr Philip R S Speare is Senior Lecturer in the Department of Civil Engineering at City University London. His research interests include durability of concrete, the use of recycled materials as aggregate, cement replacement materials, high performance materials, bond and anchorages.

Ms Kalliopi Eleftheriou, formerly a postgraduate student at City University London, is now working as a consulting engineer in Athens. Her particular interest is in concrete construction.

Mr Surarith Siludom, formerly of City University London, is now working in the construction industry in Thailand following postgraduate study.

INTRODUCTION

It is well established that many organic ashes have a high silica content [1] and amongst those with the highest silica content is rice husk ash. Vast quantities of rice husk are produced world-wide: more than 400 million tonnes annually from which around 16 million tonnes of pozzolanic ash could potentially be produced. Relative to other organic materials, rice husk yields a high percentage of ash (22%) with a very high silica content, typically 93%. Under controlled combustion conditions the form of the ash varies depending on the combustion conditions. From 400 to 500 °C the silica is still amorphous, at around 600 °C quarts may occur and above this temperature other crystalline forms are produced, such as crystobalite. Combustion time also influences the final form. Under certain conditions an active form of silica can be produced and the silica activity index has been shown to be related to the amorphousness of the silica in the ash [2]. Values in excess of 70% may be obtained.

Such ashes, suitably ground, have been used as cement replacement materials. In view of the large quantities of rice husk ask produced such use has naturally been widely investigated. More recently there has been investigation of the potential for the production of high strength mixes incorporating finely ground ash [3].

The emphasis of the work reported in this paper is on the use of the ash in a finely ground state and used as an additive rather than replacement material. The objective was to determine whether such use results in an enhancement of mechanical or durability properties of concrete in comparison with concrete without additives and mixes containing silica fume. The tests included in the programme were chosen to study a wide variety of performance characteristics for a restricted range of mixes.

MIXES STUDIED

Materials

The rice husk ash (rha) used was prepared by controlled low temperature combustion in a specially designed prototype furnace. The ash was then milled to increase the fineness to an economically limiting value. The relevant properties of the ground ash were:

    silica activity index   76.8%
    loss on ignition   7.6%
    BET surface area $14.15 \text{ m}^2 \text{ g}^{-1}$.

Other materials used were Portland cement, Thames Valley gravel course aggregate and natural sand fines. The silica fume was used in slurry form.

Mixes

A high performance material specification was used as the basis of this study so a moderately high cement content of 400 kg m^{-3} was adopted for all mixes and the water content adjusted to give values of slump in the range 30 - 40 mm. A superplasticiser, Cormix SP6, was used.

The range of mixes studied was limited to concentrate on the principal additive parameter and hence all mixes had a coarse aggregate content of 1200 kg m^{-3}, 646 kg m^{-3} of fine aggregate and 1.29 l m^{-3} of superplasticiser. The only variations between mixes were the additive type and content and the necessary adjusted water/cement ratio. These are summarised in Table 1.

Table 1 Mixes used

MIX	ADDITIVE	WATER/CEMENT RATIO
A	none	0.32
B	5 % silica fume	0.33
C	10 % silica fume	0.32
D	10 % rha	0.37
E	5 % rha	0.34

All specimens were demoulded after 24 hours and water cured at 19 ± 2 °C until either tested or transferred to the specific curing regimes required for particular tests.

MECHANICAL PROPERTIES

Compressive Strength

The 28 day compressive strength obtained from 100 mm cubes are shown in Table 2.

Table 2 Compressive strength

MIX	COMPRESSIVE STRENGTH AT 28 DAYS, N/mm^2
A no additive	58.0
B 5 % silica fume	63.5
C 10 % silica fume	60.0
D 10 % rha	61.0
E 5 % rha	61.0

The values show that, for these mixes, the 28 day strength values are all very close with small increases over the control for all mixes with additives.

Bond Tests

Bond strength was determined from simple pull-out tests using plain, 12 mm diameter bars, embedded to a depth of 150 mm.

This procedure was adopted so that the principal bond action would be by adhesion. The average ultimate bond strength, f_{bu}, is shown in Table 3 together with values normalised for concrete strength, f_{cu}.

Table 3 Bond strength

MIX	f_{bu} N mm^{-2}	$f_{bu}/\sqrt{f_{cu}}$
A no additive	2.45	0.30
B 5 % silica fume	2.95	0.36
C 10 % silica fume	3.65	0.47
D 10 % rha	5.95	0.76
E 5 % rha	5.35	0.63

Enhanced bond performance is apparent for all mixes containing additives, very significantly so in the case of rha, increasing with additive content for both silica fume and rha mixes.

ABSORPTION PROPERTIES

Three tests of absorption were carried out: absorption by immersion, surface absorption (ISAT) and sorptivity.

Absorption by Immersion

100 mm cubes were used, three per mix. They were cured in water for 56 days, dried to constant mass at 105 ± 5 °C and then stored in air-tight containers until testing. The specimens were then immersed in water so that they were covered to a depth of 25 ± 5 mm. The samples were weighed before immersion and after immersion for 30 minutes, 1, 2 and 4 hours. Surplus water was lightly wiped from the surfaces before weighing. The percentage by weight of water absorbed for each mix and immersion time is shown in Table 4.

Table 4 Absorption by immersion

	ABSORPTION, %, AFTER IMMERSION TIME			
MIX	30 min	1 h	2 h	4 h
A no additive	1.53	2.06	2.72	2.85
B 5 % silica fume	1.47	1.89	2.25	2.55
C 10 % silica fume	1.44	1.94	2.15	2.30
D 10 % rha	1.68	1.96	2.10	2.59
E 5 % rha	1.60	2.01	2.28	2.57

It is evident that both silica fume and rha addition lowers the absorption, this reduction usually increasing with additive content.

Initial Surface Absorption Test

The ISAT tests were carried out on 150 mm cubes, water cured for 56 days, oven dried to constant mass at 105 ± 5 °C and then stored in air-tight containers until testing. The results of the tests conducted using a pressure head of 200 mm of water are shown in Table 5 and are the averages of tests on three specimens.

Table 5 Initial surface absorption

MIX	ABSORPTION, ml m^{-2} s^{-1} AFTER TIME		
	10 min	30 min	1 h
A no additive	0.265	0.138	0.083
B 5 % silica fume	0.221	0.088	0.039
C 10 % silica fume	0.216	0.088	0.050
D 10 % rha	0.238	0.105	0.044
E 5 % rha	0.309	0.116	0.050

At 10 minutes the two silica fume results are below that for the control whilst the rha results are either side of it. After a longer period a consistent pattern emerges of all results with additive showing lower values than the control, with the rha absorption lying between those for the control and those for the silica fume mixes.

Sorptivity

Sorptivity tests have the advantage that there is effectively negligible head of water as driving force so that absorption is purely by capillary suction. They were carried out on 100 mm cubes water cured for 56 days and then oven dried to constant mass at 105 ± 5 °C. The four faces which were vertical in the test configuration were sealed with chlorinated rubber paint and the inflow face and the face opposite it left unpainted. The face opposite the trowelled face was used as the inflow face in all cases for consistency.

The cubes were placed in water so that the water level was 5 ± 1 mm above the inflow face. The cubes were weighed at the start of the test and subsequently after intervals of 10, 30 minutes, 1, 2, and 4 hours. The inflow face was lightly wiped to remove excess water before weighing. At the conclusion of the test each specimen was split across the inflow face so that the height of water rise could be measured. The average results for three specimens for each mix are shown in Table 6. The values for sorptivity were found to be rather variable between specimens of the same mix, possibly due to the difficulty of measuring accurately changes in mass which are very small in relation to the mass of the cube, and these values are not included.

Table 6 Results of sorptivity tests

MIX	HEIGHT OF WATER RISE, mm
A no additive	36
B 5 % silica fume	24
C 10 % silica fume	21
D 10 % rha	27
E 5 % rha	35

OXYGEN PERMEABILITY

The oxygen permeability determination was carried out on 150 mm diameter, 50 mm thick specimens, cured in water for 28 days and then stored at 20 ± 2 °C and 65 ± 5% relative humidity to an age of 37 days before testing. The tests were at gauge pressures of 1.5 and 2.5 bar and the results given for permeability are an average of the values at these two pressures. The tests were carried out in accordance with Cembureau recommendations [4]. The values of permeability coefficients determined are shown in Table 7.

Table 7 Oxygen permeability coefficient, K

MIX	K, m^2
A no additive	4.62×10^{-17}
B 5 % silica fume	2.50×10^{-17}
C 10 % silica fume	2.38×10^{-17}
D 10 % rha	4.36×10^{-17}
E 5 % rha	4.10×10^{-17}

The permeability coefficients are low for all mixes, as would be expected with the higher cement content selected for the basic mix. However, it can be seen that whilst silica fume leads to a significant reduction in permeability, the presence of rha has only a small effect.

FREEZE THAW DE-ICING SCALING RESISTANCE

This test was carried out using 150 mm cubes cured in water for 14 days and then in air for a further 14 days before testing. A dyke was formed around the perimeter of the test face using 10 mm thick plastic strips to allow the face to be ponded with chloride solution.

A depth of 2 - 3 mm of 3% sodium chloride solution was maintained on the test face and the specimens, two per mix, were then subjected to 50 freeze-thaw cycles consisting of 16 hours of storage at -20 °C followed by 8 hours of thawing at room temperature.

The specimens were kept in the -20 °C environment at any times, for example during weekends, when it was not possible to follow the cycling regime. At the conclusion of the cycling the scaled particles were filtered off, dried and weighed and a visual inspection of the surface carried out. The mass of scaled particles and a description of the test surface are shown in Table 8.

Table 8 Freeze-thaw surface scaling

MIX	MASS OF SCALED PARTICLES, kg m^{-2}	SCALED AREA
A no additive	2.02	Large area with shallow scaling
B 5 % silica fume	0.43	Small patches of shallow scaling
C 10 % silica fume	1.71	Moderate area; some deep pits
D 10 % rha	1.72	Moderate area; shallow scaling
E 5 % rha	0.40	Small patches; a few deep pits

All specimens with additive show a consistent pattern of reduction in scaling, most pronounced at the 5% additive level. At 10 % addition of either silica fume or rha, the mass scaled is only a little less than for the control mix. The scaling patterns show some dramatic differences; in some cases the scaling is localised but in the form of deep pits, whilst in others it is more extensive but shallow.

CHLORIDE EXPOSURE

The compressive strength of cubes stored in 10% sodium chloride solution for nine months was measured and compared with corresponding cubes stored in fresh water throughout. The strength of the cubes exposed to chlorides was found in each case to be 95% or more of that of the corresponding cubes kept in freshwater. The loss of strength was minimal and no differences were detected between the control mixes and those with either additive.

CONCLUSIONS

1. The addition of finely ground rice husk ash leads to a small increase in compressive strength at 28 days. A more significant increase might be expected at later stages due to delayed pozzolanic action during hydration.

2. Bond strength due to adhesion is markedly increased by the addition of rha.

3. Absorption characteristics are enhanced by the presence of rha, though not to the same extent as corresponding mixes containing silica fume.

4. Oxygen permeability is reduced very slightly with the inclusion of rha. This is in contrast to equivalent silica fume mixes where more significant reduction occurs.

5. The incorporation of additives modifies the extent and pattern of surface scaling due to freeze-thaw cycling in the presence of deicing slats; the effect is particularly pronounced at the 5% content.

6. There is no perceptible change in strength under prolonged exposure to chlorides.

ACKNOWLEDGEMENTS

The authors would like to express their appreciation to Kingsway Technology Ltd for their support for the work reported in this paper, in particular for the processing of the raw husk material.

REFERENCES

1. UNIDO/ESCAP/RCTT. Proceedings of the workshop on rice husk ash cements, Peshawar, Pakistan, 1979. Published by the Regional Centre for Technology Transfer, Bangalore, India.

2. MEHTA, P.K. The chemistry and technology of cements made from rice husk ash. Published in reference [1].

3. CHATVEERA, B. and NIMITYONGSKUL P. High performance concrete containing modified rice husk ash. Approrpiate Concrete Technology, Edited R. K. Dhir and M. J. McCarthy. E. & F. N. Spon, 1996.

4. KOLLEK J. J. The determination of the permeability of concrete to oxygen by the Cembureau method - a recommendation. Materials and Structures, 1989, 22, 225-230.

PROPERTIES OF PORTLAND CEMENT MORTARS CONTAINING FBC FLY ASH

N Ghafoori

S Kassel

Southern Illinois University

United States of America

ABSTRACT. The fresh and hardened properties of several mortars, made with fly ash obtained from a Fluidized Bed Combustion (FBC) unit, were determined. Variables included four different fly ash addition rates as a partial replacement of portland cement and two distinct levels of cementitious materials factor. FBC fly ash mortars, prepared at a uniform flow, were tested for bleeding, time for setting, adiabatic temperature, compression, drying shrinkage, and resistance to internal sulfate attack.

Test results show that the time of setting, unit weight, and compressive strength decreases as the FBC fly ash replacement of portland cement increases. The adiabatic temperature reaches its peak at a rate similar to that of control mortars, whereas the bleeding elevates with increases in the FBC fly ash content of the test matrices. The expansion (due to internal sulfate attack) and drying shrinkage strains of FBC fly ash mortars are directly proportional to the cement content and the amount of fly ash substituted for portland cement.

Keywords: FBC fly ash, Mortars, Wastes, Fresh properties, Cementitious materials, Compression, Shrinkage, Ettringite, and expansion.

Dr Ghafoori is currently a professor of Civil Engineering at the Southern Illinois University, Carbondale, U.S.A. His special interests are in the areas of durability, strength, and behavior of concrete systems, and utilization of industrial by-products for construction applications. He is a member of various technical committees of ACI, ASCE, and TMS.

Mr S Kassel is currently working on his MS in civil engineering at the Southern Illinois University, Carbondale. He holds a BS in civil engineering form the same institution. His research interests are in concrete materials and structures.

INTRODUCTION

The State of Illinois is one of the largest coal producing areas in the United States. Most of the coal produced and utilized in and around the Illinois coal basin is high sulfur coal. In the light of the 1990 Clean Air Act Amendment requirements to reduce sulfur and nitrogen oxides emissions, electric utility and co-generation plants are forced to a adopt a system from various available clean combustion technologies. To date, the utilization of fluidized bed combustion by the power industry has and will continue to emerge as one of the most viable option for burning high sulfur coal. This is mainly because of its economical and environmental advantages over other available conventional methods (Tavoulareas, et al. 1987). The main idea behind the FBC system is to burn crushed coal at the bottom part (combustion zone) of the boiler where a sorbent, usually limestone or dolomite, is added (Bland, et al. 1987). Both coal and sorbent are kept suspended during combustion on the bed by an upward stream of hot air producing a liquid-like mixture having characteristics very similar to those of a fluid, coining the name fluidized bed boilers (EPRI 1987).

Although there has been considerable research on pulverized coal combustion (PCC) fly ash, there has been little investigation dealing with the use of fluidized bed combustion (FBC) fly ash for the concrete industry. This paper reports on a laboratory study that attempted to examine the extent of utilization of FBC fly ash, derived from burning high-sulfur coal, in portland cement mortars. Ten different mixes containing various proportions of portland cement, FBC fly ash, and natural fine aggregate were blended to produce mortars of similar consistency. The fresh properties and hardened characteristics of the FBC fly ash mortars are compared to those of the equivalent reference mixes in order to ascertain the suitability of the FBC fly ash in portland cement concretes.

EXPERIMENTAL METHODOLOGY

Materials

The fly ash as obtained from a fluidized bed combustion co-generation plant from the State of Illinois. Chemical and Physical properties of the FBC fly ash were: SiO_2 = 22.10%, Al_2O_3 = 6.80%, Fe_2O_3 = 6.67%, SO_3 = 15.67%, CaO = 38.70%, MgO = 1.29%, LOI = 5.46%, Na_2O = 0.5%, K_2O = 1.12%, specific gravity = 2.61, #325 seive fineness = 40.20%, autoclaved expansion = 0.08%, water requirement = 106.60%, and 7-day compressive strength = 61.20% of that of the control sample. The average pH value for the FBC fly ash was 12.2, indicating a passive environment when reinforcing steel is used. A type V portland cement, conforming to the requirements of ASTM C 150, was used. Its chemical components were: SiO_2 = 22.60%, Al_2O_3 = 4.0%, Fe_2O_3 = 4.0%, CaO = 64.70%, C_3S = 54.0%, C_2S = 24%, C_3A = 3.8%, C_4AF = 12%, and LOI = 1.0%. This cement had an average specific gravity of 3.15.

The natural siliceous fine aggregate had well graded particles with a fineness modules value of 2.8. The oven and saturated surface dry specific gravity were 2.65 and 2.66, respectively. Its SSD absorption value was 0.5%. Tests on organic impurities failed to indicate the presence of any organic matters. The percentage of combined clay lumps and friable particles was measured at 1.75%. The fine aggregate particles were dense, smooth in texture, and had a well-rounded shape.

Mixture Proportions

Two levels of cementitious materials factor, 15 and 30% by mass of total dry solids, were used. Various proportions of FBC fly ash mortars were investigated by batching each mixture with 10, 20, 30, and 40 percent of total cementitious materials by mass of fly ash to replace a portion of portland cement. A mortar with zero percent FBC fly ash was the control. A constant flow of 100 ± 3% was used for all selected mixtures. The matrix constituents and proportions as well as the corresponding water-cementitous materials ratio, resulting from a uniform flow, are shown in Table 1.

Table 1 Mixture Proportions of FBC Fly Ash Mortars

SAMPLE DESIGNATION	PORTLAND CEMENT (%)	FBC FLY ASH* (%)	CEMENT SUBSTITUTION (%)	NATURAL FINE AGGREGATE (%)	W/C**
A-15-0	15	0	0	85	0.90
A-15-10	13.5	1.5	10	85	0.90
A-15-20	12	3	20	85	0.91
A-15-30	10.5	4.5	30	85	0.92
A-15-40	9	6	40	85	0.93
A-30-0	30	0	0	70	0.39
A-30-10	27	3	10	70	0.41
A-30-20	24	6	20	70	0.43
A-30-30	21	9	30	70	0.44
A-30-40	18	12	40	70	0.45

* % by mass of total dry solids
** water-cementitious materials ratio generating flow of 100 ± 3%

Preparation of Test Samples and Testing Procedures

Mortar cylinders and prisms cast in molds with internal dimensions of 100 x 200 mm and 25 x 25 x 285 mm, respectively, were used. After being removed from their molds, all cylinders and prisms were cured in a tank filled with lime-saturated water, at temperature of 23 ± 1 °C. The cylinders were tested for compressive strength after curing periods of 28, 90, and 180 days. Samples tested for evaluation of internal sulfate attack were kept in lime-saturated water for a period of one year. Shrinkage specimens were moist-cured for 28 days and then placed indoors for the remainder of testing period (23 ± 1 °C and 50 ± 5% R.H.).

Tests used for the evaluation of the mortar samples included: ASTM C 109 (flowability), ASTM C 237 (bleeding), ASTM C 807 (time of setting), ASTM C 39 (compressive strength), ASTM C 452 with modifications (internal sulfate attack), and ASTM C 157 (drying shrinkage). For the measurement of adiabatic temperature, the fresh mortar was placed in a 130 mm diameter by 130 mm in height cylindrical insulated plastic container lined with a plastic bag. A Type J thermocouple wire was inserted into the fresh mortar to a depth of 100 mm. The container was then securely sealed into a tightly sealed insulated 670 x 670 x 730 mm wooden box. The thermocouple wire was plugged into a data recorder which registered and logged the mortar temperature at 1 hour intervals over a period of 100 hours.

RESULTS AND DISCUSSION

Mix unit weight at one day, right after demolding, is shown in Table 2. Due to the lower specific gravity of FBC fly ash, all FBC fly ash mortars exhibited lower unit weights compared to that of the control mixes. The lower unit weight of the FBC fly ash mortars can also be attributed to the higher water-cementitious materials ratio which increased the pore size and, thus, resulted in a lower unit weight of the hardened mortars. On average, 1.5 and 2% reduction in density can be expected for every 10% replacement of portland cement by FBC fly ash for the mortars containing 15 and 30% cementitious materials content by mass of total dry solids, respectively.

Table 2 Fresh Properties of FBC Fly Ash Mortars

SAMPLE DESIGNATION	1-DAY UNIT WEIGHT (kg/m^3)	BLEEDING (%)	TIME OF SETTING (MIN)	FRESH MORTAR TEMP (°C)	PEAK ADIABATIC TEMP (°C) (PEAK TIME)
A-15-0	2222	17.23	317	21	34 (21 hrs)
A-15-10	2210	20.2	298	17	34 (24 hrs)
A-15-20	2191	21.8	288	18	32 (25 hrs)
A-15-30	2189	23.0	278	18	30 (25 hrs)
A-15-40	2164	23.3	275	19	30 (26 hrs)
A-30-0	2312	5.7	258	16	45 (22 hrs)
A-30-10	2301	7.2	239	22	53 (22 hrs)
A-30-20	2282	7.9	236	21	48 (22 hrs)
A-30-30	2249	8.5	228	22	46 (22 hrs)
A-30-40	2240	8.5	224	21	43 (22hrs)

Table 2 also presents setting times measured by a modified Vicat apparatus for the mortar samples with various percentages of portland cement replaced by FBC fly ash. As can be deduced from the test results, the setting time of the FBC fly ash mortars decreased, to a larger extent, with increases in cement content and, to a lesser extent, with higher amounts of FBC fly ash added. The setting times for the mortar containing 15% cementitious materials with various percentages of FBC fly ash ranged, in an approximately linear fashion, from 4 hr:58 min to 4 hr:35 min, which is lower than the setting time of 5 hr:17 min for the reference mortar. For the mixtures containing 30% cementitious materials content, the setting time of the FBC fly ash mortars also decreased with increases in cement replacement and were lower than that of the reference mix by an average of 26 minutes (10%). On the whole, test results indicate that the use of FBC fly ash, as a partial replacement of portland cement, does not exhibit any adverse effects for the parameters tested, but it does reduce the setting time of the mortar.

Because of the increased demand for mixing water, the FBC fly ash mortars displayed a higher degree of bleeding compared to that of control mixes. As shown in Table 2, the amount of bleeding for the FBC fly ash mortars containing 15% cementitious materials content, varied almost linearly from 20.2 to 23.3%, accounting for an increase of 28% over that of control mortar of similar cementitious materials content (no FBC fly ash). When cementitious materials content of the mortar increased to 30% by mass of total dry solids, the accumulated bleeding water of both reference and FBC fly ash mortars were drastically reduced. For these mixes the bleeding water ranged from 7.2 to 8.5% in a nearly linear variation. This accounted for an increase in the amount of bleeding, over that of control mortar, by about 40%. No particle segregation was observed for all mortar samples with various percentage of FBC fly ash used. The visual inspections of internal surfaces of crushed mortar cylinders also did not reveal any segregation.

Table 2 also gives the peak adiabatic temperature rise and the corresponding elapsed time for both FBC fly ash and reference mixes. Test results show that mortars with various percentages of FBC fly ash generate a maximum temperature rise at a rate similar to that of the companion control samples. The number of hours elapsed to reach the maximum temperature was also similar for FBC fly ash and control mixtures containing 15 and 30% cementitious materials contents.

Table 3 shows the compressive strengths for the mortar cylinders with various percentages of FBC fly ash at different curing ages for the mixtures containing 15 and 30% cementitious materials content, respectively. After 90 and 180 days of curing the 28-day compressive strength of FBC fly ash mortars having 15% cementitious materials content increased by an average of 58 and 70%, respectively. Increases for the FBC fly ash mixes containing 30% portland cement were 18 and 31%, respectively. When compared with the reference mix having 15% cement, the 28-day compressive strength decreased by 25, 40, 45, and 56% for 10, 20, 30, and 40 percent of FBC fly ash, respectively. At the end of 180 days, the reduction in the compressive strength of the FBC fly ash mortars, over that of the reference mix, were 8, 13, 19, and 33% for 10, 20, 30 and 40% FBC fly ash blended to produce 15% cementitious materials content. For the 30% cementitious materials content of the mortars, the compressive strength of FBC fly ash cylinders were also below that of the control mix by nearly 35, 32, and 30% at curing ages of 28, 90, and 180 days.

Table 3 Compressive Strength of Mortars Containing 15 and 30% (by Mass of Total Dry Solids) Cementitious Materials Content

SAMPLE DESIGNATION	COMPRESSIVE STRENGTH (MPA) CURING AGES (DAYS)		
	28	90	180
A-15-0	7.10	8.70	9.40
A-15-10	5.40	7.75	8.25
A-15-20	4.40	7.10	8.00
A-15-30	3.80	6.25	7.50
A-15-40	3.00	4.90	6.25
A-30-0	32.0	37.25	38.80
A-30-10	29.0	32.5	37.0
A-30-20	22.0	28.0	31.25
A-30-30	18.30	20.0	23.0
A-30-40	10.90	14.20	19.0

Figures 1 and 2 display the drying shrinkage curves as a function of specimen age for both control and FBC fly ash mortars.

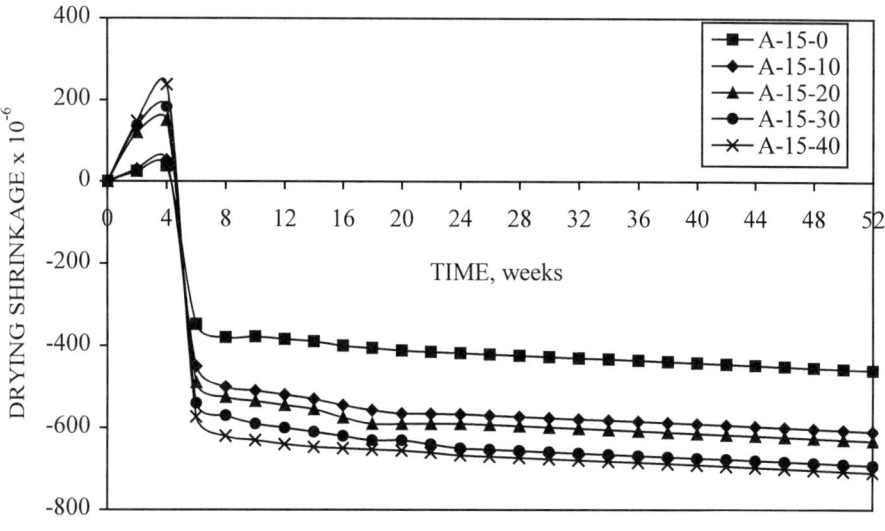

Figure 1 Dry shrinkage strain of mortars containing 15% (by mass of total dry solids) cementitious materials content

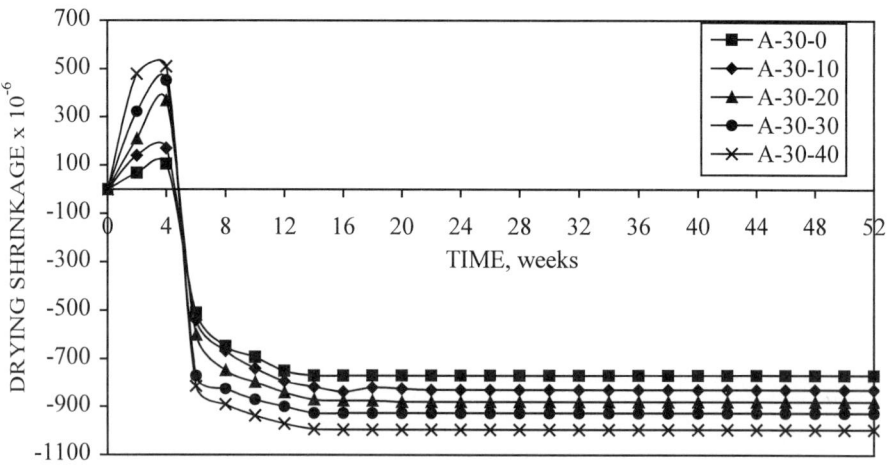

Figure 2 Dry shrinkage strain of mortars containing 30% (by mass of total dry solids) cementitious materials content

The drying shrinkage of all mortar prisms increased with time and the shrinkage strains increased with an increase in percentage of FBC fly ash. The one-year drying shrinkage of the mortars having 15% cementitious materials content increased form 460×10^{-6} for reference sample with 0% FBC fly ash to 610×10^{-6}, 630×10^{-6}, 690×10^{-6}, and 710×10^{-6} for 10, 20, 30, and 40% FBC fly ash, respectively. For the mortars containing 30% cementitious materials content, the one-year drying shrinkage also exhibited similar results, ranging from 770×10^{-6} for reference prism with 0% fly ash to 1000×10^{-6} for 40% FBC fly ash. On average, the inclusion of FBC fly ash resulted in an increase of 44 and 18% in drying shrinkage for the mortars containing 15 and 30% cementitious materials content, respectively. This behavior can be easily explained through higher water-cementitious materials ratios resulting from increases in FBC fly ash content.

The chemical reactions between tricalcium aluminate of portland cement and gypsum-contained FBC fly ash, potentially results in a chemical compound known as ettringite. Formation of ettringite has a deteriorative effect on the mortar strength and, in addition, causes expansion which may result in microcracking of the cement paste. The influence of the FBC fly ash content and the amount of cement on resistance to internal sulfate attach, expressed in terms of length change, of the FBC fly ash and control mortars are presented in Table 4 for the mixes containing 15 and 30% cementitious materials content. In all mortars, the expansion strain steadily increased with increases in the FBC fly ash content. The one-year expansion strain of the mixes of low cementitious materials content (15%) were 0.016, and 0.023, 0.0267, and 0.0579% for 10, 20, 30, and 40% FBC fly ash, indicating that the increases in expansions are 167, 284, 345, and 865%, over that of reference mixture, respectively. The availability of greater quantity of C_3A (by mass) in the mortars containing 30% cementitious materials content resulted in higher one-year expansions of 0.028, 0.12, 0.13, and 0.14%; an increase of 40, 500, 550, and 600% over that of the control mixture, respectively.

Table 4 Expansion of mortars containing 15 and 30% (by mass of total dry solids) cementitious materials content

SAMPLE DESIGNATION	EXPANSION STRAIN (%) IMMERSION AGE (WEEKS)						
	4	8	12	16	26	40	52
A-15-0	0.002	0.002	0.0025	0.004	0.0065	0.0065	0.006
A-15-10	0.01	0.012	0.014	0.015	0.016	0.017	0.0175
A-15-20	0.016	0.018	0.021	0.022	0.023	0.023	0.0235
A-15-30	0.02	0.023	0.024	0.026	0.027	0.027	0.0275
A-15-40	0.036	0.05	0.054	0.056	0.0580	0.0585	0.0580
A-30-0	0.008	0.01	0.012	0.016	0.020	0.020	0.022
A-30-10	0.018	0.024	0.022	0.023	0.028	0.028	0.03
A-30-20	0.036	0.054	0.074	0.082	0.108	0.12	0.122
A-30-30	0.05	0.03	0.10	0.104	0.124	0.132	0.132
A-30-40	0.064	0.086	0.16	0.125	0.138	0.142	0.142

In addition to expansion, loss of mass resulting from internal sulfate attack was observed. However, mass residue was not found throughout the experiment.

CONCLUSIONS

Based on the experimental results conducted in this study, the following conclusions can be drawn.

1. Setting time of the FBC fly ash mortars were marginally (10%) lower than that of the reference mixtures.

2. Because of the increased demand for mixing water required to achieve an equivalent consistency, the FBC fly ash mortars displayed a higher degree of bleeding than the control mixtures.

3. The mean 28, 90, and 180-day compressive strength of reference mortars were 70, 46, and 38%, respectively, higher than those of companion FBC fly ash mortar cylinders.

4. FBC fly ash mortar exhibited an increased drying shrinkage (44 and 18% for mixtures of 15 and 30% cementitious materials content, respectively) in comparison with that of the control samples.

5. The expansion (due to internal sulfate attack) of the FBC fly ah mortars was significantly higher than the volume increase exhibited by the equivalent reference mixtures.

ACKNOWLEDGEMENTS

This work was supported by grants made possible by the Illinois Department of Energy and Natural Resources through its Coal Development Board and Illinois Clean Coal Institute, and by the U.S. Department of Energy (Grant Number DE-FB22-91PC913340). However, any opinions, findings, conclusions, or recommendations expressed herein are those of the authors and do not necessarily reflect the views of IDENR, ICCI, and the DOE.

REFERENCES

1. BLAND, A E, JONES, C E, JG, AND JARETT, M N. Production of Non-Cement Concretes Utilizing Fluidized Bed Combustion Waste and Power Plant Fly Ash. Proceedings: 9th International Conference on Fluidized Bed Combustion, Boston, Mass., 1987, pp 947-953.

2. ELECTRIC POWER RESEARCH INSTITUTE. Utilization Potential of Advanced SO_2 Control ByProducts. EPRI CS-5269, Project No. 2708-1, Prepared by ICF Northwest, 1987.

3. GHAFOORI, N, AND SAMI, S. A Simple and Practical Approach for Effective Prehydration of Fluidized Bed Combustion Residues. Annual Report, Materials Technology Center, Southern Illinois University at Carbondale, August, 1992, 12 pp.

4. TRAVOULAREAS, S, HOWE, W, GOLDEN, D, AND EKLUND, G. EPRI'S Research on AFBC By-Product Management. Proceedings: 9th International Conference on Fluidized Bed Combustion, Boston, Mass., 1987, pp 954-959.

CHARACTERISATION OF FILLER SANDCRETES WITH RICE HUSK ASHES ADDITIONS – STUDY APPLIED TO SENEGAL

I K Cissé

Polytechnic School of Thiès

Senegal

M Laquerbe

INSA of Rennes

France

ABSTRACT. In order to capitalise on the local materials of Senegal (agricultural and industrial wastes, residual fines from crushing process, sands from dunes, etc.), rice husk ashes and residues of industrial and agricultural wastes have been used as additions in sandcretes.

The mechanical resistance of sandcrete blocks obtained when unground ash (and notably the ground ash) is added reveals that there is an increase in performance over the classic mortar blocks. Moreover the use of unground rice husk ash enables the production of a light weight sandcrete, with insulating properties, at a reduced cost. The ash's pozzolanic activity is its main characteristic which explains the high strengths obtained.

Keywords: Sandcrete, Rice husk ash, Agricultural wastes, Ground ash, Blockworks, Pozzolanicity

Dr I K Cissé is a Research/Teaching Fellow in Concrete Technology in the department of civil engineering of Polytechnic School of Thiès (Senegal) .He has specialised in the mechanical characterisation of building materials for habitat and roads. He is also an ancient researcher of the Geomechanics, Thermic and Materials laboratory, INSA of Rennes (France). He has made some publications on building local materials in Senegal

Professor Michel Laquerbe is Director the Geomechanics, Thermic and Materials laboratory, INSA of Rennes (France). He specialises in habitat building materials, particularly on the cold extrusion. He has developed a process of block stabilization to cold called the SBF process.

INTRODUCTION

The exploitation of local resources and the development of innovative techniques, the use of sandcrete has been the object of studies [1],[2],[3] and production [4],[5] in Senegal. However, the additions used was exclusively residual filled sands resulting from the crushing of limestone, sandstone, chert and basalt. The accumulation of agricultural wastes, such as rice husk, has posed environmental problems, thus it is judicious to recycle these products to resolve its.

The employment, promotion and exploitation of agricultural and industrial by-products, with the aim of minimising production costs or producing new products, would therefore represent an interesting proposition, thanks to saving in raw materials.

The objective of the study of filler sandcretes using rice husk ashes additions is viewed from three points of view : preservation of the environment, to generate an increase in the value added in the construction of buildings, while ascribing physical and mechanical properties to the material.

ORIGIN OF THE MATERIALS

Sand

Red dune sand exists in " inexhaustible " quantities, covering approximately 70 % of the national territory. The exploitation of these sands have been carried out close to suburbs of Dakar (North Foire).

Ashes

The ashes are supplied by SO.NA.COS.(Marketing National Society of Oleaginous) who use the rice husk as a supplement to combustion, the average annual production of rice husk ash is of the order of 2 300 tonnes.

Cement

The cement is produced by the SO.CO.CIM. (Cement Marketing Society) and is type CEM II/A32,5 conforming to the standard NS-02 currently extant in Senegal.

DESCRIPTION OF MATERIALS

Dune Sand

Observation of the sand grains using a binocular optical microscope, revealed that the grains were essentially quartz (99 %) with a red staining. The staining was due to a thin film of iron shale that covered the grains [6].

The characteristics of the sand are set out on Table 1. It is noticed that the sand is very fine and " well graded " according to the ATTERBURG classification .

Table 1 Geotechnical characteristics of used materials

MATERIALS	% OF FILLERS BY VOLUME (m³)	γ_s	U_r	F_m	S_s (cm²/g)
Portland Cement	84	3,14	---	---	2935
Unground ashes	7	2,35	4	1,7	---
Ground ashes	85	2,35	---	---	6960

γ_s = Specific density $\qquad U_r$ = Uniformity ratio = Hazen ratio
F_m = Fineness modulus $\qquad S_s$ = Specific surface

Examination, using a scanning electron microscope (SEM) at a 50 magnification, shows grains of regular form, well rounded which is attributed to their aeolian transportation mechanism in dunes. A magnification of 2000 reveals that the land particles are covered with a coating of iron shale, this being responsible for the red colour (see Figures 1 and 2). The efficiency of the sands as a substitute for beach sand, the exploitation of which is forbidden, has been previously proven [6].

Figure 1 Form of the dune sands examined by SEM

Portland Cement

The principal characteristics of the cement are shown in Table 1. The values found are in reasonable agreement with those supplied by the producer, a compressive resistance at 28 days of 42,5 N/mm² (σ_{c28} = 42,5 N/mm²). The results of the chemical analysis (% by weight) of the cement are shown on Table 2.

Table 2 Chemical analysis results of cement and ash

OXIDES	Fe$_2$O$_3$	CaO	Na$_2$O	K$_2$O	Al$_2$O$_3$	SiO$_2$	LOSS ON IGNITION
Cement	8,9	60,1	0,3	0,3	4,0	19,8	6,3
Ash	2,6	6,0	6,3	5,5	1,6	79,2	0,5

Figure 2 Image of the land particles of dune sand

Rice Husk Ashes

Two types of ashes have been used, ground and unground. The use of unground rice husk ash has enabled both the production of a light weight insulating concrete and the reduction cost due to energy saving by the elimination the grinding process.

Chemical analysis (see Table 2) reveals highly siliceous nature of the ash which give the ash a pozzolanic quality, as described by DIOUF (1994) [7]. The pozzolanic quality is also linked with amorphous structure of silica.

A grading analysis of the ground rice husk ash, of which 85 % by volume of the particles are smaller than 80 microns, was carried out by laser. Results indicate that 34,46 % by volume of particles are smaller that 7 microns [9].; these particles appear to be responsible for the pozzolanic reactivity of the amorphous silica [8]. The grading analysis, by sieving, of the unground ash shows the presence of 7 % fines with a low fines modulus, and a well graded size distribution (Cu =4 ; Fm =1,7)[9]. The characteristics of the material are shown in table 1 and confirm the low dry density of the ash (γ_d = 0,30 g/cm^3 and 0,72 g/cm^3 for, respectively, unground and ground ash) and the extreme fineness of the ground ash (Blaine specific surface, Ss = 6960 cm^2/g).

The analysis using SEM shows Unground ash particles in a tubular form split longitudinally with a presence of small bristles distributed over an undulated surface (see Figure 3). Figure 4 confirms the presence of hydroscopic pores.

Figure 5 shows there cellular structure. The pozzolanicity test following norm NF P 15-462 shows that the rice husk is very reactive in comparison with other products, which is reputed to be pozzolanic such as basalt, volcanic slags and the ash of groundnut shells [9].

CHARACTERISTICS OF THE SANDCRETES

Sample of filler sandcretes (4cmx4cmx16 cm prisms) were produced in accordance to the method described by CISSE (1996) [9].The details of the constituents of the mix are set out in Table 3 ; E' represents the actual quantity of water added to the mix to attain 10 cm to 12 cm spreading of the mix when subjected to 15 seconds on a vibrating table ; C' represents the total filler (cement + ash fillers) E' takes into account the rate of absorption of the water by the ash and the dune sand.

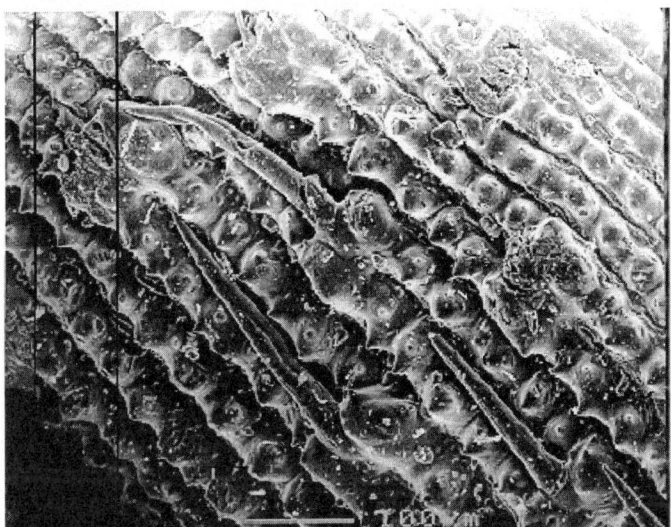

Figure 3 Texture of unground rice husk ash particles

Table 3 Batch weights (kg/m^3) of the filler sandcretes

BATCHING PARAMETERS	UNGROUND SANDCRETE	GROUND SANDCRETE
Cement	250	200
Dunes sand	913	1864
Addition	885	279
E'	330	196
E'/C	1,32	0,98
E'/C'	1,10	0,65

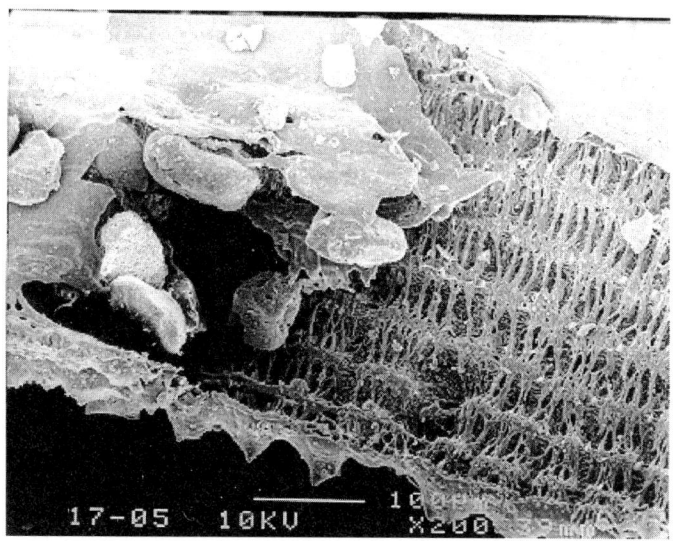

Figure 4 Porous unground rice husk ash

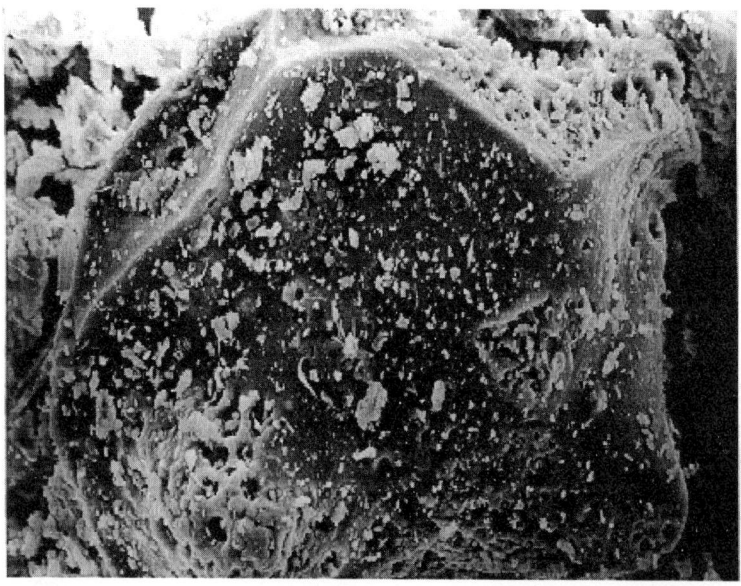

Figure 5 SEM image of the cellular structure of the ground rice husk ash

Characteristics and Mechanical Properties

Table 4 summarises characteristics. Examination of the data lead to the following:

1. It is noted that there is loss of weight for samples cured in air, therefore water, between 7 and 28 days ; this loss is slightly more the unground ash than the ground one. On the contrary the gains of weight of samples cured in water by the absorption of water is almost double in the same period. That is to say that the ash absorbs more water than liberated, which is illustrated by their morphology as demonstrated by observation by SEM (Figures 4 and 5)

2. The mechanical strength increases through the period of cure , more rapidly for the samples cured in water ; it is well known that humid conditions enhance the hydration process .The ratio of the age coefficients demonstrates the importance of water curing.

3. The mechanical strength of the sandcretes with ground ash is on average twice of the one with unground ash whatever the curing method and age. This is due to the pozzolanic nature of the ash, which will develop more rapidly in the case of the fine particles than the coarse particles which have the same mineralogical and chemical compositions [10]. The percentage of particles smaller than 7 microns, higher for the ground ash, is a characteristic that accentuates the pozzolanicity as already demonstrated JARRIGE [8].

4. If the cumulative effect of density, grinding - which reduces the water cement ratio from 1.1 to 0.65 - and the pozzolanicity explain the highly significant increase in strength of concrete made with ground ash, only a increased cement content (250 Kg/m^3 in place of 200 Kg/m^3) and the pozzolanicity would enable the use of unground ash. In the case using unground ash, two factors contribute to the reduction in mechanical strength of the concrete ; firstly the lower density of the ground ash and, secondly, at the time of mixing there is a reduction in the maximum aggregate size (D) which would increase a minimum content of fines (cement + addition) greater than that predicted by the empirical formulae (in $\sqrt[5]{D}$).This factor has not been taken into account as it would be necessary to know the dimension of D. Additionally one must consider that water content of the unground ash is higher than the ground ash ; this will similarly lead to a reduction in strength.

Table 4 Mechanical parameters of the different sandcretes

PARAMETERS MEASURED	UNGROUND ASHES SANDCRETE		GROUND ASHES SANDCRETE	
	7 Days	28 Days	7 Days	28 Days
$(\sigma_{c28})_{water}$ (Mpa)	3,69	9,56	8,31	18,59
$(\sigma_{c28})_{air}$ (Mpa)	3,31	4,98	6,50	10,37
Loss of weight (%)	12,60	13,81	7,30	7,60
Gain in weight (%)	0,84	1,61	0,85	1,59
Bulk density (Kg/m^3)	1880		2050	
$(\sigma_{c28}/\sigma_{c7})_{water}$	2,59		2,24	
$(\sigma_{c28}/\sigma_{c7})_{air}$	1,50		1,59	
$\{(\sigma_{c28}/\sigma_{c7})_{water}\}/\{(\sigma_{c28}/\sigma_{c7})_{air}\}$	1,72		1,40	

The fact that the grinding of rice husk ash demands a higher water content in the mix was found by PATEHA in 1991[11]. Also the use of plasticizers which reduce water content would allow the increase in the efficiency of these fillers which are hydraulically reactive. The influence of the characteristics of fillers on sandcretes was studied at INSA, Lyon [12].

Finally, even it is true that the strength found from the 4cmx4cmx16cm samples is not significant, it was noted that the sandcretes made with rice ash husk demonstrated a superior mechanical strength when compared with sandcretes made with filled limestone or chert [9]and having the same or higher cement contents (Table 5).

Table 5 Comparison with other types of sandcretes

ADDITIONS	CEMENT CONTENT (Kg/m^3)	$\sigma_{28\,days}$ (N/mm^2)
Unground ashes	250	9,56
Ground ashes	200	18,59
Filled no sandy limestone 0/3	250	6,78
Filled sandy limestone 0/3	250	9,10
Filled cherts 0/3	265	7,47

$\sigma_{28\,days}$ = compressive stress at 28 days

One sees therefore, that there is an interest in promoting filler sandcretes with rice husk ash addition, especially ground, since mechanical strengths are greater than those obtained using other additions. In effect, even volcanic materials (tuffs, scorias , basalt) that are reputed to be pozzolanic do not attain values as high as those produced by ground ash (Table 6).

Table 6 Comparison with sandcretes using pozzolanic fillers

ADDITIONS	CEMENT CONTENT (Kg/m^3)	$\sigma_{28\,jours}$ (N/mm^2)
Ground ashes	200	18,59
Filled basalt 0/3	250	12,93
Ground black volcanic slags	200	16,68
Ground volcanic tuffs	200	10,25

CONCLUSION

The introduction of rice husk ash as addition to sandcretes has therefore allowed to improve the physico-mechanical performance of this material. In fact, with the use of ground ash one achieve unexpectedly high strengths , while the unground ash permits the production of a robust material. These results are explained respectively by the chemical nature and morphology of this material of organic origin. Equally the financial competitiveness of the material should be pointed out, since sandcrete blockwork with rice husk addition has a lower cost compared with those using other types of additions (Table 7).

Table 7 Cost comparative study of the different Sandcretes (f CFA = Senegalese currency)

TYPE OF CONCRETE	COST OF CUBIC METRES (f CFA)
Sandcrete with filled limestone addition	13 522
Sandcrete concrete with filled basalt addition	14 454
Sandcrete concrete with filled sandstone addition	12 808
Sandcrete concrete with filled cherts addition	13404
Mortar of sand and cement	13 505
Sandcrete with unground ashes addition	12 965
Sandcrete with ground ashes addition	11 051

In summary, the use of these agricultural wastes is highly justified and confirms the conclusions of previous studies [7],[11].

REFERENCES

1. NDIAYE, K H. Optimisation des formulations de bétons de sable. Mémoire de fin d'études d'ingénieur en génie civil de l'E.P.T., Université Cheikh Anta Diop, Sénégal, 1991

2. EUSEUBIO, A G. Etude du revêtement des canaux à ciel ouvert par du béton de sable : Application au Canal du Cayor (Sénégal). Mémoire de fin d'études d'ingénieur de génie civil de l'E.P.T., Université Cheikh Anta Diop, Sénégal, 1991

3. DIASSÉ, B. Les bétons de sable routiers au Sénégal : Proposition de formulation, caractérisation - dimensionnement . Mémoire de fin d'études d'ingénieur - géologue de l'I.S.T., Université Cheikh Anta Diop, Sénégal, 1996

4. DIOP, P M B, THIOUNE, S L. Rapport d'évaluation expérimentale de stabilisation de trottoirs en béton de sable à la Médina, Dakar, Sénégal, 1991

5. THOMAS, F. Application de la technique du béton de sable à la réalisation de voirie . Commune de Saint Louis, Sor , Sénégal, 1996

6. LAQUERBE, M, CISSÉ, I, AHOUANSOU, G. Pour une utilisation rationnelle des graveleux latéritiques et des sables des dunes comme granulats à béton : Application au cas du Sénégal,. Materials and structures , Vol 28, Dec., 1995, pp.604-610.

7. DIOUF, B. Caractérisation d'un ciment Portland à ajout de cendres de balles de riz, Mémoire de fin d'études d'ingénieur - géologue de l'I.S.T., Université Cheikh Anta Diop de Dakar, Sénégal, 1995

8. JARRIGE, A. Les cendres volantes : Propriétés - Applications industrielles, Edtions Eyrolles, Paris 1971.

9. CISSÉ, I. Contribution à la valorisation des matériaux locaux au Sénégal: Application aux bétons de sable, Thèse de doctorat de génie civil de l'INSA de Rennes, 1996

10. COSTA, U, MASSAZZA, F. Factors affecting the reaction with lime pozzolanas, Communication supplémentaire au $6^{\text{ème}}$ congrès International de la chimie du ciment, Moscou, 1974

11. PATEHA, K M. Contribution à la valorisation des sous - produits industriels et agricoles dans l'industrie du ciment, Thèse de doctorat en génie civil et sciences de la conception, I.N.S.A. de Lyon, 1991

12. AMBROISE, J, PERA, J. Relations entre les caractéristiques des fillers et les bétons de sable dans lesquels ils sont employés. Etude sur onze fillers : Etude de la porosité avec six fillers, Fév.1993 - Janvier 1993, Rapport d'étude,.

RICE HUSK ASH: A FILLER FOR SAND CONCRETE

I K Cissé
Ecole Supérieure Polytechniqu
Sénégal

R Jauberthie   M Temimi
Institut National des Sciences Appliquées

J P Camps
Université de Rennes 1
France

ABSTRACT. Concrete is usually made of gravel, sand, cement and water. In some world areas the natural resource in gravel may be low and then expensive. The sand concrete can be substituted to traditional concrete for a number of uses. In the rice producing countries, the rice husk is usually used as a power plant combustible. This by-product can be substituted to a part of cement in concrete. This paper discusses physical and chemical properties of this concrete. The performances of the ground ash concrete are higher than those of the reference concrete. This is the consequence of the filler effect combined with the pozzolanic effect of ashes. Concrete with rice husk ash has a lower cost than the same concrete containing sandstone filler, a part of cement can be replaced with ground ash without reducing the performances of the concrete. It can be used to make concrete blocks or roads.

Keywords: Waste, Filler, Sand, Concrete, Ash, Rice husk.

Dr I Cissé is Maître de Conférences, Département Génie Civil, Ecole Supérieure Polytechnique, Thiès, Sénégal. He specialises on the properties of sand concretes.

Dr R Jauberthie is Maître de Conférences, Département Génie Civil, Institut National des Sciences Appliquées, Rennes, France. His main research interests include the properties of different phases in hydrated cements, the durability and the protection of concretes.

Dr M Temimi is a researcher at the "Géomécanique, Thermique et Matériaux" laboratory of the Civil Engineering Department, Institut National des Sciences Appliquées, Rennes, France. He has completed a thesis on the processing of low cost new materials using natural resources.

Dr J P Camps is Maître de Conférences, teaching at the Institut Universitaire de Technologie, Département de Génie Civil, Université de Rennes 1, France. His research area developed at INSA's GTMa lab involves the durability and protection of building materials and the behaviour of clay cement mixes.

INTRODUCTION

The concrete is usually made of gravel, sand, cement and water. In some world areas the natural resource in gravel may be low and then expensive. The sand concrete can be substituted to traditional concrete for a number of uses [1]. Furthermore, the cement content of a concrete can be lowered by using fillers while the performances of the final product are not reduced [2]. The aim of this work is to study if Senegal rice husk ash can be used as an efficient filler to make sand concrete. This material, usual in this country, was studied using X ray diffraction, scanning electron microscopy (SEM) and chemical analysis. Its mechanical behaviour was compared to the behaviour of a similar concrete but containing a sandstone filler instead of rice husk ash.

The performances of both sorts of concrete are similar. This means that, a concrete containing rice husk ash as a filler presents a lower cost than the same concrete made with a sandstone filler [3]. Making rubble stone or road concrete using it is quite possible.

FILLER IDENTIFICATION

Sandstone Filler

The deposit is located at Toglou, 55 km from Dakar, in the N-W direction. The chemical analysis is reported on table 1

Table 1 Chemical analysis of the sandstone filler (mass percent)

SiO_2	Fe_2O_3	Na_2O	K_2O	Al_2O_3	CaO	TOTAL
77.6	9.4	5.5	4.3	3.1	0.2	100

The X ray diffraction analysis shows that silica is present under its quartz form as a large amount and it masks the other crystals. The X ray pattern (figure 1) does not show any amorphous phase, indeed no rise of the background noise and no halo can be detected on this diagram.

The SEM examination displays sandstone grains which the most size is lower than 100 µm (figure 2). Their surface is spiked with very fine particles (figure 3) probably issue from the quartz binder of the original rock. They are mainly composed of a clayey mineral from the illite family. The granulometric distribution, established by means of a laser granulometre COULTER LS 130, is represented figure 4. It confirms that :

- the maximum particles size is 100 µm
- 50 % of the particles have a size smaller than 10 µm
- the volume percentage of particles smaller than 7 µm is 47 %.

The specific area measured by the Blaine's method is 4,800 cm²/g.

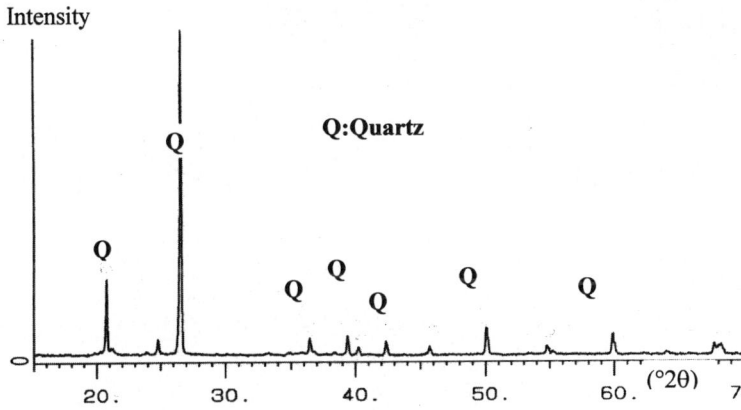

Figure 1 X ray diffraction pattern of sandstone filler (Cu Kα)

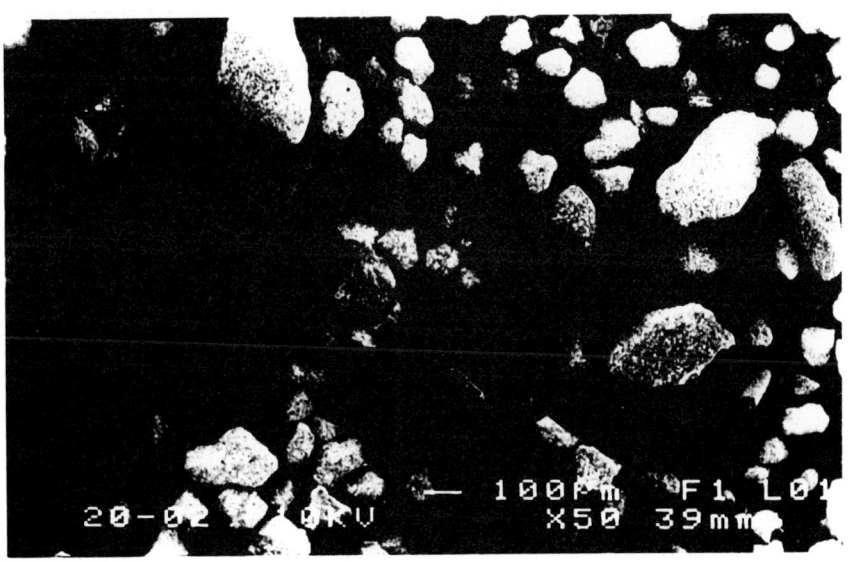

Figure 2 SEM micrograph of sandstone particles (lower than 100 μ)

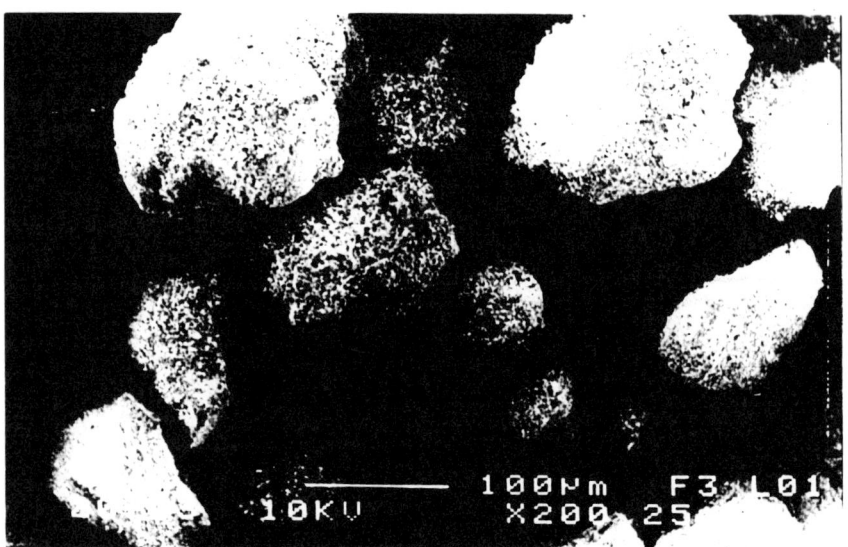

Figure 3 SEM micrograph of sandstone particles (surface)

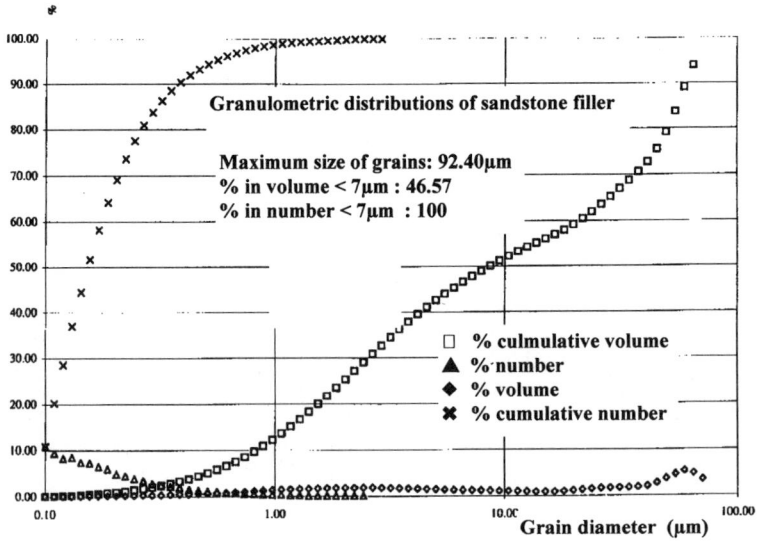

Figure 4 Granulometric distribution of sandstone filler

Rice Husk Ash Filler

The rice husk is a complementary combustible burned in electric power plants. It is used as a substitute to save the fuel oil or the coal. The ash produced by this combustion is presently used only in an agricultural purpose. The table 2 gives its chemical analysis.

Table 2 Chemical analysis of the rice husk ash (mass percent)

SiO_2	Fe_2O_3	Na_2O	K_2O	Al_2O_3	CaO	TOTAL
78.2	2.6	6.2	5.5	1.6	5.9	100

Silica makes up the major part of the ash like it does for the sandstone filler. The main difference with the sandstone can be seen on the X Ray pattern (figure 5) where it appears that silica is present under its quartz form but also and particularly it is present as a large quantity of cristobalite form which is absent in sandstone.

A large part of amorphous component is emphasised by the halo centered at 22° (22°2θ). Before grinding, the burnt rice husk has a fibrous aspect (figure 6). It contents very small spherical particles (from 1 to about 10 µm) probably quartz or cristobalite (figure 7).

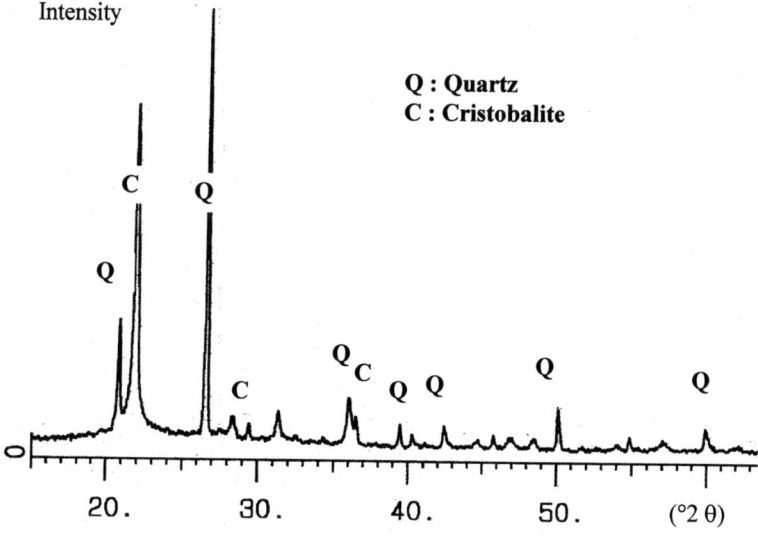

Figure 5 X ray diffraction pattern of rice husk ash (Cu Kα)

Figure 6 SEM micrograph of rice husk ash (fibrous aspect)

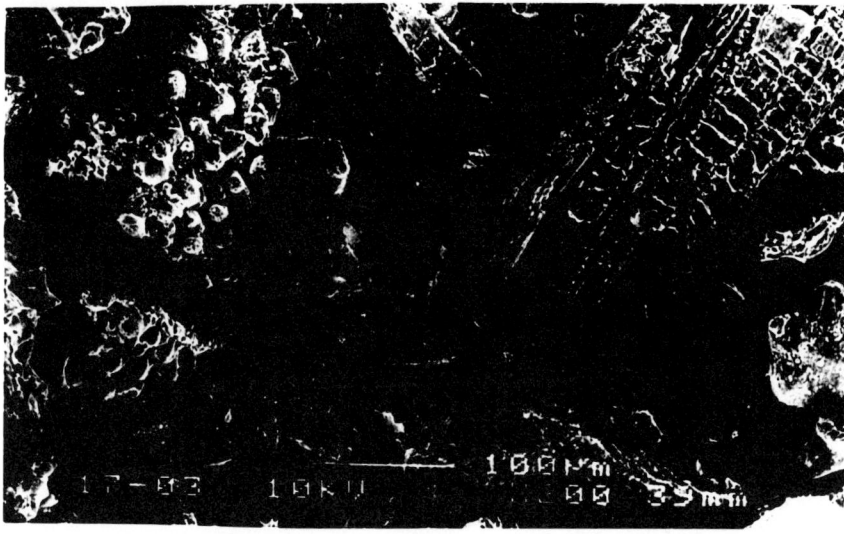

Figure 7 SEM micrograph of rice husk ash (spherical particles)

The granulometric distribution of ground rice husk ash (figure 8) underlines the analogy with the sandstone filler :

- the maximum particles size is 100 μm
- 50 % of the particles have a size smaller than 10 μm
- the volume percentage of particles smaller than 7 μm is 35 %.

The specific area is 6,960 cm²/g.

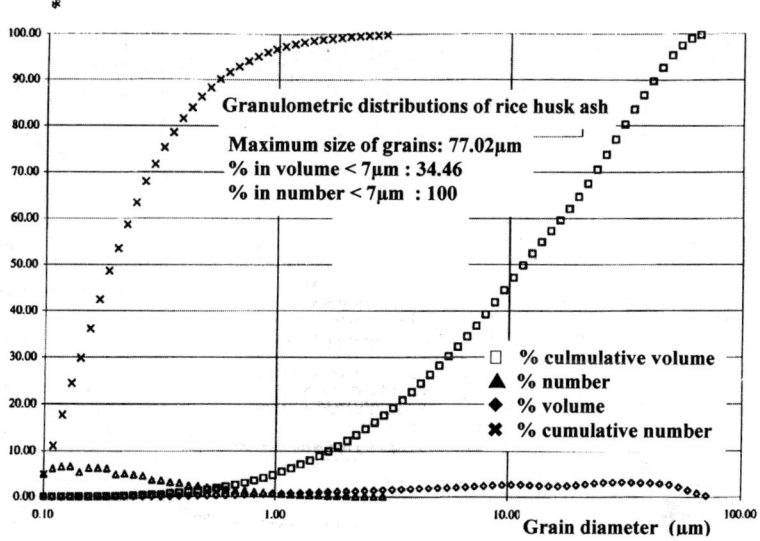

Figure 8 Granulometric distribution of rice husk ash

EXPERIMENTS AND RESULTS

The tests were carried on 3 concrete mixes containing: sandstone filler, raw rice husk ash, ground rice husk ash. The composition of the mixes is given table 3 where W/C is the ratio mass of water to mass of cement. The water content is fitted in order to make mixes having the same workability.

Table 3 Composition of the different concrete mixes

FILLER NATURE	FILLER CONTENT kg/m^3	CEMENT CONTENT kg/m^3	W/C
Sandstone	58.2	250	0.75
Raw rice husk ash	57.5	250	1.28
Ground rice husk ash	234	200	0.92

The cement is an ordinary Portland cement CPA-CEM I. The samples are moulded in a parallelepiped shape 4 cm × 4 cm ×16 cm. The concrete set takes place in a wet room (temperature 20°C, Relative humidity 100 %) for 24 hours. Then a first set of samples is immersed into water at 20°C and a second set is stored in an air conditioned room (temperature 20°C, Relative humidity 50%).

The mechanical resistance is measured at 7 and 28 days : the tensile strength by means of a 4 point bending test, the compressive strength is measured with the two pieces resulting from the bending test. The data are given table 4 and 5.

Table 4 Mechanical resistance of samples stored in the air conditioned room

FILLER NATURE	TENSILE STRENGTH (MPa)		COMPRESSIVE STRENGTH (MPa)	
	7 days	28 days	7 days	28 days
Sandstone	3.20	2.85	5.64	6.18
Raw rice husk	1.50	1.40	3.31	4.98
Ground rice husk	2.34	2.32	6.50	10.34

Table 5 Mechanical resistance of samples immersed into water

FILLER NATURE	TENSILE STRENGTH (MPa)		COMPRESSIVE STRENGTH (MPa)	
	7 Days	28 days	7 days	28 days
Sandstone	2.80	3.37	5.88	11.53
Raw rice husk ash	1.70	2.81	3.69	9.56
Ground rice husk ash	2.62	4.55	8.31	18.59

These results reveal that the behaviour of the rice husk ash is different according to its ground or raw condition. Using the sandstone concrete as a reference, the raw rice husk has a lower mechanical resistance than the reference in spite of a more or less similar composition of the mix. On the other hand, when the size of the ash is reduced by grinding, the performances of the concrete are clearly improved even if with a lower cement content. The cement can be partly replaced with ground ash but when the amount cement + ash increases a larger water content is needed to keep a good workability of the mix. The use of some admixtures could probably reduce the necessary water quantity and so improve the good resistance.

CONCLUSIONS

In the rice producing countries, as Senegal is, the rice husk is usually used as a power plant combustible. The produced ash contains a large amount of silica. When it is substituted to a part of cement in the sand concrete the performances of the material can be saved and, even, improved.

This is probably a consequence of the both effects : filler effect and pozzolanic reactions. Its behaviour can be compared with fly ash.

ACKNOWLEDGEMENTS

The authors are grateful to Colonel P.M.Diop, Conseiller en Génie Civil à la Présidence de la République du Sénégal, for his encouragement and interest in this study.

REFERENCES

1. SABLOCRETE. Bétons de sable : caractéristiques et pratiques d'utilisation. Presses de l'E.N.P.C., 1994, Paris, France.

2. Laboratoire Régional des Ponts et Chaussées. Etudes des fillers. Octobre 1992, France.

3. CISSE, I.K. Contribution à la valorisation des matériaux locaux au Senegal : Application aux bétons de sable. Thèse de doctorat, octobre 1996, INSA, Rennes, France.

ASSESSING THE PROPERTIES OF MORTARS CONTAINING MUNICIPAL SOLID WASTE INCINERATION FLY ASH

S Rémond **P Pimienta**
CSTB

N Rodrigues **J P Bournazel**
Université de Marne-la-Vallée LERM

France

ABSTRACT. This study comes within a research project aiming to develop a method for evaluating concrete containing waste. This method is drawn up by studying mortars containing experimentation waste: Municipal Solid Waste Incineration (MSWI) fly ash. In this report we are presenting the work relating to the study of the performance characteristics of these materials. The setting times, the workability, and the compressive strengths are presented. Incorporating MSWI fly ash into mortars increases their setting time and reduces their fluidity. For ash contents of less than 10%, the compressive strengths of the mortars are improved.

Keywords: Mortars, MSWI fly ash, Soluble salts, Chlorides, Performance characteristics.

Sébastien Rémond is a construction engineer, studying for a PhD at the CSTB, France. He is working on evaluating concrete made from waste and more specifically is studying the changes in the microstructure of this concrete during leaching.

Dr P Pimienta is the mineral materials laboratory manager in the Structures Division of the Centre Scientifique et Technique du Bâtiment, France. One of his main fields of interest concerns the use of recycled materials in concrete.

Nelson Rodrigues is a DEA student at the Université de Marne-la-Vallée, France.

Dr Jean Pierre Bournazel is the scientific director of LERM. His interest lies in the link between physical, chemical and mechanical properties in relation to the durability of concrete.

INTRODUCTION

In France, legislation stipulates that from 2002, only final waste (which cannot be technically or economically reused) can be stored in landfill sites. The various waste producers should therefore, in the years to come, develop solutions for reusing their waste. The construction industry, which uses a large amount of materials, has for many years been using by-products from other industrial sectors as substitution materials. Incorporating waste into concrete might provide a satisfactory solution for managing certain types of waste. However, before using waste in concrete, the properties of the concrete made in this way must be studied to ensure that the new material can be used as a construction material.

The main objective of this research project is to develop a method for evaluating concrete containing waste. This method is drawn up by studying mortars containing"experimentation" waste: Municipal Solid Waste Incineration (MSWI) fly ash. This waste is very irregular and contains large amounts of heavy metals and soluble salts, which may lead to problems when incorporating waste into concrete. The properties of the mortars studied are: the performance characteristics, the durability and the environmental impact. Given here is the study relating to the evaluation of the performance characteristics of these materials. The short term properties (setting time and fluidity) of mortars containing increasing amounts of MSWI ash were studied first. The properties of the hardened mortars (compressive strength, dimensional variations…) were then analysed.

EXPERIMENTAL PROCEDURES

Materials

The cement used is a CPA-CEM I 52.5 PM-ES with a low C_3A content (4.2 %). The sand is a normalised sand complying with standard EN 196-1 [1]. Fluidity tests were performed with a naphthalene sulphonate based superplasticizer (Rhéobuild 1000 from M.B.T.).

The MSWI ash comes from the incineration plant in Lagny (France, 77) which operates with a wet scrubber. The physico-chemical and mineralogical characteristics of this ash are given in [2]. The main components of the ash are given in Table 1. Analysing the ash using X-ray Diffraction showed the presence of quartz, sylvite (KCl), halite (NaCl), anhydrite ($CaSO_4$), calcite ($CaCO_3$) and quick lime (CaO).

Table 1 Main components of the MSWI ash

COMPONENT	CONTENT (%)	COMPONENT	CONTENT (%)
LOI (975°C)	13.00	MgO	2.52
SiO_2	27.23	Fe_2O_3	1.80
CaO	16.42	Cl	7.20
Al_2O_3	11.72	SO_3	3
Na_2O	5.86	Zn	0.11
K_2O	5.80	Pb	0.4

Mortar Composition

Table 2 gives the compositions of the mortars used.

Table 2 Composition of the mortars used expressed in grams

	M0	M5	M10	M15	M20
Cement	450	450	450	450	450
Sand	1350	1327.5	1305	1282.5	1260
MSWI ash	0	22.5	45	67.5	90
Water	225	225	225	225	225

Test Methods

The initial and final setting times of the mortars were worked out using the Vicat needle in accordance with French standard NF P18-356 [3]. However, the mixing times of the mortars were increased to 5 minutes so as to limit the segregation observed during the preliminary tests performed with a mixing time of three minutes.

The fluidity of the mortars and the change in the fluidity with time (during the first 45 minutes) were studied by measuring the flow times of the mortars using a March cone, in accordance with the experimental procedure of the new AFREM method [4]. For these tests, the amount of mortar paste (cement + water + MSWI ash) was increased keeping the same W/C content so as to increase fluidity and to allow the flow times to be measured using the Marsh cone. The composition of the reference mortar was as follows: 650 g of cement, 1350 g of sand and 325 g of water (to be compared with test mortar M0 of table 2). MSWI ash was added to the mortar in place of the sand. To study the effect of the soluble fraction of ash, fluidity tests were also performed with mortars containing washed ash or wash water. Ash was washed with water for 3 minutes. The wash water was then filtered and the solid residue dried at 33°C until a constant mass was obtained. Two washes (referred to as L10 and L20) were performed (L10: 65 g of ash in 325 g of water, L20: 130 g of ash in 325 g of water). The ash/water ratios used in these washes correspond to the ash/water ratios of mortars M10 and M20, given in Table 2.

The compressive strengths of mortars M0, M5, M10, M15 and M20 were worked out after 7, 28, 90 and 565 days in accordance with standard EN 196-1 [1]. The mortars were stored in sealed bags at 20°C. The 28 day compressive strengths of the mortars made with wash water and with washed ash (wash L10) were also worked out.

RESULTS AND DISCUSSION

Setting Times

Figure 1 shows the initial and final setting times of the mortars as a function of MSWI fly ash content.

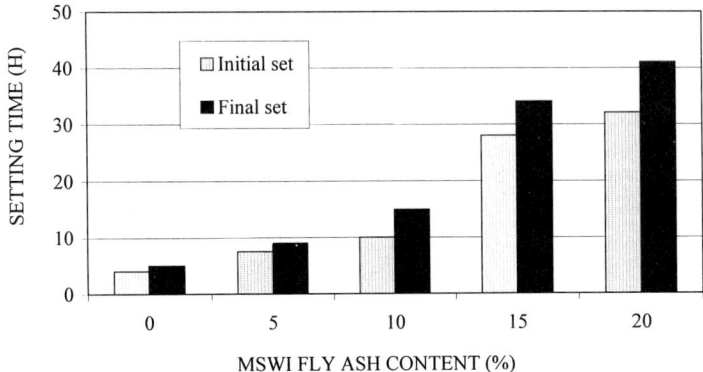

Figure 1 Initial and final setting times of mortars M0, M5, M10, M15 and M20

It can be seen in Figure 1 that the setting times increase with MSWI ash content in the mortar. In particular, for ash contents above 15%, the increase is setting times is very significant. The heavy metals in the MSWI ash, especially zinc and lead, are well known setting retarders. These metals are certainly responsible for slowing down the setting times seen.

Fluidity of the Mortars

Figure 2 shows the logarithm of the flow times of mortars containing 0, 10 and 20 % MSWI ash (instead of sand) as a function of the superplasticizer content expressed as a percentage of dry extract with respect to the mass of cement. The vertical lines show the saturation doses of superplasticizer, worked out in accordance with the new AFREM method [4].

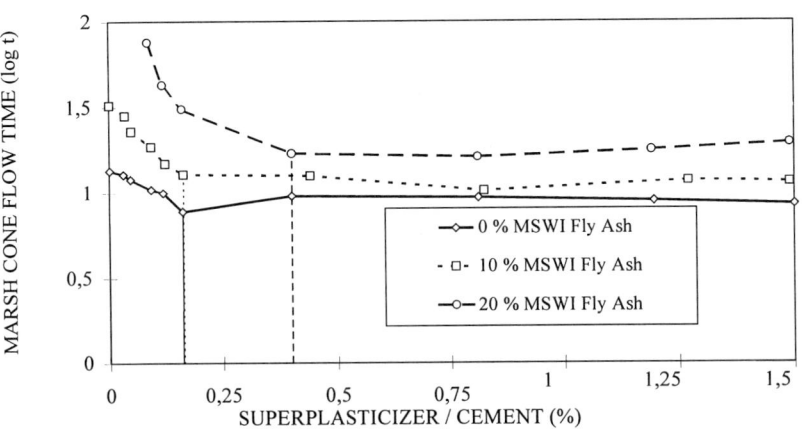

Figure 2 Fluidity of mortars containing 0, 10 and 20 % MSWI fly ash as a function of their superplasticizer content

Given these results, it essentially seems that incorporating MSWI fly ash into mortars without admixtures reduces their fluidity. Up to 20% ash, the greater the ash content, the longer the flow times. This effect of the ash continues when the superplasticizer is added. The curves obtained for the two mortars containing ash, although offset, are very similar in appearance to the very traditional curve for the reference mortar without ash. The flow times initially decrease quite quickly when superplasticizer is added but are then practically constant beyond the saturation dose.

Figure 3 shows the change with time of the fluidity of mortars containing 0, 10 and 20% MSWI ash up to 45 minutes. The amounts of superplasticiser used for these tests were equal to the saturation doses worked out from Figure 2. The change in fluidity of the mortars containing 0 and 10% MSWI ash without added superplasticiser is also shown.

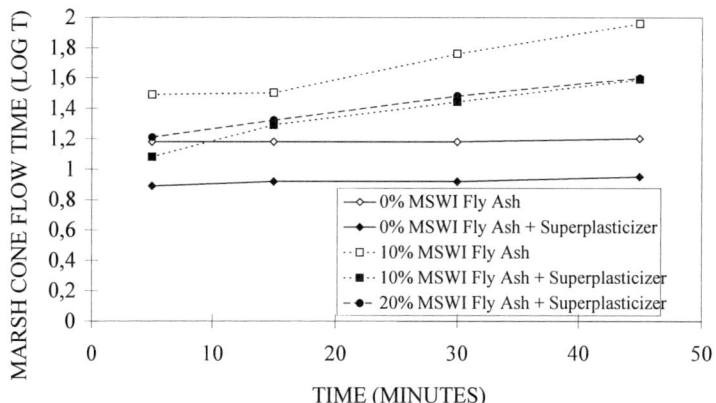

Figure 3 Change in fluidity of mortars with time

These results show that the fluidity of mortars without MSWI ash (with or without superplasticiser) remains constant with time whilst that of mortars with ash (with or without superplasticiser) decreases. The curves obtained with or without superplasticiser are practically parallel.

Figures 2 and 3 therefore show that adding MSWI ash to admixture-free mortars greatly effects their fluidity. Ash leads to an instant loss of workability as well as a reduction in workability with time. Figure 2 shows that the gains in fluidity of mortars with or without ash, caused by adding superplasticiser, are the same. Comparing the curves represented by squares (solid and not solid) in Figure 3 shows that the variations in fluidity with time of the mortar with ash M10, with and without superplasticizer, are also the same. These two results tend to show that the superplasticiser used is not incompatible with the ash in the study.

Figure 4 provides a comparison of the change with time of the fluidity of mortars without ash made with pure water and wash water (wash L20) and mortars containing 20% ash and 20% washed ash (wash L20).

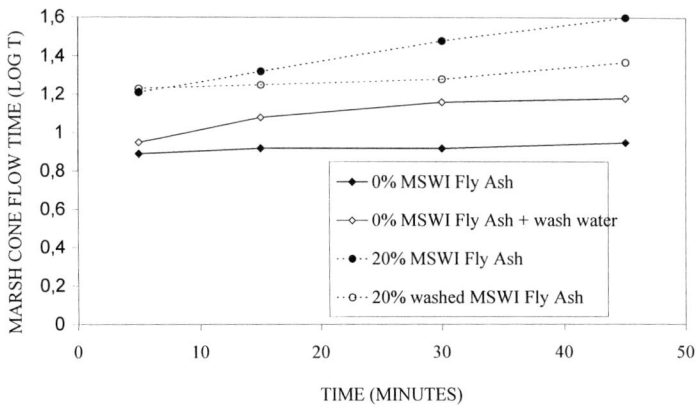

Figure 4 Changes in fluidity of the various mortars with time (0% ash, 0% ash + wash water, 20% ash and 20% washed ash)

Adding wash water (wash L20) to mortars without ash leads to a reduction in fluidity over time. The fluidity of mortars containing 20% ash decreases more slowly when the ash has been washed (wash L20).

These results seem to indicate that the soluble salts contained in the ash are partly responsible for the loss of fluidity of the mortars with time. The wash water was analysed. The results show that practically all the chlorides were extracted during washing as well as a smaller amount of sulphates. The stiffening seen might therefore be due to the formation of chloroaluminates right from the start of mixing.

It should be noted that the ash wash water had little effect on the fluidity of the mortars in the first few minutes. The flow times after 5 minutes of mortars without ash containing pure water and wash water are in fact practically the same. Likewise, replacing the ash with washed ash does not affect the initial flow times of the mortars. The soluble salts are therefore not the cause of the reduction in initial fluidity of the mortars seen in figure 2. The fact that the ash absorbs the water might explain this result. MSWI ash is essentially finer than the sand used. Also, observations using a Scanning Electron Microscope (SEM) performed on this ash showed that many particles are porous and have a very high specific surface area [2]. Partially replacing the sand with ash therefore leads to an increase in the specific surface area of the mixing grains and therefore greater water absorption. The latter phenomenon is surely responsible for the reduction in the fluidity of the mortars containing ash.

Compressive Strengths

Figure 5 shows the compressive strengths of mortars M0, M5, M10, M15 and M20 after 7, 28, 90 and 565 days.

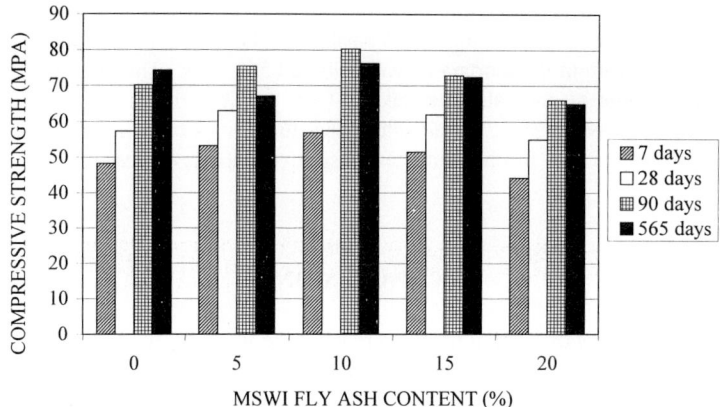

Figure 5 Compressive strengths of mortars after 7, 28, 90 and 565 days

In general, it appears on this graph that the compressive strengths of the mortars are at an optimum for a MSWI ash content of around 10%. The strength increases when the ash content varies from 0 to 10% and then decreases. After 90 days, mortar M10 was approximately 15% stronger than the reference mortar M0. However, the strength of mortar M20 is approximately 5% less. The strength of the mortars containing MSWI ash increases with age up to 90 days although the strengths obtained at 565 days are lower (the strength of mortar M5 dropped by around 10%).

The 28 day compressive strengths of mortars without ash made with wash water (wash L10) and mortars containing 10% washed ash were also worked out and compared with the 28 day strength of mortar M10. Using wash water gives a 35% increase in strength with respect to mortar M10, although washing the ash leads to a 20% reduction in strength. These results show that the soluble fraction of ash leads to a clear improvement in the compressive strengths of the mortars. The presence of large amounts of chlorides in the ash wash water is surely the cause of these increases in strength. Chlorides are effectively well known hardening accelerators. The non soluble (or slightly soluble) fraction of ash seems on the contrary to have an unfavourable effect on the compressive strength. In effect, the compressive strength of mortar M10 is lower than that of the mortar made with the ash wash water. The amount of chlorides contained in these two mortars is however very similar since washing leads to a total extraction of chlorides. Furthermore, the compressive strength of the mortar containing 10% washed ash is less than that of mortar M10.

Our research cannot categorically explain the effect of the non soluble part of the ash on the reduction in compressive strength. One explanation that we can however give is the following: the MSWI ash particles might form weak points in the mortars. Kessler et al [5] effectively showed using an SEM study of cement pastes containing MSWI ash that the bond between the cement paste and the ash seemed to be weak.

The competition between the positive effect of the chlorides and the negative effect of the non soluble part might explain why the optimum strength was obtained for an ash content of 10%. Above 10% ash, the negative effect of the non soluble part might be greater thus leading to a reduction in mechanical strength.

Neither of these two assumptions can however explain the decrease in 565 day compressive strength seen in mortars containing MSWI ash. It should be noted that no increase in cracking could be seen in the mortars between 90 and 565 days. The presence in the MSWI ash of high amounts of sulphates and alkalis (and possibly the presence of certain trace elements such as heavy metals) might reduce the durability of the mortars. However, our study does not allow the mechanisms responsible for these decreases in strength to be identified.

CONCLUSIONS

MSWI fly ash increases the setting time and reduces the workability of mortars. The delay in setting is probably due to the retarding action of the zinc and lead in the ash. Instantaneous falls in mortar workability may be partly attributed to the absorption of water by the ash. The ash is effectively porous and has a high specific surface area. However, we have shown that the reduction in fluidity of the mortar over time is mostly due to the action of the soluble salts in the ash. In particular, the early formation of chloroaluminates might explain this result. Adding a superplasticizer to mortars containing ash provides a gain in workability which is comparable to that in mortar without ash. The superplasticizer tested did not seem to show any incompatibilities with the ash, at least not during the first 45 minutes.

The properties of the hardened mortars were also studied. For contents less than 10%, MSWI ash increases the strength of mortars up to 90 days. However, the long term strength decreases when ash is used. Most of the increase in the short and medium term strengths of mortars containing 10% ash may be due to the hardening accelerator effect of the chlorides. The cause of the decrease in long term strength could not be explained.

REFERENCES

1. AFNOR. NF EN 196-1. Méthodes d'essais des ciments - Détermination des résistances mécaniques, Août 1995.

2. REMOND, S, BENTZ, D P, PIMIENTA, P AND BOURNAZEL, J P. Cement hydration in the presence of Municipal Solid Waste Incineration Fly Ash. Proceedings of the first International Meeting "Material Science and Concrete Properties", Toulouse, France, March 5-6 1998, pp 63-70.

3. AFNOR. NF P18-356. Adjuvants pour bétons, mortier et coulis - Détermination des temps de prise sur mortier, Août 1985.

4. DE LARRARD, F, BOSC, F, CATHERINE, C AND DEFLORENNE, F. La nouvelle méthode des coulis de l'AFREM pour la formulation des bétons à hautes performances. Bulletin des Laboratoires des Ponts et Chaussées, N°202, Mars-Avril 1996, pp 61-69.

5. KESSLER, B, ROLLET, M AND SORRENTINO, F. Microstructure of cement pastes as incinerator ash host. Proceedings of the First International Symposium on Cement Industry, Solution to Waste Managment, Ed R W Piggot, Calgary, pp 235-251, 1992.

ELABORATION OF A MSWI FLY ASH SOLIDIFICATION STABILISATION PROCESS: USE OF STATISTICAL DESIGN OF EXPERIMENTS

S Morel-Braymand
P Clastres
URGC-Structures of INSA
A Pellequer
Electricité de France
France

ABSTRACT. Ashes produced from the Incineration of Municipal Solid Wastes (MSWI fly ashes) require to be solidified and stabilised by an hydraulic binder before being landfilled in a waste disposal site. Ground granulated blast furnace slag was adapted to stabilise these ashes. This study proposes a new approach for the development of a solidification-stabilisation process based on specific planning and exploitation of experiments : the Statistical Design of Experiment. First, empirical laws have been determined which describe the physical and mechanical behaviour of the mixture of MSWI ash with slag. One model has been found for a set of tests. Next, using these calculated models, an optimised formulation that complies with the French regulation has been found. Then, experiments on the optimised composition found have been led to validate the models.

Keywords: Blast furnace slag, Municipal solid waste incinerator ashes, Fly ash, Hazardous materials, Experimental design, Solidification, Stabilisation, Leachability, Swelling, Durability.

S Morel-Braymand is a graduate of Civil Engineering of ENSAIS, Strasbourg, France (in 1993), Doctor, Researcher in URGC-Structures of INSA (Lyon, France). Her research focused the use of Statistical Design of Experiments for the use of hydraulic binders and the formulation of mixtures for wastes solidification.

P Clastres is an ACI member, Doctor es Sciences, Senior Researcher in URGC-Structures Laboratory of INSA (Lyon, France). His research interest is in the hardening and deformation of materials using hydraulic binders.

A Pellequer is the Head of Cemete Department at Electricité de France, graduate of Civil Engineering of ENSAIS (Strasbourg, France) and of ISBA, (Marseille, France). He worked in the field of civil engineering works and in the field of industrial waste valorisation.

INTRODUCTION

The increasing production of MSW (Municipal Solid Waste) in France has required increasing and improving methodology for their treatment. Incineration of these MSW is the most common process used in France. However, it is considered only as an intermediate waste processing technology because the ash derived from the incineration of MSW (MSWI fly ash) still requires further treatment before it can be landfilled in hazardous waste disposal site. This ash is a powdery material comprising mainly chlorides, lime and sulphates. It also contains alkaline salts, alumina, silica and toxic heavy metals.

One treatment method commonly used prior to landfilling is the solidification/stabilisation process utilising a hydraulic binder. In this study, ground granulated blast furnace slag was used. This binder is well adapted to the solidification and stabilisation of heavy metals . It also offers good durability with regards to chlorides. In addition, the components of MSWI ash act as a catalyst for the blast furnace slag hydration.

The methodology used to work out this process is based on the following points :

- Using the Statistical Design of Experiments in order to master the experimental variance influence and reduce the number of experiments. Moreover, this method makes it possible to calculate the effects of the studied factors and the interactions that may exist between these factors. It is also possible to establish empirical laws that concern the material behaviour in function of the formulations parameters.

- Optimising the process as regards multicriteria characteristics in order to meet technical, economical and industrial needs. At the end the material, whose composition had been optimised was put through different tests first to assure its lasting feature as regards particular thermohygrometric conditions and then to characterise its mechanical behaviour under compressive strength.

METHODOLOGY AND EXPERIMENTAL RESULTS

A mixture of blast furnace slag with industrial MSWI ash has been studied, using the parameters : A/S : Ash/Slag, W/(A+S) : Water/(Ash + Slag), and water temperature.

Table 1 Variation ranges of mixture of industrial MSWI ash and blast furnace slag

VARIATION RANGES OF FACTORS	A/(A+S) %	W/(A+S)	WATER TEMPERATURE °C
Minimal value -1	50	0.4	5
Maximum value +1	100	0.6	45

Methodology

A particular terminology is used in this paper, parameters are called 'factors', influences of parameters are called 'factors effects' or 'roles of factors'.

The compositions which were prepared have been chosen to facilitate the calculation of the role of the factors defined previously and the potential effects of their interactions effects. Empirical laws based on these calculations will be derived. The choice of the compositions is also made to obtain the best accuracy in the calculations, i.e. to decrease the influence of the experimental variance on the calculated effects.

The set of tests to be carried out with one composition is called one 'experiment'. All the experiments are represented schematically with an 'experimental design'. Each line of the table corresponds to one experiment. Each column of the table is related to the values of the factors for each experiment. The same design is used for the different tests. Experimental results for each test are called 'responses'. The values of factors are codified as follow : the lower value for a factor is generally codified as 'level –1', its upper value is codified as 'level +1', with a linear progression of the factor in between. The lower and upper limits of a factor determine its range of variation. The 'experimental domain' is the area defined by the variation ranges of the factors.

In this study, three factors were considered, their variation and their codification are given in Table 1. A three variables second degree polynomial models was postulated for the following responses : initial setting time, mechanical strength , leachability, workability and swelling. The interaction between the three factors (X_1 = A/S, X_2 = W/(A+S) and X_3 = water temperature) will be added to this polynomial model, y : response (initial setting time, mechanical strength, etc).

$$y = b_0 + \sum_{i=1 \text{ à } 3} b_i X_i + \frac{1}{2} \sum_{i,j=1 \text{ à } 3, i \neq j} b_{ij} X_i X_j + \sum_{i=1 \text{ à } 3} b_{ii} X_i^2 + b_{123} X_1 X_2 X_3$$

The b_i and b_{ij} coefficients are calculated thanks to the experiments worked out by a multilinear regression calculation using the least square method. Eleven coefficients have to be found for each model, so a minima of eleven compositions have to be chosen among workable points in the experimental domain. The experimental design is calculated in order to obtain the most accurate estimation of the models coefficients with the minimum number of experiments.

Thanks to an iterative calculation, we decided to realise thirteen experiments, spread over the whole experimental domain, and one more (repeated three times), at the centre of the experimental domain.

The influence of the variation of each factor on the responses changes was then analysed according to the values of the two other factors, i.e. according to the calculated interactions effects. This interpretation is called 'effects analysis'.

To optimise the solidification process formulation, a second application of the Statistical Design of Experiments has been used : a 'multi-criteria' optimisation using empirical modelling of phenomena.

A classic approach to assess the optimum composition would consist in setting up a network over the experimental domain and choosing the best composition.

With the models calculated previously, an optimised formulation can be developed, which satisfies the different criteria prescribed by the regulation for mechanical and physico-chemical properties of the solidified ash : a good mechanical strength, an initial setting time which would allow satisfactory placement, a good resistance to water attack and a minimum of pollutants leachability. This optimised composition is chosen amongst all the possible compositions of the experimental domain and is not necessarily one of those carried out .

The chemical characteristics and reactivities of MSWI fly ash and blast furnace slag had been studied separately during preliminary experiments.

Chemical Characteristics and Reactivity of Studied Materials

Blast furnace slag

A haematite ground granulated blast furnace slag containing 98% of glass, was used. The mineral compositions of this slag is given in Table 2. Blast furnace slag requires a catalyst to set, and the catalyst can either be calcic, alkaline, or sulfatic.

MSWI ash

This MSWI ash derived from an incineration plant which uses dry flue gas treatment, has a hydraulic activity.

Chemical composition of MSWI ash: The chemical elements were identified by chemical analysis (see Table2). $Ca(OH)_2$, $CaClOH$, $CaSO_4$, $NaCl$, and KCl minerals were identified by XRD (X Ray Diffraction) analysis. Al_2O_3 and SiO_2 exist in their non crystalline form, they are amorphous. The polluting agents are mainly PbO, $PbCl_2$, ZnO.

Table 2 Chemical Analysis of studied materials

CHEMICAL ANALYSIS, %	BFS	MSWI ASH
SiO_2	36.04	13.3
Fe_2O_3	1.21	1.7
Al_2O_3	10.22	8.3
CaO	42.57	50.9
MgO	6.46	1.6
Na_2O	0.38	1.9
K_2O	0.51	1.5
Mn_2O_3	0.31	0.2
SO_3	*	2.6
Zn	*	0.255
Pb	*	0.266
Chlorides	*	14.1

MSWI ash reactivity: When the ash were mixed with water, at a W/A = 0.5, an initial setting time of one hour was observed, the mechanical strengths obtained for this specimen were :

Compressive strength :   1 day : 5 MPa ;   28 days : 6 MPa
Tensile strength :   1 day : 2.1 MPa ;   28 days : 2.7 MPa

With these mechanical properties the solidified material could be landfilled in the hazardous waste disposal site (a minimum of 1 MPa of strength is required). However, at 28 days of age its water resistance is insufficient. The material breaks up after 24 hours of soaking in water. As a consequence of this and the potential leachability of its polluting agents it cannot be landfilled. Furthermore, during its hydration, this material swells up until it sets. This swelling which can reach up to 10% of the specimen height decreases its mechanical strength and increases its permeability. This is due to a reaction between metallic aluminium (Al) and lime contained in the MSWI ash ; which leads to the production of hydrogen. An additive which acts through electrical attraction on the Al or $Ca(OH)_2$ particles was used to prevent this swelling.

Empirical Laws Determination

With these 16 materials defined previously, which all have a different composition, we have made the following experiments :

Initial setting time, Swelling, Workability, Mechanical strength, Immersion with analysis of the filtrate, Leachability with analysis of leachate.

The last three experiment categories have been realised with samples that had a 24 hours, 7 days, 28 days, 90 days and one year of maturation.

For each model (each testing) 11 coefficients have been calculated. Then the experimental and calculated results of the 16 compositions were compared with a statistical test. If the results don't fit, it's possible to realise a transformation over this model.

All the models calculated for all the experiments are not detailed in this paper. For example, we obtained following models :

Log(initial setting time)
$2.024 - 0.444x_1 + 0.362x_2 + 0.267 x_3 + 0.291x_1^2 + 0.069x_2^2 - 0.170x_3^2 - 0.324x_1x_2 + 0.103x_1x_3 - 0.088x_2x_3 + 0.026x_1x_2x_3$

Compressive strength, 28 days
$12.9 - 3.8x_1 - 8.4x_2 + 1.1x_3 - 0.5x_1^2 + 0.8x_2^2 + 0.5x_3^2 + 2.3x_1x_2 - 1.5x_1x_3 - 0.5x_2x_3 + 1.9x_1x_2x_3$

Conductivity of filtrate, 28 days
$7.96 + 5.57x_1 + 0.27x_2 - 0.39x_3 + 0.25x_1^2 - 0.50x_2^2 + 0.02x_3^2 + 0.30x_1x_2 - 0.40x_1x_3 + 0.06x_2x_3 + 0.08x_1x_2x_3$

A detailed analysis of the different models lead to the following conclusion

1. The more one increases the MSWI ash content (or the more one decrease the slag content), the more is shortened the setting time.

2. The increase in water content and water temperature lengthen the initial setting time. The effect of this last factor results from the water temperature action on the MWSI ash solubility and more particularly on the CaClOH solubility.

3. As regards the mechanical strength experiments, important interactions exist between the MSWI ash content and the water ratio. Moreover these interactions depend on the age of the material. With a 28 days old material, it appears that the fact to increase the MSWI ash content is the more harmful to mechanical strength, less the water content is important.

4. As regards the leachate analysis, it clearly appears that increasing the MSWI ash content is bad to the material stabilisation and this, however old the material is.

Multicriteria Optimisation

With all the models obtained for the different tests, the multicriteria optimised composition can be calculated. Moreover, a satisfaction criterion for each testing has been defined (si). The general satisfaction can then be calculated : $S = \Pi s_i$ (i=1 to n : number of testings). An iterative algorithm will enable the best compromise to be established. With these models, an estimation of the expected values for the different tests can be calculated.

Then, the following composition has been obtained:

$$X1 = -0.469 \Rightarrow \text{Ash/(Ash + Slag)} = 63.2862 \%$$
$$X2 = -0.840 \Rightarrow \text{Water/(Ash + Slag)} = 0.4160$$
$$X3 = 0.247 \Rightarrow \text{Water temperature} = 29.9437 \text{ °C}$$

With this composition, the response results are for example : see Table 3

Table 3 Optimised composition validation

TEST	CALCULATED VALUE	EXPERIMENTAL VALUE
Workability	50%	45%
Compressive strength 1 day	4 N/mm²	1.2 N/mm²
Compressive strength 28 days	23.9 N/mm²	19 N/mm²
Conductivity of filtrate 28 days	5.05 mS/cm	5.1 mS/cm

Optimised Composition Validation

Validation of the model

An experiment with this composition determined previously has then been lead. These are the results obtained with the optimised material compared with the calculated values.

The difference between the calculated and the experimental value is due to the unlinearity of the material behaviour in function of the MSWI ash content as regards the 24 hours mechanical strength.. The estimations that concern the other experiments are very satisfying. The whole results of the testings (which are not presented there) show that the experimental values always fit the calculated ones, except for the 24 hours mechanical strength and the initial setting time.

Additional testing

The solidified material has been put through the following conserving conditions:

- Ten 50°C and –25°C thermal cycles day/night
- 96 hours long immersions in neutral or aggressive baths.

At the end of these cycles, mechanical strength tests have been realised on the test specimen. Some materials composed of 100% MSWI ash and other composed of 90% slag+10% Ca(OH)$_2$ (the other factors remain the same) have been put through the same conditions. Moreover, the whole results are compared with these obtained with reference test specimen that haven't been put through any bath or immersion. The results are given Table 4.

Table 4 Compressive strength of materials after baths and immersions cycles

THERMAL CYCLE	OPTIMISED COMPOSITION	100% MSWI	90% SLAG+10% LIME
Alternate	18.5	2.3	19
50°C	20	7	17
-25°C	21	12.5	15.5
Reference	20	14.5	15
IMMERSION CYCLE	OPTIMISED COMPOSITION	100% MSWI	90% SLAG+10% LIME
ph7	22	2	17.5
ph4.5	19	2.2	16
ph12.5	22.2	2	16
Reference	21.5	13.2	15

These last results show that examining the compressive strength and the analysis of the 28 days old leachate are not sufficient to accept a material as a wasted material. Additional tests such as thermal cycles can bring to the force the material weaknesses that can be harmful to its behaviour.

CONCLUSIONS

A general method that permits to elaborate a waste solidification process has been explained in the previous pages. This method takes the different aspects of the problem into account, that is to say not only the MSWI ash behaviour but also the determination of an optimised composition for which several parameters intervene and which must satisfy technical, industrial and economic criteria.

The interest of the statistical designs methodology is demonstrated when several parameters have to be studied to formulate a material composition. The study using empirical models (study of polynomial factors or models) made the characterisation and comprehension of the material hydration and hardening easier. In particular, this technique permits to study the interactions between the composition factors of the mixing. The swelling due to the aluminium presence in a metallic form in the MSWI ash is strongly harmful to these criteria. The use of a specific admixture helped to solve partly this problem. It is also important to take into account the hydraulic activity of the MSWI ash that must be seen as a material and not only as a waste. Moreover, nowadays the norms do not aim to put the solidified materials through such essays as the ones presented in the last part of this document. But it has been proved that a material capacity to guarantee a minimal compressive strength ruled by the norms does not assure its durability.

Thermal cycles may strongly affect the mechanical characteristics of the materials that contain a lot of MSWI ash; these materials then don't meet the current normalisation criteria that permit to be put in a waste disposal site anymore.

ACKNOWLEDGEMENT

The authors gratefully acknowledge the support of Mrs. Florence Kraus, engineer at « Electricité de France », Research and Development Division, for her technical help in the environmental science; and Mr. Joël Olivier, engineer at « Electricité de France », Engineering and Construction Division, for having initiated this project. The authors acknowledge also Mr. Didier Mathieu, engineer-chemist, Doctor es Sciences and Professor at the University of the Mediterranean.

REFERENCES

1. MATHIEU, D, AND PHAN TAN LUU, R. NEMROD Software, LPRAI, Marseille, France.

2. MARAVAL, S., « Etude comparative de divers liants hydrauliques et de la vitrification, cas des cendres d'incinération d'ordures ménagères », Thèse INSA Lyon, 1994, 327p.

3. YANG, GORDON C C, CHEN, SHIUANN-YIH. « Statistical analyses of control parameters for physicochemical properties of solidified incinerator fly ash of municipal solid wastes », Journal of Hazardous Materials, Vol 39, 1994 pp 317-333 and Vol 459, 1996 pp 149-173

4. MOREL-BRAYMAND, S. "Elaboration d'un procédé de solidification/stabilisation de REFIOM. Apport de la méthodologie des Plans d'Expériences., Thèse INSA de Lyon, 1999, 560p.

5. MOREL, S, CLASTRES, P, MATHIEU, D, PELLEQUER, A., "Optimised solidification of mixtures of MSWI ashes with blast furnace slag; use of the statistical design of experiments" , 6th CANMET / ACI International Conference on fly ash, silica fume, slag and natural pozzolans in concrete. Bangkok 1998 – Vol II pp. 975-996.

UTILIZATION OF VITRIFIED CONTAMINATED SOIL MATERIALS IN CONCRETE MIXTURES

S Amirkhanian
Clemson University
United States of America

ABSTRACT. The United States is facing a major crisis regarding its waste materials. One major area of concern is the landfilling of hazardous or contaminated waste materials. This research project was established to determine the effectiveness of using one type of contaminated soil material in highway construction. The activities of a naval base in South Carolina caused some soil contamination. The materials were vitrified in a laboratory and the final product was used as a source of an aggregate in various concrete mixtures. The results indicated, in general, that the use of vitrified contaminated soil materials in various concrete mixtures does not adversely effect the strength of the mixtures. In some cases, the compressive strength of mixtures containing these materials was the same as the control mixture.

Keywords: Hazardous waste, Vitrification, Contaminated soil, Concrete, Vitrified glass, Recycling, Flexible pavements.

Professor Serji N Amirkhanian joined Clemson University in 1987. He teaches undergraduate and graduate level courses in the construction area involving construction materials, pavement design and maintenance, construction practices, and construction productivity. He has been a consultant in the materials area for many firms in the state. Professor Amirkhanian has published over 100 refereed papers, articles, and technical reports. His work has been published in various technical publications and journals around the country. In addition, he has given over 125 technical presentations around the country regarding various topics associated with the paving industry. He has been involved with several research projects sponsored by the South Carolina Department of Transportation and Federal Highway Administration involving rubberized asphalt mixtures and the use of waste materials in highway construction materials. In addition, he has been involved with some private firms to reduce or eliminate their by-product materials. He is a member of American Society of Civil Engineers (ASCE), American Society for Testing and Materials (ASTM), Transportation Research Board (TRB) and many other associations.

INTRODUCTION

Over 4.1 billion metric tons of non-hazardous solid wastes (including agricultural) are generated in the United States annually. This equals to about 16 metric tons of waste per person, per year. The majority of this waste is being disposed in landfills despite efforts to recycle, reuse, or reprocess some of these materials. Disposal of this waste is a national concern due to many factors such as environmental issues, costs, and public's views. Of the 4.1 billion tons, 540 million tons are municipal and industrial hazardous waste. Since 1980, only 423 of the 1200 Superfund sites (32%) on the National Priorities List (NPL) have been cleaned up. The NPL is expected to increase to 2000 in the next 5 to 10 years. It is estimated the cost of clean up programs for Superfund sites is $750 billion over the next 30 years.

The Environmental Protection Agency (EPA) estimates that approximately 80% of existing landfills will close within next 10 to 15 years. In addition, due to regulations and design modifications, the cost of opening and operating a new landfill has increased. Therefore, there is a need to reuse or reprocess as many products or by-products as possible [1, 2].

Due to many reasons (e.g., geology) in some parts of the country (e.g., Charleston, South Carolina), a quality aggregate source cannot be found. The aggregate must be imported from other locations and this increases the cost of construction. Therefore, it is important, in such cases, for engineers and researchers to determine an alternative solution to the problem.

BACKGROUND

South Carolina, like many other states, used to be the home of many military bases which were ordered to be closed due to many factors (e.g., budget cuts, end of cold war, etc.). One of these facilities (called the Naval Complex) was located in Charleston, South Carolina, and was closed in 1996. An engineering company contracted Clemson University to determine the feasibility of using the contaminated soil at site in the construction industry (e.g., concrete mixes). The contaminated soil was vitrified and then tested for suitability in the mixes.

VITRIFICATION PROCESS

Vitrification technology is not new. Taking silica, alumina, and oxides of alkali and alkaline earth elements, such as sodium, calcium, barium and potassium, and heating to relatively high temperatures, a solid, tightly bound chemical structure is formed. The result is a fairly rigid material (glass) with low porosity with rigidity ranges up to a Mohs scale hardness of 6.0.

In this project, a new vitrification technology was used to remove hazardous constituents from the contaminated materials. The cleansed molten glass in the arc furnace is tapped and cast into billets followed by production of aggregate by crushing techniques.

Due to the confidentiality of the process, the details of the work cannot be described in great detail at this point. The vitrified product was subject to TCLP testing (toxic characteristic leaching procedure) to fully assess the safety of the product to the environment (Table 1).

Table 1 Chemical Analysis of Waste Samples Total and TCLP Metals

CONSTIT-UENT	SWMU 5 TCLP, mg/l	SWMU 5 TOTAL, mg/kg	SWMU 25 TCLP, mg/l	SWMU 25 TOTAL, mg/kg	SPOILS TCLP, mg/l	SPOILS Total, mg/kg
Ag	U / <0.1	0.28 / <10	U / <0.1	7.04 / <10	<0.1	0.0
As	U / <0.2	2.1 / <100	0.0658 / <0.2	1.23 / <100	0.2	0.0
Ba	0.13 / <0.5	32 / 120	0.104 / 1.1	71.6 / <100	1.2	0.0
Cd	U / <0.1	0.11 / <10	0.0675 / <0.1	34.5 / <10	<0.1	0.0
Cr	U / <0.1	4.9 / 80	249 / 126	12200 / 1950	<0.1	90
Pb	143 / 74	10500 / 3610	U / <0.1	555 / 70	0.1	30
Se	U / <0.2	0.55 / <500	U / <0.2	2.14 / <100	<0.2	0.0
Hg	0.00152/<0.1	0.23 / 0.4	0.00058 / <0.1	0.0318 / <0.1	<0.1	0.0
Al		174 / 17500		/ 4710		59250
Be		0.11 / -				
Ca		531 / 1307		/ 41800		12090
Cu		15.8 / 50		/ <10		0.0
Fe		2040 / 59750		/ 10300		45400
Mg		38.3 / 830		/ 930		6810
Mn		13.7 / 100		/ 50		260
Ni		1.4 / 30		/ 20		10
K		232 / 5250		/ 1570		10520
Na		60.6 / 2920		/ 280		11560
Tl		0.55 / -				
V		1.7 / -				
Zn		21.2 / 160		/ 30		100
Sn		7.9 / <100		/ <100		0.0
CN		0.35 / -		0.294 / -		
Cr+6				4250 / -		
P		/ 260		/ 250		1090
Si		/ 398210		/ 394000		305440
Ti		/ 890		/ 770		4030
Zr		/ 30		/ 10		80

Vitrification is the process of using heat to fuse or melt materials forming oxide polymers commonly known as glasses. This provides for a material, which can be tested to certify its safety for use in many products (e.g., concrete mixes). The process of vitrification involves chemical reactions and produces a product in which waste constituents are no longer recognizable and thereby cannot be separated in a physical process from that product. It employs very high temperatures (e.g., greater than 1550°C) which allow for destruction of organic constituents, converting them to carbon dioxide and water vapor.

In general, the vitrification process includes the "locking" of hazardous metals or inorganics in the glass polymer rendering these materials non-accessible or non-leachable into the environment.

However, the process utilized in this project caused metals and inorganic volatilization due to the extreme heat available in the electric arc process, thereby separating the hazardous materials from the glass polymer such that the glass meets leaching and land application goals in recycling. The developed glass formed in two distinct states: primarily amorphous (rapid cooling) or primarily crystalline (slow cooling). The rate of cooling of the molten glass dictated the change in state. The molten glass was collected in molds, each holding approximately 360 kg of material.

Lead and zinc, which are considered toxic metals by Environmental Protection Agency (EPA), were successfully removed from the raw material, and thus absent from the glass product. Table 2 shows the elemental composition of the resulting glass product from the two waste streams: the combined Charleston Naval Complex and the dredge spoils. A comparison of the above analyses with results shown in Table 1 indicates the reduction in constituents of interest including lead and shows the chromium reduced to the trivalent, chemically stable state.

In 1997, two drums (approximately 270 kg each) of vitrified glass were delivered to the Civil Engineering Materials Laboratory at Clemson University. One of the drums contained aggregates produced from Naval Shipyard and the second drum contained aggregates from Charleston area dredge spoils. In this report, the aggregates produced from Naval Shipyard and Charleston area dredge spoils are designated as Sample A and Sample B, respectively. The physical properties of these aggregates were tested and the results shown in Table 3.

OBJECTIVES

The specific objectives of this part of the project included:

a) Performing various physical tests (e.g., Los Angeles Abrasion, bulk specific gravity, etc.) on Samples A and B.

b) Performing ACI-211 concrete mix design procedures to obtain the optimum ingredients of control mixtures prepared using a local aggregate source as well as concrete mixtures containing Samples A and B.

c) Determining the unit weight, air void, slump, elastic modulus, poisson's ratio, and compressive and bending strengths of all concrete mixtures.

METHODOLOGY

The initial phase of this part of the research included the determination of the optimum percentage of Samples A or B in a concrete mix. Several percentages of samples A and B were used to make concrete mixes. The strength of these mixtures was measured to determine the optimum percentages.

After the initial laboratory testing, it was concluded that three percentages of each aggregate source would be used for further testing. These percentages included 5%, 10%, or 15% (by weight) of vitrified aggregate being substituted for the coarse aggregate.

The ACI-211 mix design procedures were used to obtain the initial quantities of the ingredients for various mixtures. Cylinders were made according to ASTM C192 procedures.

Table 2 Glass product total metals analysis

CONSTITUENT	DREDGE SPOILS TOTAL, mg/kg	NAVAL COMPLEX TOTAL, mg/kg	TOTAL GLASS[1] mg/kg
Ag	<2	<2	<2
Al	62460	31480	40760
As	<50	20	10
Ba	40	50	30
Ca	145340	159950	177460
Cd	<10	<10	<10
Cr	3760	2150	2080
Cu	20	3	10
Fe+2	20130	15610	14580
Fe+3	<100	<100	<100
Metallic Fe	500	290	230
Hg	<0.1	<0.1	<0.1
K	5140	5620	4910
Mg	69770	59690	49060
Mn	180	70	170
Na	3380	1060	3410
Ni	<20	<20	4
P	900	70	250
Pb	<20	20	10
Se	<50	10	10
Si	235950	266280	251940
Sn	<50	<50	<50
Ti	3210	450	1740
Zn	50	4	20
Zr	240	20	140

[1] - average of spoils and Complex glass

The results of the compressive strength and bending strength of mixtures containing Samples A or B were compared to a control mixture (compressive strength of 27.6 MPa or 4,000 psi mix). A total of 70 concrete cylinders (15.2 cm by 30.5 cm), some containing Samples A or B and an air-entrained agent were made and tested.

The compressive strengths of these mixtures were measured after 28 days of wet curing. In addition, 70 beams (7.62 cm x 7.62 cm x 35.56 cm) were made and tested for bending strength.

EXPERIMENTAL DESIGN AND RESULTS

The experimental design for this part of the research is shown in Figure 1. The physical properties of samples A and B are shown in Table 3. The quantity of each ingredient (e.g., cement, fine aggregate, etc.) for non-air-entrained and air-entrained mixtures is shown in Tables 4.

The %air void, unit weight, compressive strength, and bending strength of mixtures containing non-air-entrained and air-entrained agents are shown in Tables 5 and 6, respectively. Table 7 shows the elastic modulus and poisson's ratio of all mixtures used for this research work.

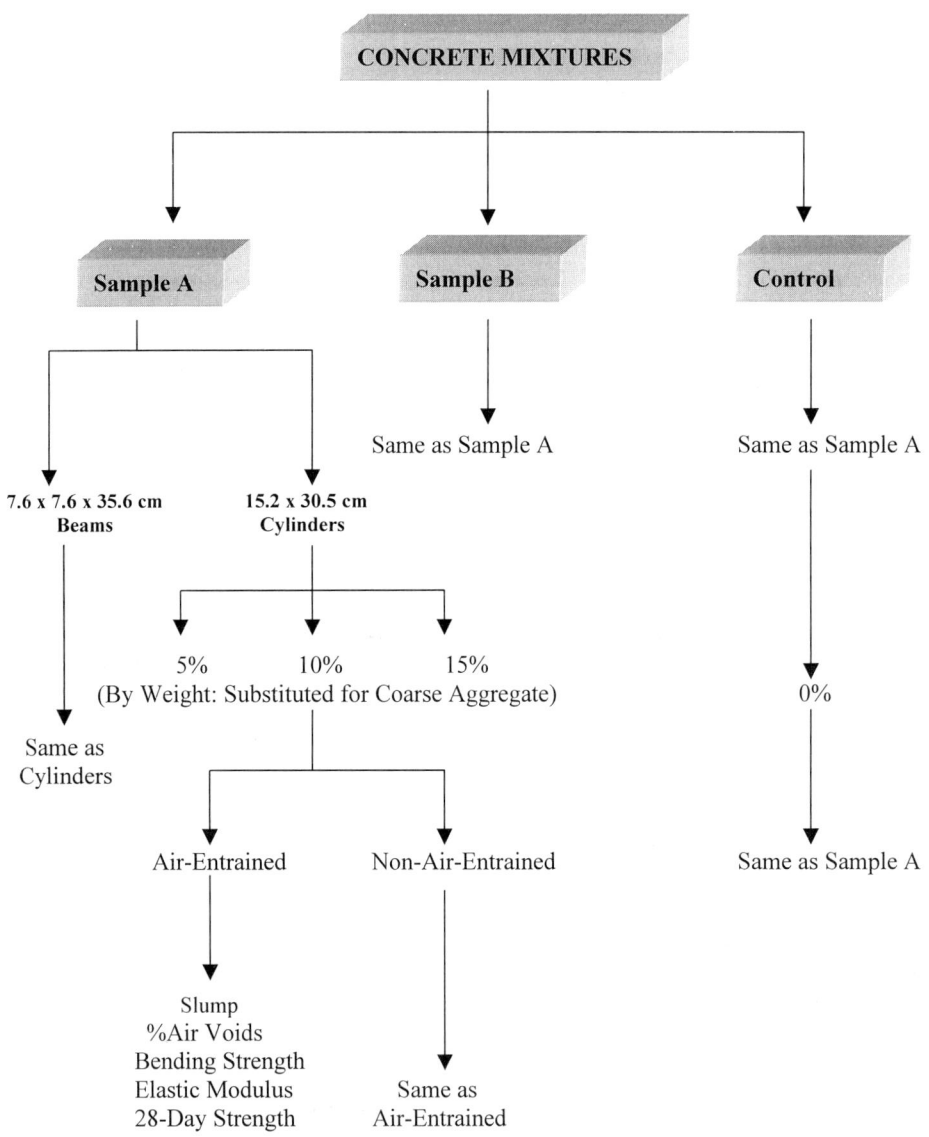

Figure 1 Experimental Design for Concrete Mixtures (n=5)

Table 3 Physical properties of samples A and B

PROPERTY	SAMPLE A	SAMPLE B
Bulk Specific Gravity	2.669	2.707
Bulk Spec. Gravity (SSD)	2.689	2.753
Apparent Specific Gravity	2.722	2.814
Percent Absorption	1.23	1.79
Fineness Modulus	5.41	4.93
Los Angeles Abrasion	40.4	45.3
Sieve Size	Sample A	Sample B
37.5 mm or 1.5"	100	100
25.4 mm or 1"	100	99.5
19 mm or ¾"	96.2	96.4
12.5 mm or ½"	63.6	78.7
4.76 mm or No. 4	41.6	55.6
2.36 mm or No. 8	26.6	36.9
1.18 mm or No. 16	15.4	21.5
0.6 mm or No. 30	8.3	11.0
0.3 mm or No. 50	4.8	5.6
0.15 mm or No. 100	2.5	1.6

Table 4 Quantity of each ingredient for a batch of concrete mixture (3 cylinders, one beam, and %air void determination): slump = 7.6 – 10.2 cm

MIX	CEMENT kg (lb)	AIR-ENTRAINED * (gm)	FINE AGG. kg (lb)	COARSE AGG. kg (lb)	SAMPLE A kg (lb)	SAMPLE B kg (lb)
Control	9.9 (21.7)	9.2	15.5 (34.2)	22.8 (50.1)	---	---
5% Sample A	9.9 (21.7)	9.2	15.5 (34.2)	21.6 (47.6)	1.1 (2.5)	---
10% Sample A	9.9 (21.7)	9.2	15.5 (34.2)	20.5 (45.1)	2.3 (5.0)	---
15% Sample A	9.9 (21.7)	9.2	15.5 (34.2)	19.4 (42.6)	3.4 (7.5)	---
5% Sample B	9.9 (21.7)	9.2	15.5 (34.2)	21.6 (47.6)	---	1.1 (2.5)
10% Sample B	9.9 (21.7)	9.2	15.5 (34.2)	20.5 (45.1)	---	2.3 (5.0)
15% Sample B	9.9 (21.7)	9.2	15.5 (34.2)	19.4 (42.6)	---	3.4 (7.5)

* Used only for mixtures containing air-entrained agent.

Table 5 Average %air voids, unit weight, compressive and bending strengths of mixtures containing non-air-entrained agent

MIX	%AIR VOID	UNIT WEIGHT kg/m^3 (lb/ft^3)	COMP. STRENGTH MPA (psi)	STAN. DEV. MPA (psi)	BENDING STRENGTH MPA (psi)	STAN. DEV. MPA (psi)
Control	1.7	2523 (157.4)	50.3 (7294)	0.29 (39)	6.16 (893)	0.42 (61)
Sample A, 5%	1.9	2448 (152.7)	48.9 (7088)	1.36 (197)	6.14 (891)	0.31 (45)
Sample A, 10%	2.3	2438 (152.1)	43.7 (6338)	5.03 (729)	6.07 (880)	0.36 (52)
Sample A, 15%	2.4	2430 (151.6)	48.9 (7085)	1.89 (274)	6.01 (872)	0.48 (69)
Sample B, 5%	1.7	2459 (153.4)	45.4 (6583)	1.32 (191)	6.20 (899)	0.32 (46)
Sample B, 10%	1.9	2451 (152.9)	44.5 (6448)	3.72 (539)	6.71 (973)	0.32 (47)
Sample B, 15%	1.9	2446 (152.6)	45.3 (6575)	9.01 (1307)	6.27 (910)	0.26 (37)

Table 6 Average %air voids, unit weight, compressive and bending strengths of mixtures containing air-entrained agent

MIX	%AIR VOID	UNIT WEIGHT kg/m^3 (lb/ft^3)	COMP. STRENGTH MPA (psi)	STAN. DEV. MPA (psi)	BENDING STRENGTH MPA (psi)	STAN. DEV. MPA (psi)
Control	3.0	2443 (152.4)	41.4 (6001)	1.91 (277)	6.30 (914)	0.57 (82)
Sample A, 5%	3.3	2435 (151.9)	40.9 (5937)	3.34 (484)	6.05 (877)	0.38 (55)
Sample A, 10%	3.6	2424 (151.2)	32.9 (4775)	1.87 (271)	5.57 (808)	0.29 (42)
Sample A, 15%	3.9	2406 (150.1)	35.3 (5121)	0.83 (121)	5.52 (800)	0.32 (47)
Sample B, 5%	3.4	2440 (152.2)	39.9 (5792)	2.31 (335)	6.12 (888)	0.36 (52)
Sample B, 10%	3.6	2432 (151.7)	42.7 (6198)	2.60 (377)	6.14 (891)	0.24 (35)
Sample B, 15%	3.9	2422 (151.1)	42.6 (6183)	0.97 (140)	5.72 (829)	0.34 (50)

Table 7 Elastic modulus and poisson's ratio of various concrete mixtures

NON-AIR ENTRAINED CONCRETE MIXTURES

Mix	Elastic Modulus MPa (psi)	Poisson's Ratio
Control	33,470 (4,850,677)	0.15
5% Sample A	26,920 (3,901,252)	0.17
10% Sample A	18,300 (2,652,643)	0.27
15% Sample A	21,175 (3,068,764)	0.18
5% Sample B	21,630 (3,134,750)	0.20
10% Sample B	15,990 (2,317,498)	0.22
15% Sample B	17,353 (2,514,936)	0.28

Air-Entrained Concrete Mixtures

Mix	Elastic Modulus MPa (psi)	Poisson's Ratio
Control	26,930 (3,902,586)	0.21
5% Sample A	18,375 (2,662,952)	0.23
10% Sample A	15,360 (2,226,320)	0.33
15% Sample A	16,490 (2,390,180)	0.28
5% Sample B	17,355 (2,515,000)	0.25
10% Sample B	23,520 (3,408,886)	0.13
15% Sample B	20,520 (2,974,376)	0.15

CONCLUSIONS AND RECOMMENDATIONS

The following are some of the conclusions, observations, and recommendations regarding this portion of the research project:

1. In most cases, the control mixture produced higher compressive and bending strengths compared to mixtures containing Samples A or B. However, the mixtures containing Samples A or B produced strengths that are much higher than the initial mix design criteria (i.e., 27.6 MPa or 4,000 psi).

2. The compressive and bending strengths of mixtures containing an air-entrained agent were lower than mixtures without the agent.

3. The compressive and bending strengths of mixtures containing no air-entrained agent and made with Sample A were similar to those mixtures containing Sample B. This indicates, based on this limited work, that both mixtures could be used in the field.

4. The compressive and bending strengths of mixtures containing an air-entrained agent and made with Sample B, in general, were higher than strengths of mixtures containing Sample A. However, the mixtures containing Sample A still could be used in the field based on the minimum criteria set for the mix design (minimum compressive strength of 27.6 MPa or 4,000 psi).

5. The elastic modulus of all treated mixtures (non-air and air-entrained) was lower than the control mixtures. In some cases the control mixtures had elastic modulus values which were twice of those of treated mixtures.

6. The use of Samples A and B in concrete mixtures is recommended in several field applications such as parking lots, sidewalks, and secondary roads. It is recommended, at this point, to construct several field experimental sections of these types of mixtures and monitor them for at least 18 months to determine their short-term performance in the field.

7. It is also recommended to perform more testing on other uses of admixtures such as retarders and water reducers to determine the affects of these admixtures on the performance of concrete mixtures containing Samples A or B.

8. Several other aggregate sources should also be used in the next phase of the project to determine the compatibility of aggregate sources used throughout the state with Samples A or B.

ACKNOWELDGEMENTS

The author would like to express his appreciation to the SCANA Corporation for financial support of this research work. In addition, the assistance of Dr. Jeong Ki Min, a post-doctoral Fellow from Research Center for Advanced Mineral Aggregate Composite Products of Kangwon National University of South Korea, is appreciated and acknowledged.

REFERENCES

1. AMIRKHANIAN, S. "UTILIZATION OF WASTE MATERIALS IN HIGHWAY INDUSTRY – A LITERATURE REVIEW," The Journal of Solid Waste Technology and Management, Volume 24, Number 2, May 1997, pp 94-103.

2. AMIRKHANIAN, S AND MANUGIAN, D. "A FEASIBILITY STUDY OF THE USE OF WASTE MATERIALS IN HIGHWAY CONSTRUCTION," Report FHWA-SC-94-01, Clemson University, Civil Engineering Department, Clemson, SC 29534-0911.

INDEX OF AUTHORS

Aba, M	121-130	Nagaraj, T S	95-100
Amirkhanian, S N	337-346	Naik, T R	23-36
Baguant, B K	81-94	Olorunsogo, F T	163-170
Benmalek, M L	243-250	Pavlenko, S I	101-108
Bijen, J M J M	209-216		251-260
Blockmans, S	139-150	Payá, J	47-56
Bonner, D G	189-198	Pellequer, A	327-336
Borrachero, M V	47-56	Pimienta, P	319-236
Bouguerra, A	217-226	Queneudec, M	217-226
	243-250		243-250
Bournazel, J P	319-236	Rad, T	189-198
Camps, J P	309-318	Rémond, S	319-236
Chia-Ming, L	37-46	Robl, T L	57-66
Cisse, I K	299-308	Rodriques, N	319-236
	309-318	Ruby, M	151-162
Clastres, P	327-336	Ryan, C A	233-242
Desmyter, J	139-150	Samarin, A	1-22
Dheilly, R M	217-226	Sato, Y	37-46
Dhir, R K	67-80	Shameem, M	109-120
Eleftheriou, K	283-290	Shoya, M	121-130
Fleischer, W	151-162	Siludom, S	283-290
Gailius, A	275-282	Speare, P R S	283-290
	261-274	Splittgerber, F	199-208
Ghafoori, N	299-298	Sugita, S	121-130
Girniene, I	275-282	Szwabowski, J	261-274
Goual, M S	243-250	Taguchi, S	37-46
Groppo, J C	57-66	Takeda, Y	37-46
Hansen, H	261-274	Temimi, M	309-318
Hobbs, A	57-66	Thompson, A	233-242
Ibrahim, M	109-120	Tittle, P A J	67-80
Ishikawa, T	95-100	Tokuhasi, K	121-130
Jauberthie, R	309-318	Tolchin, S M	131-138
Kassel, S	299-298	Tsukinaga, Y	121-130
Khan, M N	109-120	van Dessel, J	139-150
Kiyohara, C	37-46	West, R P	233-242
Kraus, R N	23-36	Wild, S	261-274
Laquerbe, M	227-232	Yakushiji, T	37-46
	299-308	Zaichenko, N M	131-138
Ledhem, A	217-226	Zakaria, M	179-188
Malyshkin, V I	101-108		
Maslehuddin, M	109-120		
Matviyenko, V A	131-138		
Maultzsch, M	171-178		
McCarthy, M J	67-80		
Meinhold, U	171-178		
Mellmann, G	171-178		
Mimoune, F Z	227-232		
Mimoune, M	227-232		
Monzó, J	47-56		
Morrel-Braymand, S	327-336		
Mueller, A	199-208		

SUBJECT INDEX

This index has been compiled from the keywords assigned to the papers, edited and extended as appropriate. The page references are to the first page of the relevant paper.

Abrasion resistance 163
Absorption 283
Additions 209
Aggregate 131, 163, 171
    artificial... 179
    natural... 179
Aggregates 109, 199, 209
Agricultural wastes 299
Air-entrained
    fine-grained concrete 251
    admixtures 67
Alkali-aggregate reaction 151
Alkali-silica reaction 275
Ash 309
ASTM C277 test 275

Basic and acidic slags from foundry 101
Blast furnace slag 327
Bleeding 81
Blockworks 299
Bottom ash 23
Brickdust 275
Broad definition of concrete 1
Burnt spoils 251

C&D waste aggregates 139
California bearing ratio (CBR) 81
Cement 227, 275
    paste 199
    replacement 261
    stabilisation 81
Cementitious materials 291
Chlorides 319
Clay brick 261
Clayey concretes 217
Coal combustion by-products 23
Compound 227
Compression 291
Compressive
    strength 81, 163, 189
    and tensile strength 179
Concrete 23, 109, 179, 275, 309, 337
    fine-grained cementless slag... 101
    fine-grained... 131, 251
    lightweight... 243
    mixer requirements 67

    pavements 151
    properties 199
    self compacting... 121
    sludge 37
Conditioned pulverized-fly ash 67
Contaminated soil 337
Controlled low strength materials 23
Creep 171, 233
Crushed brick aggregate 179
Crushing 151
Crushing and grinding process 199

Drying shrinkage 81
Durability 109, 121, 171, 209, 261, 283, 327

Economics of wastes utilisation 1
Effect of ponding 57
Effects on the greenhouse gas emissions 1
Elasticity 171, 179
Electro-surface properties 131
Enrichment 199
Environment 209
Environmental pollution 1
Ettringite 291
Expansion 275, 291
Experimental design 327

FBC fly ash 291
Filler 309
Fine-grained
    cementless slag concrete 101
    concrete 131, 251
Flexible pavements 337
Flexural strength 163
Flowable slurry 81
Fly ash 23, 47, 227, 327
Freeze-thaw resistance 283
Fresh properties 291
Frost resistance 151

Froth flotation 57

Ground ash 299

Hazardous
>materials   327
>waste   337
Hydraulic classification   57
Hydro-removed ash   251

Industrial
>and domestic wastes   1
>waste   243
Kyoto protocol   1

Landfill   37
Landfilled and ponded fly ash   57
Leachability   327
Liaison   227
Lightweight concrete   243
Limestone powder   121
Load-bearing and enclosing structures   251
LOI reduction   57

Mechanical
>properties   109
>strength   47
Modulus
>of elasticity   81, 179
>of rupture   179
Moisture content   243
Mortar   189, 261, 275, 291, 319
MSWI fly ash   319
Municipal solid waste incinerator ashes   327

Natural aggregate   179
New applications for wastes   1

Particle size distribution   57, 199
Pelletized fly ash aggregate   95
Performance characteristics   319
Permeability   283
Porosity   171
Portland cement (PC) 47
Pozzolanic activity   47
Pozzolanicity   299
Pozzolans   261
Processed
>and unprocessed wastes   1
>building rubble   171
Processing and grinding of rocks   251
Pulverized dry sludge   37

Ready-mixed concrete   37
Recyclability   209
Recycled   189
>aggregate   163
>aggregates   151
>concrete   139
>concrete   233

paper   233
Recycling   23, 199, 337
>concrete   171
Replacement   275
Returned and surplus concrete   37
Rice husk   309
>ash   283, 299
Rockdust   81
Roller-compacted concrete   233

Sand   309
Sandcrete   299
Sawdust   227
Screening   151
Self compacting concrete   121
Separation   199
>-purification technologies   139
Setting   47
Sewage sludge ash (SSA)   47
Shrinkage   171, 233, 291
Size variation reductions   217
Slag fine aggregate   121
Slipform paver   151
Sludge cake   37
Sludge water   37
Solidification   327
Soluble salts   319
Soundness   47
Stabilisation   327
Stabilized bases   151
Steel slag   109
Strength   131, 171, 227, 261
>index   57
Structural lightweight concrete   95
Structure formation   131
Super-plasticizing admixtures   67
Sustainability   209
Sustainable development   1
Swelling   327

Technology of processing slags   101
Temperature   243
Thermal conductivity   243
Two component composite concrete   95

Unbound roadbases   151
Uniformity   67

Vitrification   337
Vitrified glass   337

Waste   261, 309 management   47
>material   109
>products   131
>water   37

Wastes 209, 291
    as aggregates 1
    as binders 1
Wood 227
    aggregate treatments 217
    ash 23
Workability 47, 163, 189